Springer Optimization and Its Applications

Volume 168

Aims and Scope

Optimization has continued to expand in all directions at an astonishing rate. New algorithmic and theoretical techniques are continually developing and the diffusion into other disciplines is proceeding at a rapid pace, with a spot light on machine learning, artificial intelligence, and quantum computing. Our knowledge of all aspects of the field has grown even more profound. At the same time, one of the most striking trends in optimization is the constantly increasing emphasis on the interdisciplinary nature of the field. Optimization has been a basic tool in areas not limited to applied mathematics, engineering, medicine, economics, computer science, operations research, and other sciences.

The series **Springer Optimization and Its Applications (SOIA)** aims to publish state-of-the-art expository works (monographs, contributed volumes, textbooks, handbooks) that focus on theory, methods, and applications of optimization. Topics covered include, but are not limited to, nonlinear optimization, combinatorial optimization, continuous optimization, stochastic optimization, Bayesian optimization, optimal control, discrete optimization, multi-objective optimization, and more. New to the series portfolio include Works at the intersection of optimization and machine learning, artificial intelligence, and quantum computing.

Volumes from this series are indexed by Web of Science, zbMATH, Mathematical Reviews, and SCOPUS.

More information about this series at http://www.springer.com/series/7393

Michael Th. Rassias
Editor

Harmonic Analysis and Applications

Springer

Editor
Michael Th. Rassias
Institute of Mathematics
University of Zurich
Zurich, Switzerland

Moscow Institute of Physics and
Technology Dolgoprudny, Russia

Institute for Advanced Study
Program in Interdisciplinary Studies
Princeton, NJ, USA

ISSN 1931-6828 ISSN 1931-6836 (electronic)
Springer Optimization and Its Applications
ISBN 978-3-030-61889-6 ISBN 978-3-030-61887-2 (eBook)
https://doi.org/10.1007/978-3-030-61887-2

Mathematics Subject Classification: 32A50, 42Axx, 42B35, 42B37, 42Cxx, 43Axx, 11K70, 37A45, 65Txx, 26D05, 33B10, 35C09, 41A10

This Springer imprint is published by the registered company Springer Nature Switzerland AG
The registered company address is: Gewerbestrasse 11, 6330 Cham, Switzerland

Preface

Harmonic Analysis and Applications publishes high-quality works devoted to a broad spectrum of areas in which Harmonic Analysis plays a central role. This branch of Mathematics is known for its applicability in a plethora of diverse areas, such as Differential Equations, Number Theory, Optimization Theory, Representation Theory, Quantum Mechanics, Neuroscience, and Signal Processing, to name just a few, in the interplay of Mathematics, Physics, Finance, Electrical Engineering, Computer Science, and other branches.

The goal of the book in hand is to present essential developments in various areas in which Harmonic Analysis is applied. Particularly, the contributed chapters discuss topics on structure and optimization in computational harmonic analysis, sampling and approximation in shift invariant subspaces of $L_2(\mathbb{R})$, optimal rank one matrix decomposition, the Riemann Hypothesis, large sets avoiding rough patterns, Hardy Littlewood series, Navier-Stokes equations, sleep dynamics exploration and automatic annotation by combining modern harmonic analysis tools, harmonic functions in slabs and half-spaces, Andoni-Krauthgamer-Razenshteyn characterization of sketchable norms, random matrix theory, and multiplicative completion of redundant systems in Hilbert and Banach function spaces.

The chapters within this book have been chosen to represent a variety of different topics and have been contributed by eminent experts, presenting the state of the art in the corresponding topics and problems treated. Effort has been made for the content of the book to constitute a valuable resource for graduate students but also senior researchers working on Harmonic Analysis and its various interconnections with related areas.

We would like to express our sincere thanks to all contributors of book chapters who have participated in this publication, especially under the very difficult circumstances caused by the current unprecedented Coronavirus global crisis. Last but not least, we would like to warmly thank the staff of Springer for their valuable help throughout the publication process of this book.

Zurich, Switzerland   Michael Th. Rassias

Contents

Sampling and Approximation in Shift Invariant Subspaces of $L_2(\mathbb{R})$

Nikolaos Atreas

Abstract Let ϕ be a continuous function in $L_2(\mathbb{R})$ with a certain decay at infinity and a non-vanishing property in a neighborhood of the origin for the periodization of its Fourier transform $\widehat{\phi}$. Under the above assumptions on ϕ, we derive uniform and non-uniform sampling expansions in shift invariant spaces $V_\phi \subset L_2(\mathbb{R})$. We also produce local (finite) sampling formulas, approximating elements of V_ϕ in bounded intervals of $\mathbb{R}$, and we provide estimates for the corresponding approximation error, namely, the truncation error. Our main tools to obtain these results are the finite section method and the Wiener's lemma for operator algebras.

1 Introduction

Sampling theory allows the recovery of a function f from a pre-determined sequence of linear functionals (measurements) $c_f = \big(c_n(f)\big)$ on a sampling set $\tau = (\tau_n)$; see [11, 14, 25, 31, 40] for an overview on sampling theory and applications. Before we state the classical sampling theorem, we provide some notation. Throughout this text, $L_p(K)$ $(1 \le p \le \infty)$ denotes the space of all p-integrable functions on a Lebesgue measurable set $K \subseteq \mathbb{R}$ with usual norm $\|\cdot\|_{L_p(K)}$ (or $\|\cdot\|_{L_p}$ for brevity, in case $K = \mathbb{R}$). If $p = 2$, we denote by $\langle\cdot,\cdot\rangle_{L_2}$ the usual inner product on the Hilbert space L_2. We define the Fourier transform of a function $f \in L_1$ by

$$\widehat{f}(\xi) = \int_{\mathbb{R}} f(x) e^{-2\pi i \xi x} dx, \quad \xi \in \mathbb{R}$$

and we recall the Plancherel theorem extending the definition of the Fourier transform from the space $L_1 \cap L_2$ to an isometric isomorphism onto L_2. For any

N. Atreas (✉)
School of Electrical and Computer Engineering, Faculty of Engineering, Aristotle University of Thessaloniki, Thessaloniki, Greece
e-mail: natreas@ece.auth.gr

M. Th. Rassias (ed.), *Harmonic Analysis and Applications*, Springer Optimization and Its Applications 168, https://doi.org/10.1007/978-3-030-61887-2_1

$\Omega > 0$, we say that a square-integrable function g is *Ω-bandlimited*, if $\widehat{g}(\xi) = 0$ for all $|\xi| > \Omega$, and we denote by

$$\mathscr{B}_\Omega = \{g \in L_2 : \ g \text{ is } \Omega\text{- bandlimited}\},$$

the corresponding well-known *Paley Wiener* subspace of L_2. Now we have:

Theorem 1 ((Classical sampling theorem) [12]) *Let $T, \Omega > 0$ be such that $2T\Omega \leq 1$. Then, every function $f \in \mathscr{B}_\Omega$ can be reconstructed from the formula*

$$f(x) = T \sum_{n\in\mathbb{Z}} f(nT) \frac{sin(2\pi\Omega(x - nT))}{\pi(x - nT)}, \ x \in \mathbb{R}, \tag{1}$$

where the convergence is uniform on $\mathbb{R}$ and in the L_2-sense.

The function $sinc_\Omega(x) = \frac{\sin(2\pi\Omega x)}{\pi x}$ is a *sampling function* for $\mathscr{B}_\Omega$. If $2T\Omega < 1$, then the set $\{sinc_\Omega(\cdot - nT) : \ n \in \mathbb{Z}\}$ is overcomplete in $\mathscr{B}_\Omega$; hence the above sampling expansion is not unique. On the other hand, if $2T\Omega = 1$, then the set $\{sinc_\Omega(\cdot - nT) : \ n \in \mathbb{Z}\}$ is an orthonormal basis for $\mathscr{B}_\Omega$, and Theorem 1 becomes the well-known *Shannon-Whittaker-Kotelnikov* sampling theorem, providing uniqueness of the reconstruction formula (1). Notice that in (1), the *sampling period* is equal to T, the sampling set is $\tau = \{nT : \ n \in \mathbb{Z}\}$ and the sampling theorem is called *uniform or regular*, because τ is equispaced. The sampling *rate* is equal to the number of samples per second, i.e., it is equal to $\frac{1}{T}$. Hence, the minimum sampling rate for perfect reconstruction in (1), called *Nyquist rate,* is equal to $2\,\Omega$.
Theorem 1 is important because it is a prototype of a digital to analog reconstruction formula and vice versa. Roughly speaking, an Ω-bandlimited (analog) signal f is associated with a discrete set $c(f) = \{f(nT)\}_{n\in\mathbb{Z}}$, and formula (1) provides the means for perfect reconstruction of f on $\mathbb{R}$. In order to model this process, e.g. for $2T\Omega = 1$, we could say that

(i) the *sampling operator* $S : \mathscr{B}_\Omega \ni f \longmapsto c_f = \{f(nT)\} \in \ell_2(\mathbb{Z})$ is bounded and invertible and
(ii) *the reconstruction operator* $R : \ell_2(\mathbb{Z}) \longmapsto \mathscr{B}_\Omega$ is the adjoint of the sampling operator, i.e., $RS(f) = S^*Sf = f$ for all $f \in \mathscr{B}_\Omega$.

Here and hereafter, $\ell_2 = \ell_2(\mathbb{Z})$ denotes the Hilbert space of all square summable sequences $c : \mathbb{Z} \to \mathbb{C}$ with usual inner product $\langle\cdot,\cdot\rangle_{\ell_2}$ and norm $\|\cdot\|_{\ell_2}$.
Theorem 1 can be extended to cover the case of non-square integrable bandlimited functions (see [13, Theorem 3.1]) or to cover the d-dimensional bandlimited case (see [11]). Moreover, the sinc function in (1) can be replaced with another sampling function with a more rapid decrease at infinity, to enable *local* approximation of f in bounded intervals of $\mathbb{R}$. To this direction, we have:

Theorem 2 ([9]) *Let $T, \Omega > 0$ be such that $2T\Omega \leq 1$. Then we may find a rapidly decreasing function s in $\mathscr{B}_{\frac{1}{2T}}$, such that every function $f \in \mathscr{B}_\Omega$ can be* *reconstructed from the formula*

$$f(x) = T \sum_{n \in \mathbb{Z}} f(nT) s(x - nT), \tag{2}$$

where the convergence is uniform on $\mathbb{R}$ *and in the* L_2*-sense.*

Theorem 2 can be extended in order to cover the case of non-necessarily bandlimited functions as well. This generalization is motivated, because in practice there are no bandlimited functions. Hence, we may select the sampling function in (2) to be a non-bandlimited function. Even more generally, we could design the whole sampling process so that information about f is gathered not on the sequence of its sampled values $\{f(nT)\}_{n\in\mathbb{Z}}$ but on more general averaging measurements, obtained, for example, from a convolution process of the original function f with a suitable kernel function, say $\phi \in L_2(\mathbb{R})$, on a sampling set τ, i.e.,

$$L_2(\mathbb{R}) \ni f \longmapsto f * \overline{\phi(-\cdot)} \in L_2(\mathbb{R}) \underset{\text{sampling}}{\longmapsto} c_f = \left\{\langle f, \phi(\cdot - \tau_n)\rangle_{L_2}\right\}_{n\in\mathbb{Z}}. \tag{3}$$

Notice that the above *generalized sampling scheme* (3) is ill-posed on $L_2(\mathbb{R})$ [30, 32]. For well-posedness, we need to determine a certain subspace of L_2 where f lives in and require a space for the sample vector c_f to stay. For more about *average sampling expansions*, we refer to [1, 3, 4, 7, 9, 30, 36, 37] and related references therein.

Below we deal with classical (rather than average) sampling expansions on a class of (non-necessarily bandlimited) subspaces of L_2, namely, the shift invariant spaces.

2 Uniform Sampling in Shift Invariant Spaces of L_2

The original motivation to select a shift invariant space V of $L_2(\mathbb{R})$ originates from Theorems 1 and 2. Another think is that the theory of shift invariant space fits perfectly with Fourier Analysis. Indeed, the representation (2) or (3) for any Ω-bandlimited function implies that $\mathscr{B}_\Omega$ is a *T-shift invariant subspace* of L_2. More generally, given a *generator* function $\phi \in L_2$, let

$$V_\phi^0 = \operatorname{span}\{\phi(\cdot - k) : k \in \mathbb{Z}\}$$

be a subspace of L_2 containing all *finite* linear combinations of the integer shifts of ϕ. Notice here that when we talk about a *finite* sequence, we mean a sequence where at most finitely many entries are non-zero. Consider the L_2-closure of V_ϕ^0 to be

$$V_\phi = \overline{\operatorname{span}} V_\phi^0.$$

Definition 1 ([17, Thm 3.6.6]) We say that the set $\{\phi(\cdot - k) : k \in \mathbb{Z}\}$ forms a *Riesz basis* for V_ϕ, if there exist two positive constants A and B such that for any finite scalar sequence $\{c_k\}$, we have

$$A \sum_k |c_k|^2 \leq \Big\| \sum_k c_k \phi(\cdot - k) \Big\|_{L_2}^2 \leq B \sum_k |c_k|^2.$$

For an excellent review about the theory of Riesz bases and the theory of frames, we refer to [17]. Now, we may prove the following:

Theorem 3 (Sampling theorem for shift invariant spaces) *Let ϕ be a continuous function on $\mathbb{R}$ such that*

$$\|\phi\|_{W_p(L_\infty, u_\alpha)} = \Big\| \left\{ \|u_\alpha(\cdot - n)\phi(\cdot - n)\|_{L_\infty[-\frac{1}{2},\frac{1}{2}]} \right\}_{n\in\mathbb{Z}} \Big\|_{\ell_p} < \infty, \tag{4}$$

where $1 \leq p \leq +\infty$, $u_\alpha(x) = (1 + |x|)^\alpha$ for some $\alpha > 1 - \frac{1}{p}$ and

$$\Phi^\dagger(\xi) = \sum_{n\in\mathbb{Z}} \phi(n) e^{-2\pi i n\xi} \neq 0 \text{ for any } \xi \in \Big[-\frac{1}{2}, \frac{1}{2}\Big]. \tag{5}$$

Then the set $\{\phi(\cdot - k) : k \in \mathbb{Z}\}$ is a Riesz basis for its closed linear span $V_\phi = \overline{span}\{\phi(\cdot - k) : k \in \mathbb{Z}\}$, and moreover, V_ϕ is a sampling subspace of L_2, i.e., any function $f \in V_\phi$ is uniquely and stably reconstructed from its sample set $\mathscr{L}(f) = \{f(n)\}_{n\in\mathbb{Z}}$ by the formula

$$f = \sum_{n\in\mathbb{Z}} f(n) S(\cdot - n) \tag{6}$$

in the L_2-sense and uniformly on compact intervals of $\mathbb{R}$, for some sampling function S whose Fourier transform is determined by the following equality

$$\widehat{S}(\xi) = \frac{\widehat{\phi}(\xi)}{\Phi^\dagger(\xi)}, \quad \xi \in \mathbb{R}.$$

Proof The assumption $\alpha > 1 - \frac{1}{p}$ on the exponent of the polynomial weight u_α in condition (4) ensures that $\phi \in L_1 \cap L_2$. In order to prove this claim, it suffices to prove that $\|\phi\|_{W_1(L_\infty, u_\alpha)} < \infty$. Indeed, let q be the conjugate exponent of p (i.e., $\frac{1}{p} + \frac{1}{q} = 1$). Then we have

$$\|\phi\|_{W_1(L_\infty, u_\alpha)} = \Big\| \left\{ \|\phi(\cdot - n)\|_{L_\infty[-\frac{1}{2},\frac{1}{2}]} \right\}_{n\in\mathbb{Z}} \Big\|_{\ell_1} = \|(\phi u_\alpha) u_\alpha^{-1}\|_{W_1(L_\infty, u_0)}$$

$$\leq \|\phi\|_{W_p(L_\infty, u_\alpha)} \Big(\sum_{n\in\mathbb{Z}} \frac{1}{(|n| + 1/2)^{\alpha q}} \Big)^{1/q} \leq C \|\phi\|_{W_p(L_\infty, u_\alpha)} < \infty. \tag{7}$$

In addition, we observe that V_ϕ is a space of continuous square integrable functions on $\mathbb{R}$, and moreover the 2π-periodic function $|\Phi^\dagger|$ in condition (5) is continuous on $\left[-\frac{1}{2}, \frac{1}{2}\right]$ and so

$$\|\Phi^\dagger\|_0 = \min_{\gamma\in[-\frac{1}{2},\frac{1}{2}]}|\Phi^\dagger(\gamma)| > 0 \text{ and } \|\Phi^\dagger\|_\infty = \max_{\gamma\in[-\frac{1}{2},\frac{1}{2}]}|\Phi^\dagger(\gamma)| < \infty.$$

This also implies that the set $\{\phi(\cdot - k) : k \in \mathbb{Z}\}$ is a Riesz basis for its closed linear span V_ϕ; see [17, Thm 7.2.3]. We now define the infinite matrix

$$\Phi = \{\Phi_{m,n} = \phi(m-n)\}_{m,n\in\mathbb{Z}} \tag{8}$$

as an operator on ℓ_2. Then, for any $c \in \ell_2$, we apply the Parseval equality, and we obtain

$$\|\Phi c\|_{\ell_2}^2 = \int_0^1 |\Phi^\dagger(\xi)|^2 \Big|\sum_n c_n e^{-2\pi i n\xi}\Big|^2 d\xi$$

and so

$$\|\Phi^\dagger\|_0^2\|c\|_{\ell_2}^2 \le \|\Phi c\|_{\ell_2}^2 \le \|\Phi^\dagger\|_\infty^2\|c\|_2^2 \text{ for all } c \in \ell_2. \tag{9}$$

Therefore, the operator Φ is bounded on ℓ_2, and it has bounded inverse on ℓ_2. Let f be an element of V_ϕ uniquely expressed by

$$f(x) = \sum_n c_n(f)\phi(x-n),$$

in the L_2-sense and pointwise on $\mathbb{R}$. Denote by $d_f = \{f(k) : k \in \mathbb{Z}\}$ to be the sequence of sampled values of f at the integers $x = k$. Then

$$f(k) = \sum_n c_n(f)\phi(k-n) \iff d_f = \Phi c_f \iff c_f = \Phi^{-1}d_f,$$

where

$$\Phi^{-1} : \ell_2 \to \ell_2 : \left(\Phi^{-1}c\right)_n = \sum_k \Phi_{n,k}^{-1}c_k \quad n,k \in \mathbb{Z}.$$

Hence

$$f = \sum_n c_n(f)\phi(\cdot - n) = \sum_n \left(\Phi^{-1}d_f\right)_n \phi(\cdot - n) = \sum_k f(k)S(\cdot - k),$$

where $S = \sum_n \Phi_{0,n}^{-1}\phi(\cdot - n)$. Since S is a sampling function for V_ϕ (i.e., $\sum_n \widehat{S}(\xi + n) = 1$), we have

$$\begin{aligned} S &= \sum_n \Phi_{0,n}^{-1}\phi(\cdot - n) \iff \widehat{S} = \Big(\sum_n \Phi_{0,n}^{-1}e^{-2\pi i n\cdot}\Big)\widehat{\phi} \\ &\iff \sum_n \widehat{S}(\cdot + n) = \Big(\sum_k \Phi_{0,k}^{-1}e^{-2\pi i k\cdot}\Big)\sum_n \widehat{\phi}(\cdot + n) \\ &\iff 1 = \Big(\sum_k \Phi_{0,k}^{-1}e^{-2\pi i k\xi}\Big)\Phi^{\dagger}(\xi) \iff \sum_k \Phi_{0,k}^{-1}e^{-2\pi i k\xi} = \frac{1}{\Phi^{\dagger}(\xi)}. \end{aligned} \tag{10}$$

Substituting the last equality in (10), we obtain $\widehat{S} = \frac{\widehat{\phi}}{\Phi^{\dagger}}$.

Equation (6) is a well-known result of a *regular sampling expansion* for shift invariant spaces, including the above Theorems 1 and 2 and wavelet sampling expansions [38, Theorem 9.2].

3 Perturbation Sampling in Shift Invariant Spaces

Let ϕ be a continuous function on $\mathbb{R}$ satisfying the assumptions (4) and (5) and V_ϕ be its corresponding shift invariant space as in Sect. 2. In a variety of applications, the sampling process may not be uniform, but rather shifted or perturbed by a bounded sequence $\Delta = \{\delta_n\}_{n\in\mathbb{Z}}$, called *perturbation* sequence. Δ could be either known, or unknown, if it is caused from disturbances of the acquisition device or jitter. In both cases we talk about a non-uniform sampling scheme [10, 18, 29, 40], and a basic problem is to examine whether the resulting irregular sampling set $\{n + \delta_n\}_{n\in\mathbb{Z}}$ satisfies an inequality similar to (9):

$$C\|c\|_{\ell_2} \le \|\{f(n + \delta_n)\}_n\|_{\ell_2} \le D\|c\|_{\ell_2}, \tag{11}$$

for all $c \in \ell_2$ and for some positive constants C, D. If this double inequality holds, then there exists another Riesz basis $\{\psi_n^{\Delta}\}_{n\in\mathbb{Z}}$ for V_ϕ providing a unique and stable reconstruction formula for elements $f \in V_\phi$ of the form

$$f = \sum_{n\in\mathbb{Z}} f(n + \delta_n)\psi_n^{\Delta}.$$

The above equation is a *non-uniform or perturbation sampling formula.* Stable perturbed sampling sets and formulas have been studied extensively; see [1–3, 5–8, 12, 15, 24, 27, 28, 33, 39] and references therein. It is useful to detect the largest bound of the perturbation sequence Δ for which (11) holds, called *maximum perturbation.* In this section, we derive a class of perturbed sampling expansions under a maximum perturbation δ_ϕ. To do that, first we consider the bounded and boundedly invertible operator $\Phi = \{\Phi_{m,n} = \phi(m-n) : \; m, n \in \mathbb{Z}\}$ of Eq. (8), and we denote by

$$\Phi_{\tau_\delta} = \left\{ \left(\Phi_{\tau_\delta}\right)_{m,n} = \phi(\tau_m - n) \right\}_{m,n \in \mathbb{Z}} \tag{12}$$

to be a distortion of Φ, where $\tau_\delta = \{\tau_n = n + \delta_n : |\delta_n| \leq \delta\}_{n\in\mathbb{Z}}$ is a sampling set on $\mathbb{R}$ for some $\delta > 0$ and τ_δ is also an *ordered* and *ε-separated* sampling set, in the sense that for some $\varepsilon > 0$, we have

$$\tau_{m+1} - \tau_m \geq \varepsilon > 0, \;\text{ for all } m \in \mathbb{Z}.$$

Let

$$G_\phi : \mathbb{R}^+ \to \mathbb{R}^+ : \;\; G_\phi(x) = \sum_{n\in\mathbb{Z}} \sup\nolimits_{|y|\leq x} |\phi(y+n) - \phi(n)| \tag{13}$$

be a continuous, increasing and unbounded function on $\mathbb{R}^+$ with $G(0) = 0$ and

$$\delta_\phi = \inf\left\{ x > 0 : \;\; G_\phi(x) \geq \|\Phi^\dagger\|_0 \right\}, \tag{14}$$

where $\|\Phi^\dagger\|_0 > 0$ by assumption (5). Then $0 < \delta_\phi < +\infty$.

Definition 2 Let ϕ, τ_δ, Φ_{τ_δ} and δ_ϕ be as above. If $0 \leq \delta < \delta_\phi$, then we say that the operator Φ_{τ_δ} belongs in the class $\mathscr{F}_{\delta_\phi}$.

Now we are ready to prove the following:

Theorem 4 ([7] (Perturbation theorem for shift invariant subspaces of L_2)) *Let ϕ, τ_δ, δ_ϕ be as above, and the operator Φ_{τ_δ} belongs in the class $\mathscr{F}_{\delta_\phi}$ of Definition 2. Then the set τ_δ is a set of stable sampling for V_ϕ (i.e. (11) holds), and so every function $f \in V_\phi$ is uniquely reconstructed from the set of sampled values $\mathscr{L}_{\tau_\delta}(f) = \{f(\tau_n)\}_{n\in\mathbb{Z}}$ from the formula*

$$f(x) = \sum_{n\in\mathbb{Z}} f(\tau_n)\psi_n^{\tau_\delta}(x) \tag{15}$$

in the L_2-sense and uniformly on compact intervals of $\mathbb{R}$, where

$$\psi_n^{\tau_\delta}(x) = \sum_{k\in\mathbb{Z}} \left(\Phi_{\tau_\delta}\right)^{-1}_{k,n} \phi(x-k) \tag{16}$$

and $\Phi^{-1}_{\tau_\delta}$ is the inverse of the operator Φ_{τ_δ}. Furthermore, the set $\{\psi_n^{\tau_\delta}\}_{n\in\mathbb{Z}}$ is a Riesz basis for the space V_ϕ.

Proof The detailed proof is presented in [7]. Here we give a brief sketch. The crucial think is to prove that the operator $\Phi_{\tau_\delta} \in \mathscr{F}_{\delta_\phi}$ satisfies (11). We have

$$\|\Phi_{\tau_\delta} c\|_{\ell_2}\|c\|_{\ell_2} \geq \left|\langle \Phi_{\tau_\delta} c, c\rangle_{\ell_2}\right| \geq \left|\left|\langle \Phi c, c\rangle_{\ell_2}\right| - \left|\langle (\Phi_{\tau_\delta} - \Phi)c, c\rangle_{\ell_2}\right|\right|. \tag{17}$$

By using the Cauchy-Schwarz inequality and the definition (13), we obtain

$$\left|\langle (\Phi_{\tau_\delta} - \Phi)c, c\rangle_{\ell_2}\right| \leq G_\phi(\delta)\|c\|^2_{\ell_2}.$$

Substituting this bound and the lower bound of (9) into (17), we obtain

$$\|\Phi_{\tau_\delta} c\|_{\ell_2} \geq C\|c\|_{\ell_2}, \tag{18}$$

with $C = \|\Phi^\dagger\|_0 - G_\phi(\delta) > 0$, since Φ_{τ_δ} belongs in the class $\mathscr{F}_{\delta_\phi}$. Obviously, (18) holds for the adjoint operator $\Phi^*_{\tau_\delta}$ as well. On the other hand, taking into account the upper bound of (9), we can show that

$$\begin{aligned}\|\Phi_{\tau_\delta} c\|_{\ell_2} &\leq \|\Phi c\|_{\ell_2} + \|(\Phi_{\tau_\delta} - \Phi)c\|_{\ell_2} \leq \left(\|\Phi^\dagger\|_\infty + G_\phi(\delta)\right)\|c\|_{\ell_2}\\ &\leq \left(\|\Phi^\dagger\|_\infty + D_{\phi,\delta_\phi}\right)\|c\|_{\ell_2},\end{aligned}$$

where D_{ϕ,δ_ϕ} is a positive constant depending on the selection of ϕ and the number δ_ϕ. Summarizing, there exist two positive constants C, D such that

$$C\|c\|_{\ell_2} \leq \|\Phi_{\tau_\delta} c\|_{\ell_2} \leq D\|c\|_2 \text{ for all } c \in \ell_2$$

and the above lower inequality holds for $\Phi^*_{\tau_\delta}$ as well. Therefore, Φ_{τ_δ} is onto ℓ_2, and so there exists a unique sequence $c \in \ell_2$ such that $f(x) = \sum_{k\in\mathbb{Z}} c_k\phi(x-k)$ and consequently,

$$\mathscr{L}_{\tau_\delta}(f) = \Phi_{\tau_\delta} c,$$

where $\mathscr{L}_{\tau_\delta}(f) = \{f(\tau_n)\}_{n\in\mathbb{Z}}$. By following the same steps as in the Proof of Theorem 3, we obtain the result.

Below, we present some examples. Let ϕ be generator *and* sampling function of V_ϕ, i.e., $\phi(n) = \delta_{0,n}$, where $\delta_{0,n}$ is the Kronecker's delta symbol. Then $\Phi^\dagger(\xi) = \sum_{n\in\mathbb{Z}} \phi(n)e^{-2\pi i n\xi} = 1$ for all ξ in $\left[-\frac{1}{2}, \frac{1}{2}\right]$, and so by (14), we obtain

$$\delta_\phi = \inf\{x \in \mathbb{R}^+ : G_\phi(x) \geq 1\},$$

where G_ϕ is as in (13). Consider the B_2-spline $\phi(x) = 1 - |x|$ for $|x| \leq 1$ and $\phi(x) = 0$ elsewhere. Then, for any $0 \leq x \leq 1$, we observe that

$$G_\phi(x) = \sup_{|y|\leq x}(1 - |\phi(y)|) + \sup_{|y|\leq x}|\phi(y-1)| + \sup_{|y|\leq x}|\phi(y+1)| = 3x.$$

Therefore

$$\delta_\phi = \inf\{x \in [0,1] : 3x \geq 1\} = \frac{1}{3}.$$

Notice that if $\Delta = \{\delta_n\}_{n\in\mathbb{Z}}$ is a positive (or negative) sequence, then $\delta_\phi = \frac{1}{2}$, an estimate obtained in [16, 28] as well. Moreover this estimate is optimal in the sense that for $\delta_\phi = \frac{1}{2}$, the resulting sampling set is not stable [2]. Let us consider the function $\phi(x) = \left(\frac{\sin(\pi x)}{\pi x}\right)^4$, $x \in \mathbb{R}$. In this case for any $0 \leq x \leq 1/2$, we have

$$G_\phi(x) = 1 - \phi(x) + 2\sum_{n=1}^{\infty}\phi(n-x)$$

and from this equality, we obtain numerically a maximum perturbation

$$\delta_\phi = \inf\{x \in \mathbb{R}^+ : G_\phi(x) \geq 1\} \approx 0.455.$$

Consider functions of the form $\phi_c(x) = e^{-c|x|}$, $c > 0$. In this case we use the Poisson summation formula to obtain

$$\Phi^\dagger(\gamma) = 2c\sum_{n\in\mathbb{Z}}\frac{1}{c^2 + (\gamma+n)^2},\ \gamma \in \left[-\frac{1}{2}, \frac{1}{2}\right].$$

Since $\|\Phi^\dagger\|_0 = \Phi^\dagger(\frac{1}{2})$, we have $\delta_\phi = \inf\left\{x \in \mathbb{R}^+ : G_\phi(x) \geq \Phi^\dagger(\frac{1}{2})\right\}$ where

$$G_\phi(x) = 1 - \phi(x) + 2\sum_{n=1}^{\infty}(\phi(n-x) - \phi(n))$$

for $0 \leq x \leq 1/2$. If $c = 1$, we find numerically that $\delta_\phi \approx 0.21$. We work similarly for functions of the form $\phi(x) = e^{-cx^2}$, $c > 0$.

4 Local Sampling and Approximation

Besides its theoretical importance, the sampling formula (15) is difficult to be implemented numerically, because we must know an infinite number of sampled data, and we need to compute the inverse operator $\Phi_{\tau_\delta}^{-1}$. It could be desirable to establish a numerically implemented sampling reconstruction formula approximating elements of V_ϕ on compact intervals of $\mathbb{R}$ and be able to control the corresponding truncation error. So far, we considered operators (infinite matrices) Φ_{τ_δ}, produced from a function ϕ with a certain decay rate (4). It turns out that all these operators belong in the *Gröchenig-Shur* class $\mathscr{A}_{p,u_\alpha}$ [20] which contains operators $A = \{a_{m,n}\}_{m,n\in\mathbb{Z}}$ with norm

$$\begin{aligned}\|A\|_{\mathscr{A}_{p,u_\alpha}} &= \sup_{n\in\mathbb{Z}}\|\{u_\alpha(m-n)a_{m,n}\}_{m\in\mathbb{Z}}\|_{\ell_p}\\ &\quad+\sup_{m\in\mathbb{Z}}\|\{u_\alpha(m-n)a_{m,n}\}_{n\in\mathbb{Z}}\|_{\ell_p}<\infty.\end{aligned}$$

Furthermore, by assuming that $\Phi_{\tau_\delta} \in \mathscr{F}_{\delta_\phi}$ (recall Definition 2 in the previous section), then every operator Φ_{τ_δ} in this class belongs also in the space $\mathscr{B}^2$ of all bounded operators on ℓ_2 with usual norm $\|\cdot\|_{\mathscr{B}^2}$, and it has a bounded inverse $\Phi_{\tau_\delta}^{-1} \in \mathscr{B}^2$. Therefore Wiener's lemma for infinite matrices can be applied on elements of the class $\mathscr{F}_{\delta_\phi}$. Here, we mention that Wiener's lemma is a classical result in Fourier Analysis, stating that the inverse $\frac{1}{f}$ of a non-vanishing absolutely convergent Fourier series of a function f possesses again an absolutely convergent Fourier series. Therefore, instead of a direct verification that $\frac{1}{f}$ has an absolutely convergent Fourier series which requires checking whether the Fourier coefficients of $\frac{1}{f}$ are absolutely summable, Wiener's lemma forms a much easier test, by checking only that f has no zeros. Hence, a difficult condition related with invertibility can be replaced by an easier and more convenient condition. But Wiener's lemma is more than that. It is a much more deep result about the invertibility and spectrum of certain operators. From a more abstract point of view, Wiener's lemma is about invertibility in a Banach algebra. Naimark understood that the original Wiener's lemma is a result about two Banach algebras, the algebra $C(\mathbb{T})$ of continuous functions on $\mathbb{T} = [0, 1]$ and its subalgebra $\mathscr{A}(\mathbb{T})$ of functions with absolutely convergent Fourier series. In this spirit, we can give the following:

Definition 3 Let $\mathscr{A} \subseteq \mathscr{B}$ be two Banach algebras with a common identity. Then $\mathscr{A}$ is called inverse-closed in $\mathscr{B}$, if

$$a \in \mathscr{A} \text{ and } a^{-1} \in \mathscr{B} \to a^{-1} \in \mathscr{A}.$$

The inverse-closedness is often extremely useful for the study of invertibility, because the large algebra contains more invertible elements, and so our potentialities to show invertibility are broader. Since in this section we consider operators in the *Grochenig-Shur* class $\mathscr{A}_{p,u_\alpha}$, let us see how a variation of Wiener's lemma may help us understand the spectrum of these (convolution-type) operators.

Proposition 1 ([34, Theorem 4.1] (Wiener's lemma for the Grochenig-Shur class)) *The class $\mathscr{A}_{p,u_\alpha}$ is inverse-closed in the algebra $\mathscr{B}^2$ of all bounded operators from ℓ_2 to ℓ_2.*

Now we have:

Proposition 2 *Let $\mathscr{F}_{\delta_\phi} \subset \mathscr{B}^2$ be the class of operators in Definition 2. Then $\mathscr{F}_{\delta_\phi} \subset \mathscr{A}_{p,u_\alpha}$, and so every element $\Phi_{\tau_\delta} \in \mathscr{F}_{\delta_\phi}$ satisfies $\Phi_{\tau_\delta}^{-1} \in \mathscr{A}_{p,u_\alpha}$.*

Proof Let Φ_{τ_δ} be defined in (12) for some sampling set $\tau_\delta = \{\tau_n = n + \delta_n : |\delta_n| \le \delta\}_{n\in\mathbb{Z}}$. Fix an integer i. Then for any $j \in \mathbb{Z}$, we have

$$1+\delta+|\tau_i-j| \ge \begin{cases} 1+\delta+|i-j|-|\delta_i| \ge 1+|i-j|, & |\delta_i| \le |i-j| \\ 1+\delta+|\delta_i|-|i-j| \ge 1+|\delta_i| \ge 1+|i-j|, & |\delta_i| \ge |i-j| \end{cases}.$$

Therefore, if u_α is the polynomial weight related to the decay of ϕ, then for any $1 \le p < +\infty$, we have

$$\left\|\left\{u_\alpha(i-j)\left(\Phi_{\tau_\delta}\right)_{i,j}\right\}_{j\in\mathbb{Z}}\right\|_{\ell_p}^p = \sum_{j\in\mathbb{Z}} \left((1+|i-j|)^\alpha |\phi(\tau_i-j)|\right)^p$$

$$\le \sum_{j\in\mathbb{Z}} \left((1+\delta+|\tau_i-j|)^\alpha |\phi(\tau_i-j)|\right)^p = (1+\delta)^{\alpha p} \sum_{j\in\mathbb{Z}} \left(1+\frac{|\tau_i-j|}{1+\delta}\right)^{\alpha p} |\phi(\tau_i-j)|^p$$

$$< (1+\delta)^{\alpha p} \sum_{j\in\mathbb{Z}} \left((1+|\tau_i-j|)^\alpha |\phi(\tau_i-j)|\right)^p \le (1+\delta)^{\alpha p} \|\phi\|_{W_p(L_{\infty,u_\alpha})}^p < \infty$$

and for $p = +\infty$ we obtain a similar estimate. The same bound holds if we interchange the position of i's and j's in the above estimations. Therefore $\Phi_{\tau_\delta} \in \mathscr{A}_{p,u_\alpha}$. The rest follow from Proposition 1. Hence, there exists a constant C' (depending on the norms $\|\Phi_{\tau_\delta}\|_{\mathscr{A}_{1,u_\alpha}}$ and $\|\Phi_{\tau_\delta}^{-1}\|_{\mathscr{B}^2}$ and on some other constants which are affected only from the weight u_α), such that

$$\|\Phi_{\tau_\delta}^{-1}\|_{\mathscr{A}_{p,u_\alpha}} \le C' < \infty. \tag{19}$$

For more details about Wiener's lemma for infinite matrices, we refer to [20, 26, 34, 35], and for an excellent overview on Wiener's lemma and its variations, we refer to [22] and references therein. The bound (19) enables us to deduce that the Riesz basis $\{\psi_n^{\tau_\delta}\}_{n\in\mathbb{Z}}$ associated to the reconstruction formula (15) inherits the decay rate form ϕ. In fact we can show the following:

Proposition 3 *Let $\Phi_{\tau_\delta} \in \mathscr{F}_{\delta_\phi}$. Then the Riesz basis $(\psi_n^{\tau_\delta})_{n\in\mathbb{Z}}$ of V_ϕ related to the reconstruction formula (15) satisfies*

$$\|(\psi_n^{\tau_\delta})\|_{p,\infty,u_\alpha} = \left\|\left\|(u_\alpha(\cdot-\tau_n)\psi_n^{\tau_\delta}(\cdot))_{n\in\mathbb{Z}}\right\|_{\ell_p}\right\|_{L_\infty} < +\infty.$$

Proof Let $(\psi_n^{\tau_\delta})_{n\in\mathbb{Z}}$ be as in (16), $x \in \mathbb{R}$ and $Y_{n,x} = \left\{k \in \mathbb{Z} : |k - x| \le \frac{|n-x|}{2}\right\}$. Then for $1 \le p < +\infty$ and with the notation $\|\Phi_{\tau_\delta}^{-1}\|_\infty = \sup_{k,n\in\mathbb{Z}}|(\Phi_{\tau_\delta}^{-1})_{k,n}|$, we have

$$
\begin{aligned}
&\left\|\left(u_\alpha(x-\tau_n)\psi_n^{\tau_\delta}(x)\right)_n\right\|_{\ell_p}^p\\
&\le 2^{p-1}\Big(\left\| \left(\|\left(u_\alpha(x-\tau_n)(\Phi_{\tau_\delta}^{-1})_{\cdot,n}\,\phi(x-\cdot)\right)\|_{\ell_1(Y_{n,x})}\right)_n \right\|_{\ell_p}^p\\
&+\left\| \left(\|\left(u_\alpha(x-\tau_n)(\Phi_{\tau_\delta}^{-1})_{\cdot,n}\,\phi(x-\cdot)\right)\|_{\ell_1(\mathbb{R}-Y_{n,x})}\right)_n \right\|_{\ell_p}^p\Big)\\
&\le 2^{p-1}\Big(\left\| \left(\|\left(u_\alpha(x-\tau_n)(\Phi_{\tau_\delta}^{-1})_{\cdot,n}\,\phi(x-\cdot)\right)\|_{\ell_p(Y_{n,x})}^p\right)_n \right\|_{\ell_1}\\
&+\left\| \left(\|\left(u_\alpha(x-\tau_n)(\Phi_{\tau_\delta}^{-1})_{\cdot,n}\,\phi(x-\cdot)\right)\|_{\ell_p(\mathbb{R}-Y_{n,x})}^p\right)_n \right\|_{\ell_1}\Big)\\
&\le 2^{p-1}\|\phi\|_{L_\infty}^{p-1}\left\| \left(\|(u_\alpha(x-\tau_n)(\Phi_{\tau_\delta}^{-1})_{\cdot,n}\,\phi^{\frac{1}{p}}(x-\cdot))\|_{\ell_p(Y_{n,x})}^p\right)_n \right\|_{\ell_1}\\
&+2^{p-1}\|\Phi_{\tau_\delta}^{-1}\|_\infty^{p-1}\left\| \left(\|(u_\alpha(x-\tau_n)(\Phi_{\tau_\delta}^{-1})_{\cdot,n}^{\frac{1}{p}}\,\phi(x-\cdot)\}\|_{\ell_p(\mathbb{R}-Y_{n,x})}^p\right)_n \right\|_{\ell_1}\Big)
\end{aligned}
$$

$$
\begin{aligned}
&\le 2^{p-1}(1+\delta)^{\alpha p}\|\phi\|_{L_\infty}^{p-1}\left\| \left(\|(u_\alpha(x-n)(\Phi_{\tau_\delta}^{-1})_{\cdot,n}\,\phi^{\frac{1}{p}}(x-\cdot))\|_{\ell_p(Y_{n,x})}^p\right)_n \right\|_{\ell_1(\mathbb{Z})}\\
&+2^{p-1}(1+\delta)^{\alpha p}\|\Phi_{\tau_\delta}^{-1}\|_\infty^{p-1}\left\| \left(\|(u_\alpha(x-n)(\Phi_{\tau_\delta}^{-1})_{\cdot,n}^{\frac{1}{p}}\,\phi(x-\cdot))\|_{\ell_p(\mathbb{R}-Y_{n,x})}^p\right)_n \right\|_{\ell_1}\Big),
\end{aligned}
\tag{20}
$$

because $\phi \in L_\infty(\mathbb{R})$ (see (7)), $\Phi_{\tau_\delta}^{-1} \in \mathscr{A}_{p,u_\alpha} \subset \mathscr{A}_{1,u_0}$, and finally

$$
\begin{aligned}
u_\alpha(x-\tau_n) &= (1+|(x-n)+(n-\tau_n)|)^\alpha \le (1+\delta+|x-n|)^\alpha\\
&= (1+\delta)^\alpha\left(1+\frac{|x-n|}{1+\delta}\right)^\alpha \le (1+\delta)^\alpha u_\alpha(x-n).
\end{aligned}
$$

For the first term in the right-hand side of (20), i.e., for any $k \in Y_{n,x}$, we have

$$
u_\alpha(x-n) \le 2^\alpha\left(1+\frac{|x-n|}{2}\right)^\alpha \le 2^\alpha\Big(1+|n-x|-|k-x|\Big)^\alpha \le 2^\alpha u_\alpha(k-n),
$$

and by using this inequality, we immediately deduce that the first term of (20) is bounded by $2^{p-1}2^{\alpha p}(1+\delta)^{\alpha p}\|\phi\|_{L_\infty}^{p-1}\|\phi\|_{W_1(L_\infty,u_0)}\|\Phi_{\tau_\delta}^{-1}\|_{\mathscr{A}_{p,u_\alpha}}^p$. For the second term in the right-hand side of (20), we observe that for $k \notin Y_{n,x}$, we have

$$
u_\alpha(x-n) \le 2^\alpha\left(1+\frac{|x-n|}{2}\right)^\alpha \le 2^\alpha u_\alpha(x-k)
$$

and so we obtain a bound of the form

$$2^{p-1}2^{\alpha p}(1+\delta)^{\alpha p}\|\Phi_{\tau_\delta}^{-1}\|_\infty^{p-1}\|\Phi_{\tau_\delta}^{-1}\|_{\mathscr{A}_{1,u_0}}\|\phi\|^p_{W_\infty(L_{p,u_\alpha})}.$$

If $p=+\infty$ we easily obtain

$$\|(\psi_n^{\tau_\delta})\|_{\infty,\infty,u_\alpha} \leq 2^\alpha(1+\delta)^\alpha\Big(\|\phi\|_{W_\infty(L_{1,u_0})}\|\Phi_{\tau_\delta}^{-1}\|_{\mathscr{A}_{\infty,u_\alpha}} + \|\Phi_{\tau_\delta}^{-1}\|_{\mathscr{A}_{1,u_0}}\|\phi\|_{W_\infty(L_{\infty,u_\alpha})}\Big)$$

and the proof is complete.

In order to produce a numerically implementable reconstruction formula for the space V_ϕ approximating the sampling formula (15), we need to obtain a reasonable approximation of the inverse operator $\Phi_{\tau_\delta}^{-1}$ appearing in the representation of the Riesz basis functions $\psi_n^{\tau_\delta}$. To do that we employ the finite section method, [19, 23]. This method involves approximating the operator Φ_{τ_δ} with a square "section" of the infinite matrix Φ_{τ_δ} (and $\Phi_{\tau_\delta}^{-1}$ with the inverse of this square section, if it exists) and examine if the resulting new formula provides an approximation of the original sampling formula in some sense. Let us briefly describe the method. Assume that $A : U \to V$ is a linear invertible operator between two normed infinite dimensional spaces of sequences U and V and we want to solve the equation $Af = g$. For any natural number N, let

$$P_N f = (\ldots, -f_N, -f_{N-1}, \ldots, f_{N-1}, f_N, \ldots)$$

be the orthogonal projection of an element $f \in U$ (or V) onto a $2N+1$-dimensional subspace of U (or V),

$$A_N = P_N A P_N, \quad \text{and } g_N = P_N g.$$

Then, we try to solve the "finite" system $A_N x = g_N$ (for $n = -N, \ldots, N$) and examine the relation of this "approximate" solution (if it exists) with the actual solution $Ax = b$. For a detailed convergence analysis of the finite section method, we refer to [21]. We now consider a finite set X containing successive integers, and for any positive integer R, we define the R-neighborhood of X by

$$X_R = \{\min X - R, \ldots, \max X + R\}.$$

Let

$$P_{X_R} : \ell_2(\mathbb{Z}) \to \mathscr{H}_{X_R} : P_{X_R}c = \begin{cases} c_n, & n \in X_R \\ 0, & \text{elsewhere} \end{cases}$$

be the projection of a sequence $c \in \ell_2$ onto a finite dimensional subspace $\mathscr{H}_{X_R}$ and let

$$\Phi_{\tau_\delta, Y} = \begin{cases} \left(\Phi_{\tau_\delta}\right)_{m,n}, & m, n \in Y \\ 0, & \text{elsewhere} \end{cases} \tag{21}$$

be the finite section of a matrix $\Phi_{\tau_\delta} \in \mathscr{F}_{\delta_\phi}$ on $Y \times Y \subset \mathbb{Z}^2$. Then if Φ_{τ_δ, X_R} is the finite section of a matrix $\Phi_{\tau_\delta} \in \mathscr{F}_{\delta_\phi}$ as in (21), then by exploiting (18) for $P_{X_R}c$ and by recalling the boundness of Φ_{τ_δ}, we easily obtain

$$C\|P_{X_R}c\|_{\ell_2} \leq \left\|\left(\Phi_{\tau_\delta, X_R}\right)c\right\|_{\ell_2} \leq D\|P_{X_R}c\|_{\ell_2} \quad \text{for all } c \in \ell_2 \tag{22}$$

for some positive constants C, D. In addition, for the inverse matrix $\Phi^{-1}_{\tau_\delta, X_R}$, we may prove the following:

Proposition 4 *The inverse matrix $\Phi^{-1}_{\tau_\delta, X_R}$ belongs in $\mathscr{A}_{p,u_\alpha}$, and there exists a positive constant C_0 independent of the selection of the set X and the positive integer R such that*

$$sup_{m,n \in X_R}\left\{\left\|\left\{u_\alpha(m-n)\left(\Phi_{\tau_\delta}\right)^{-1}_{m,n}\right\}_m\right\|_{\ell_p(X_R)}, \left\|\left\{u_\alpha(m-n)\left(\Phi_{\tau_\delta}\right)^{-1}_{m,n}\right\}_n\right\|_{\ell_p(X_R)}\right\} \leq C_0.$$

Proof For any finite subset Y of $\mathbb{Z}$, we obviously have

$$\left\|\Phi_{\tau_\delta, Y}\right\|_{\mathscr{A}_{p,u_\alpha}} \leq \left\|\Phi_{\tau_\delta}\right\|_{\mathscr{A}_{p,u_\alpha}} < \infty. \tag{23}$$

Consider now a partition $Y_{X_R, \lambda} = \{s + \lambda|X_R| : s \in X_R\}_{\lambda \in \mathbb{Z}}$ of $\mathbb{Z}$ and determine the infinite block-diagonal operator

$$\Phi^{\dagger}_{X_R} = \sum_{\lambda \in \mathbb{Z}} \Phi_{\tau_\delta, Y_{X_R, \lambda}}.$$

Then, for any $c \in \ell_2$, we have

$$\|\Phi^{\dagger}_{X_R} c\|^2_{\ell_2} = \sum_{l \in \mathbb{Z}} \|\Phi_{\tau_\delta, Y_{X_R, l}} c\|^2_{\ell_2}$$

and so, by (22), we obtain

$$C\|c\|_2 \leq \|\Phi^{\dagger}_{X_R} c\|_2 \leq D\|c\|_2 \quad \text{for all } c \in \ell_2 \tag{24}$$

for some positive constants C, D as above. Hence $\Phi^{\dagger}_{X_R}$ is bounded on ℓ_2 and has bounded inverse determined by $\left(\Phi^{\dagger}_{X_R}\right)^{-1} = \sum_{\lambda \in \mathbb{Z}} \left(\Phi_{\tau_\delta, Y_{X_R, \lambda}}\right)^{-1}$. By applying Wiener's lemma for infinite matrices, we obtain

$$\|(\Phi^{\dagger}_{X_R})^{-1}\|_{\mathscr{A}_{p,u_\alpha}} \leq C_0$$

for some constant C_0 independent of the set X and the positive integer R (recall (19) and combine with (23) and (24)). Since $(\Phi^{\dagger}_{X_R})^{-1}_{i,j} = \left(\Phi_{\tau_\delta, X_R}\right)^{-1}_{i,j}$ for any $i, j \in X_R$, the result is proved.

Now we can state the main result of the chapter.

Theorem 5 ([7] (Local theorem for shift invariant spaces)) *Consider a shift invariant sampling space V_ϕ, whose sampling formula is determined by* (15) *with respect to an ordered and ε-separated perturbed sampling set $\tau_\delta = \{\tau_n\}_{n\in\mathbb{Z}}$. For any $f \in V_\phi$ and for any bounded interval $\mathscr{X}$, define by*

$$f^*(x) = \sum_{n\in X_R} f(\tau_n)\varphi_n^{\tau_\delta}(x)$$

the finite reconstruction approximation of f on $\mathscr{X}$, where the set X_R is the R-neighborhood of the set $X = \{n \in \mathbb{Z} : \tau_n \in \mathscr{X}\}$,

$$\varphi_n^{\tau_\delta}(x) = \sum_{m\in X_{3R}} \left(\Phi_{\tau_\delta, X_{3R}}\right)^{-1}_{m,n} \phi(x-m)$$

and $\Phi^{-1}_{\tau_\delta, X_{3R}}$ is the inverse of a square matrix $\Phi_{\tau_\delta, X_{3R}}$ as in (21)*. Then there exists a positive constant C independent of the bounded interval $\mathscr{X}$, the set X, the positive integer R, and the function f such that the error when we reconstruct f on $\mathscr{X}$ using the finite reconstruction approximation f^* is bounded by*

$$sup_{x\in\mathscr{X}}|f(x) - f^*(x)| < C\Big(\frac{\|\{f(\tau_n)\}\|_{\ell_2(X_R)}}{R^{2\alpha-\frac{3}{2q}}} + \frac{\|\{f(\tau_n)\}\|_{\ell_2(\mathbb{Z}-X_R)}}{R^{\alpha-\frac{1}{2q}}}\Big).$$

Here, the number $\alpha > 1 - \frac{1}{p}$, $(p \geq 1)$ is the exponent of the polynomial weight $u_\alpha(x) = (1 + |x|)^\alpha$ related to the decay rate of ϕ, and q is the conjugate exponent of p.

Proof A more detailed proof is demonstrated in [7]. For any $f \in V_\phi$, and for any $x \in \mathscr{X}$, where $\mathscr{X}$ is a bounded interval of $\mathbb{R}$, we have

$$f(x) - f^*(x) = \sum_{n\in X_R} f(\tau_n)(\psi_n^{\tau_\delta}(x) - \varphi_n^{\tau_\delta}(x)) + \sum_{n\notin X_R} f(\tau_n)\psi_n^{\tau_\delta}(x), \tag{25}$$

where X_R is the R-neighborhood of the set $X = \{n \in \mathbb{Z} : \tau_n \in \mathscr{X}\}$. Taking into account the definition of τ and for the above values of x, we can show that

$$|x - \tau_n| = \begin{cases} x - \tau_n \geq \varepsilon(\min X - 1 - n), & \min X > n \\ \tau_n - x \geq \varepsilon(n - \max X - 1), & \max X < n \end{cases}.$$

To bound the second term of the right-hand side of (25), we apply the Cauchy-Schwarz inequality, and we use the above estimate on $|x - \tau_n|$ and the bound $\|(\psi_n^{\tau_\delta})\|_{p,\infty,u_\alpha} < \infty$ obtained in Proposition 3 to obtain

$$\begin{aligned}
&\sup_{x\in\mathscr{X}} \sum_{n\notin X_R} |\psi_n^{\tau_\delta}(x)|^2 = \sup_{x\in\mathscr{X}} \sum_{n\notin X_R} |u_\alpha(x-\tau_n)\psi_n^{\tau_\delta}(x)|^2 u_\alpha^{-2}(x-\tau_n) \\
&\le \sup_{x\in\mathscr{X}} \left\| \left((u_\alpha(x-\tau_n)\psi_n^{\tau_\delta}(x))^2 \right)_n \right\|_{\ell_p(\mathbb{Z}-X_R)} \left\| \left(u_\alpha(\varepsilon\cdot)^{-2}\right) \right\|_{\ell_q(\mathbb{Z}-\{-R,\dots,R\})} \\
&\le \|(\psi_n^{\tau_\delta})\|_{p,\infty,u_\alpha}^2 \left\| \left(u_\alpha(\varepsilon n)^{-2}\right) \right\|_{\ell_q(\mathbb{Z}-\{-R,\dots,R\})} < C_1 \|(\psi_n^{\tau_\delta})\|_{p,\infty,u_\alpha}^2 R^{2\alpha-\frac{1}{q}},
\end{aligned} \tag{26}$$

for some positive constant C_1 depending on α, ε, q.
In order to compute an upper bound for the first term in (25), we need estimates for $sup_{x\in\mathscr{X}} \sum_{n\in X_R} |\psi_n^{\tau_\delta}(x) - \varphi_n^{\tau_\delta}(x)|^2$. By definition

$$\phi(x-l) = \sum_{n\in\mathbb{Z}} (\Phi_{\tau_\delta})_{n,l} \psi_n^{\tau_\delta}(x) = \sum_{n\in X_{3R}} \phi(\tau_n - l)\psi_n^{\tau_\delta}(x) + \sum_{n\notin X_{3R}} \phi(\tau_n - l)\psi_n^{\tau_\delta}(x).$$

Since by (22) the projection matrix $\Phi_{\tau_\delta, X_{3R}} = \{\phi(\tau_n - l) : n, l \in X_{3R}\}$ is invertible, we multiply both sides of the above equality with the inverse matrix $\Phi_{\tau_\delta, X_{3R}}^{-1}$, and we obtain

$$\begin{aligned}
&\sum_{k\in X_R} |\psi_k^{\tau_\delta}(x) - \varphi_k^{\tau_\delta}(x)|^2 = \sum_{k\in X_R} \left| \sum_{l\in X_{3R}} (\Phi_{\tau_\delta, X_{3R}})_{l,k}^{-1} \sum_{n\notin X_{3R}} (\Phi_{\tau_\delta})_{n,l} \psi_n^{\tau_\delta}(x) \right|^2 \\
&\le 2 \sum_{k\in X_R} \left| \sum_{l\in X_{2R}} (\Phi_{\tau_\delta, X_{3R}})_{l,k}^{-1} \sum_{n\notin X_{3R}} (\Phi_{\tau_\delta})_{n,l} \psi_n^{\tau_\delta}(x) \right|^2 \\
&+2 \sum_{k\in X_R} \left| \sum_{l\in X_{3R}-X_{2R}} (\Phi_{\tau_\delta, X_{3R}})_{l,k}^{-1} \sum_{n\notin X_{3R}} (\Phi_{\tau_\delta})_{n,l} \psi_n^{\tau_\delta}(x) \right|^2.
\end{aligned} \tag{27}$$

First we deal with the first term of the right-hand side of (27). Taking into account the decay estimates obtained in Proposition 4 and the above estimate (26), we compute

$$\begin{aligned}
&\sup_{x\in\mathscr{X}} \sum_{k\in X_R} \left| \sum_{l\in X_{2R}} (\Phi_{\tau_\delta, X_{3R}})_{l,k}^{-1} \sum_{n\notin X_{3R}} (\Phi_{\tau_\delta})_{n,l} \psi_n^{\tau_\delta}(x) \right|^2 \\
&\le \left\| \Phi_{\tau_\delta, X_{3R}}^{-1} \right\|_{\mathscr{A}_{1,u_0}}^2 \sup_{x\in\mathscr{X}} \sum_{l\in X_{2R}} \left| \sum_{n\notin X_{3R}} (\Phi_{\tau_\delta})_{n,l} \psi_n^{\tau_\delta}(x) \right|^2 \\
&\le \left\| \Phi_{\tau_\delta, X_{3R}}^{-1} \right\|_{\mathscr{A}_{1,u_0}}^2 \sup_{x\in\mathscr{X}} \sum_{l\in X_{2R}} \left(\sum_{n\notin X_{3R}} |(\Phi_{\tau_\delta})_{n,l}| \right) \sum_{n\notin X_{3R}} |\psi_n^{\tau_\delta}(x)|^2 |(\Phi_{\tau_\delta})_{n,l}|
\end{aligned}$$

$$\leq \|\Phi^{-1}_{\tau_\delta,X_{3R}}\|^2_{\mathscr{A}_{1,u_0}}\Big(\sup_{l\in X_{2R}}\sum_{n\notin X_{3R}}|(\Phi_{\tau_\delta})_{n,l}|\Big)\Big(\sup_{n\notin X_{3R}}\sum_{l\in X_{2R}}|(\Phi_{\tau_\delta})_{n,l}|\Big)$$

$$\Big(\sup_{x\in\mathscr{X}}\sum_{n\notin X_{3R}}|\psi_n^{\tau_\delta}(x)|^2\Big)$$

$$\leq \|\Phi^{-1}_{\tau_\delta,X_{3R}}\|^2_{\mathscr{A}_{1,u_0}}\|\Phi_{\tau_\delta}\|^2_{\mathscr{A}_{p,u_\alpha}}\Big(\sup_{l\in X_{2R}}\big\|\{u_\alpha(n-l)\}_n\big\|_{\ell_q(\mathbb{Z}-X_{3R})}\Big)$$

$$\Big(\sup_{n\notin X_{3R}}\big\|\{u_\alpha(n-l)\}_l\big\|_{\ell_q(\mathbb{Z}-X_{2R})}\Big)\Big(\sup_{x\in\mathscr{X}}\sum_{n\notin X_{3R}}|\psi_n^{\tau_\delta}(x)|^2\Big)$$

$$< \|\Phi^{-1}_{\tau_\delta,X_{3R}}\|^2_{\mathscr{A}_{1,u_0}}\|\Phi_{\tau_\delta}\|^2_{\mathscr{A}_{p,u_\alpha}}\frac{C_2}{R^{2\alpha-2/q}}\frac{C_1\|\{\psi_n^{\tau_\delta}\}\|^2_{W(\ell_p,L_\infty)}}{R^{2\alpha-1/q}}=\frac{C'}{R^{4\alpha-\frac{3}{q}}}, \tag{28}$$

where the constant C_1 is as in (26) and the constant C_2 depends only on α and q. Hence the overall constant C' depends neither on X nor on R, because the norm $\|\Phi^{-1}_{\tau_\delta,X_{3R}}\|_{\mathscr{A}_{1,u_0}}$ is independent of the set X and the positive integer R. We work similarly for the second term in the right-hand side of (27). In this case we obtain the bound

$$\sup_{x\in\mathscr{X}}\sum_{k\in X_R}\Big|\sum_{l\in X_{3R}-X_{2R}}(\Phi_{\tau_\delta,X_{3R}})^{-1}_{l,k}\sum_{n\notin X_{3R}}(\Phi_{\tau_\delta})_{n,l}\psi_n^{\tau_\delta}(x)\Big|^2$$

$$< \|\Phi^{-1}_{\tau_\delta,X_{3R}}\|^2_{\mathscr{A}_{p,u_\alpha}}\|\Phi_{\tau_\delta}\|^2_{\mathscr{A}_{1,u_0}}\frac{\|\{\psi_n^{\tau_\delta}\}\|^2_{W(\ell_p,L_\infty)}\,C_1\,C_3}{R^{4\alpha-\frac{3}{q}}}=\frac{C''}{R^{4\alpha-\frac{3}{q}}}, \tag{29}$$

where the constant C_1 is as in (26) and the constant C_3 depends on α and q. The overall constant C'' does not depend on the set X or the positive integer R for the same reasons as above. The bounds (28) and (29) are applied to (27). The resulting bound together with the bound (26) are applied to (25), and then the result is obtained.

In some cases the sampled data $f(n)$ are perturbed without our knowledge, i.e., the sequence $\Delta=\{\delta_n\}$ is unknown. Then *jitter error* appears. To be more precise, assume we are given a set of sampled data $\mathscr{L}(f)=\{f(\tau_n)\}_{n\in\mathbb{Z}}$ on a sampling set $\tau_\delta=\{n+\delta_n\}_{n\in\mathbb{Z}}$ (where the elements δ_n are unknown) of a function $f\in V_\phi$. We are interested in a bound on the L_∞-norm of the difference

$$\Big|f(x)-\sum_{n\in\mathbb{Z}}f(n+\delta_n)\psi(x-n)\Big|.$$

We have:

Proposition 5 *For any $f\in V_\phi$, we have*

$$\Big\|f-\sum_{n\in\mathbb{Z}}f(n+\delta_n)\psi(\cdot-n)\Big\|_{L_\infty}\leq\frac{\|f\|_{L_2}\|\phi\|_{W_2(L_{\infty,u_0})}}{A\|\Phi^\dagger\|_0}G_\phi(\delta),$$

where $\Phi^\dagger$ and G_ϕ are defined in (5) *and* (13), *respectively, and A is the lower Riesz bound of the set* $\{\phi(\cdot - n)\}_{n\in\mathbb{Z}}$.

Proof Let $F(x) = \{\phi(x-m)\}_{m\in\mathbb{Z}}$ and $\Phi = \{\Phi_{m,n} = \phi(m-n)\}_{n,m\in\mathbb{Z}}$ be the bounded and invertible operator of Sect. 2. If $f(\tau_n) = \sum_{m\in\mathbb{Z}} c_m\phi(\tau_n - m)$ for some unique $c \in \ell_2$ and if Φ_{τ_δ} is an infinite matrix as in (12), then

$$\Big\|f - \sum_{n\in\mathbb{Z}} f(\tau_n)\psi_n^\tau\Big\|_{L_\infty} \le \|f(\cdot+\delta_n) - f(\cdot)\|_{\ell_2}\ \Big\|\ \{\|\psi(\cdot - n)\}\|_{\ell_2}\ \Big\|_{L_\infty}$$

$$= \Big\|\big(\Phi_{\tau_\delta} - \Phi\big)c\Big\|_{\ell_2} \|\Phi^{-1}F(x)\|_{\ell_2} \le G_\phi(\delta)\|c\|_{\ell_2}\|\Phi^\dagger\|_0^{-1}\|\phi\|_{W_2(L_{\infty,u_0})}$$

$$\le \frac{G_\phi(\delta)}{A}\|f\|_{L_2}\|\Phi^\dagger\|_0^{-1}\|\phi\|_{W_2(L_{\infty,u_0})}.$$

References

1. A. Aldroubi, Non-uniform weighted average sampling and reconstruction in shift invariant and wavelet spaces. Appl. Comput. Harmon. Anal. **13**, 151–161 (2002)
2. A. Aldroubi, K. Gröchenig, Beurling-Landau-Type theorems for non-uniform sampling in shift invariant spline spaces. J. Fourier Anal. Appl. **6**, 93–103 (2000)
3. A. Aldroubi, K. Gröchenig, Non-uniform sampling and reconstruction in shift-invariant spaces. SIAM Rev. **43**, 585–620 (2001)
4. A. Aldroubi, Q. Sun, W.S. Tang, Convolution, average sampling and a Calderon resolution of the identity for shift invariant spaces. J. Fourier Anal. Appl. **11**, 215–244 (2005)
5. A. Aldroubi, C. Leonetti, Non uniform sampling and reconstruction from sampling sets with unknown jitter. Sampl. Theory Signal Image Process. **7**, 187–195 (2008)
6. N.D. Atreas, Perturbed sampling formulas and local reconstruction in shift-invariant spaces. J. Math. Anal. Appl. **377**, 841–852 (2011)
7. N. Atreas, On a class of non-uniform average sampling expansions and partial reconstruction in subspaces of $L_2(\mathbb{R})$. Adv. Comput. Math. **36**, 21–38 (2012)
8. N. Atreas, N. Bagis, C. Karanikas, The information loss error and the jitter error for regular sampling expansions. Sampl. Theory Signal Image Process. **1**, 261–276 (2002)
9. N. Atreas, J.J. Benedetto, C. Karanikas, Local sampling for regular wavelet and Gabor expansions. Sampl. Theory Signal Image Process. **2**, 1–24 (2003)
10. J.J. Benedetto, Irregular sampling and frames, in *Wavelets: A Tutorial in Theory and Applications*, ed. by C.K. Chui (Academic, Boston, 1992), pp. 445–507
11. J.J. Benedetto, P.J.S.G. Ferreira, *Modern Sampling Theory. Mathematics and Applications*, Applied and Numerical Harmonic Analysis Series (Birkhauser, Boston, 2001)
12. J.J. Benedetto, W. Heller, Irregular sampling and the theory of frames I. Note di Matematica **X**(Suppl. n. 1), 103–125 (1990)
13. P.L. Butzer, W. Splettstöβer, On quantization, truncation and jitter errors in the sampling theorem and its generalizations. Signal Process. **2**, 101–112 (1980)
14. P.L. Butzer, W. Splettstöβer, R.L. Stens, The sampling theorem and linear prediction in signal analysis. Jber d. Dt. Math.-Verein **90**, 1–70 (1988)
15. W. Chen, S. Itoh, J. Shiki, Irregular sampling theorems on wavelet subspaces. IEEE Trans. Inf. Theory, **44**, 1131–1142 (1998)
16. W. Chen, S. Itoh, J. Shiki, On sampling in shift invariant spaces. IEEE Trans. Inf. Theory, **48**, 2802–2810 (2002)

17. O. Christensen, *An Introduction to Frames and Riesz Bases* (Birkhaüser, Boston, 2003)
18. H.G. Feichtinger, K. Gröchenig, Theory and practice of irregular sampling, in *Wavelets: Mathematics and Applications*, ed. by J. Benedetto, M. Frazier (CRC Press, Boca Raton, 1994), pp. 305–365
19. I. Gohberg, I. Feldman, *Convolution Equations and Projection Methods for Their Solution* (American Mathematical Society, Providence, 1974). Translated from the Russian by F.M. Goldware, Translations of Mathematical Monographs, vol. 41
20. K. Gröchenig, M. Leinert, Symmetry and inverse closdness of matrix algebras and functional calculus for infinite matrices. Trans. Am. Math. Soc. **358**, 2695–2711 (2006)
21. K. Gröchenig, Z. Rzeszotnik, T. Strohmer, Convergence analysis of the finite section method and banach algebras of matrices. Integr. Equ. Oper. Theory **67**, 183–202 (2010)
22. K. Gröchenig, Wiener's lemma: theme and variations. An introduction to spectral invariance and its applications, in *Four Short Courses on Harmonic Analysis*, ed. by B. Forster, P. Massopust. Applied and Numerical Harmonic Analysis (ANHA book series) (Birkhäuser, Boston, 2010), pp. 175–234
23. R. Hagen, S. Roch, B. Silbermann, *C^*-Algebras and Numerical Analysis*, Monographs and Textbooks in Pure and Applied Mathematics, vol. 236 (Marcel Dekker Inc., New York, 2001)
24. H.C. Hak, C.E. Shin, Perturbation of non harmonic Fourier series and non-uniform sampling theorem. Bull. Korean Math. Soc. **44**, 351–358 (2007)
25. J.R. Higgins, *Sampling Theory in Fourier and Signal Analysis: Foundations* (Oxford University Press, Oxford, 1996)
26. S. Jaffard, *Proprietes des matrices bien localisees pres de leur diagonale et quelques applications*. Ann. Inst. Henri Poincare **7**, 461–476 (1990)
27. C. Leonetti, *Reconstruction from Error-Affecting Sampled Data in Shift Invariant Spaces*, Ph.D. Thesis, Department of Mathematics, Vanderbilt University, Nashville (2007)
28. Y. Liu, G. Walter, Irregular sampling in wavelet spaces. J. Fourier Anal. Appl. **2**, 181–189 (1995)
29. F. Marvasti, *Non-uniform Sampling: Theory and Practice* (Klewer Academic Publisher, New York, 2001)
30. M.Z. Nashed, Q. Sun, W.S. Tang, Average sampling in L^2. C. Acad. Sci. Paris, Ser I **347**, 1007–1010 (2009)
31. M.Z. Nashed, G.G. Walter, General sampling theorems for functions in reproducing kernel Hilbert Spaces. Math. Control Signals Syst. **4**, 363–390 (1991)
32. A. Olevskii, A. Ulanovskii, Almost integer translates. Do nice generator exist? J. Fourier Anal. Appl. **10**, 93–104 (2004)
33. Q. Sun, Non-uniform average sampling and reconstruction of signals with finite rate of innovation. SIAM J. Math. Anal. **38**, 1389–1422 (2006)
34. Q. Sun, Wiener's lemma for infinite matrices. Trans. Am. Math. Soc. **359**, 3099–3123 (2007)
35. Q. Sun, Wiener's lemma for infinite matrices with polynomial off diagonal decay. C. R. Acad. Sci. Paris ser.1, **340**, 567–570 (2005)
36. W. Sun, X. Zhou, Average sampling in shift invariant subspaces with symmetric average functions. J. Math. Anal. Appl. **287**, 279–295 (2003)
37. V.M. van der Mee Cornelis, M.Z. Nashed, S. Seatzu, Sampling expansions and interpolation in unitarily translation invariant reproducing kernel Hilbert spaces. Adv. Comput. Math. **19**, 355–372 (2003)
38. G.G. Walter, *Wavelets and Other Orthogonal Systems with Applications* (CRC Press, Boca Raton, 1994)
39. P. Zhao, C. Zhao, P.G. Casazza, *Perturbation of regular sampling in shift-invariant spaces for frames*. IEEE Trans. Inf. Theory. **52**, 4643–4648 (2006)
40. A. Zayed, *Advances in Shannon's Sampling Theory* (Boca Raton, CRC Press, 1993)

Optimal ℓ^1 Rank One Matrix Decomposition

Radu Balan, Kasso A. Okoudjou, Michael Rawson, Yang Wang, and Rui Zhang

Abstract In this paper, we consider the decomposition of positive semidefinite matrices as a sum of rank one matrices. We introduce and investigate the properties of various measures of optimality of such decompositions. For some classes of positive semidefinite matrices, we give explicitly these optimal decompositions. These classes include diagonally dominant matrices and certain of their generalizations, 2×2, and a class of 3×3 matrices.

2010 Mathematics Subject Classification Primary 45P05, 47B10; Secondary 42C15.

1 Introduction

The finite-dimensional matrix factorization problem that we shall investigate was partially motivated by a related infinite-dimensional problem, which we briefly recall.

Suppose that $\mathbb{H}$ is an infinite-dimensional separable Hilbert space, with norm $\|\cdot\|$ and inner product $\langle\cdot,\cdot\rangle$. Let $\mathcal{I}_1 \subset \mathcal{B}(\mathbb{H})$ be the subspace of trace-class operators. For a detailed study on trace-class operators, see [5, 9]. Consider an orthonormal basis $\{w_n\}_{n\geq 1}$ for $\mathbb{H}$, and let

$$\mathbb{H}^1 = \Big\{ f \in \mathbb{H} : |||f||| := \sum_{n=1}^{\infty} |\langle f, w_n\rangle| < \infty \Big\}.$$

R. Balan (✉) · K. A. Okoudjou · M. Rawson
Department of Mathematics, University of Maryland, College Park, MD, USA
e-mail: rvbalan@umd.edu; okoudjou@umd.edu; rawson@umd.edu

Y. Wang · R. Zhang
Hong Kong University of Science and Technology, Hong Kong, China
e-mail: yangwang@ust.hk; zhangrui112358@yeah.net

M. Th. Rassias (ed.), *Harmonic Analysis and Applications*, Springer Optimization and Its Applications 168, https://doi.org/10.1007/978-3-030-61887-2_2

For a sequence $c = (c_{mn})_{m,n=1}^{\infty} \in \ell^1$, we consider the operator $T_c : \mathbb{H} \to \mathbb{H}$ given by

$$T_c f = \sum_{m=1}^{\infty} \sum_{n=1}^{\infty} c_{mn} \langle f, w_n \rangle w_m .$$

We say that T_c is of *Type A* with respect to the orthonormal basis $\{w_n\}_{n\geq 1}$ if, for an orthogonal set of eigenvectors $\{g_n\}_{n\geq 1}$ of T_c such that $T_c = \sum_{n=1}^{\infty} g_n \otimes \overline{g_n}$, with convergence in the strong operator topology, we have that

$$\sum_{n=1}^{\infty} |||g_n|||^2 < \infty.$$

Similarly, we say that the operator T_c is of *Type B* with respect to the orthonormal basis $\{w_n\}_{n\geq 1}$ if there is some sequence of vectors $\{v_n\}_{n\geq 1}$ in $\mathbb{H}$ such that $T_c = \sum_{n=1}^{\infty} v_n \otimes \overline{v_n}$ with convergence in the strong operator topology and we have that

$$\sum_{n=1}^{\infty} |||v_n|||^2 < \infty.$$

It is easy to see that if T_c is of Type A, then it is of Type B. However, there exist finite rank positive trace-class operators which are neither of Type A nor of Type B. We refer to [7] for more details. In [1], we proved that there exist positive trace-class operators T_c of Type B which are not of Type A. Furthermore, this answers negatively a problem posed by Feichtinger [6].

Our main interest is in a finite-dimensional version of the above problem. Before stating it, we set the notations that will be used through this chapter.

For $n \geq 2$, we denote the set of all complex Hermitian $n \times n$ matrices as $S^n := S^n(\mathbb{C})$, positive semidefinite matrices as $S^n_+ := S^n_+(\mathbb{C})$, and positive definite matrices $S^n_{++} := S^n_{++}(\mathbb{C})$. It is clear that S^n_+ is a closed convex cone. Note that $S^n = S^n_+ - S^n_+$ is the (real) vector space of Hermitian matrices. We will also use the notation $U(n)$ for the set of $n \times n$ unitary matrices.

For $A \in S^n$, we let $\|A\|_{1,1} = \sum_{k,\ell=1}^{n} |A_{k,\ell}|$, and we let $\|A\|_{\mathcal{I}_1} = \sum_{k=1}^{n} |\lambda_k|$ where $\lambda_1 \leq \lambda_2 \leq \ldots \leq \lambda_n$ are the eigenvalues of A. We recall that the operator norm of $A \in S^n$ is given by $\|A\|_{\text{op}} = \max\{|\lambda_k| :, \lambda_1 \leq \lambda_2 \leq \ldots \leq \lambda_n\}$ where $\{\lambda_k\}_{k=1}^{n}$ is the set of eigenvalues of A. In addition, the Frobenius norm of A is given by $\|A\|_{\text{Fr}} = \sqrt{\text{tr} AA^*} = \sqrt{\sum_{k=1}^{n} \sum_{\ell=1}^{n} |A_{k\ell}|^2}$. One important fact that will be used implicitly throughout the paper is that all the norms defined on S^n are equivalent and thus give rise to the same topological structure on S^n.

Similarly, for a vector $x = (x_k)_{k=1}^{n} \in \mathbb{C}^n$, and $p \in (0, \infty)$, we let $\|x\|_p^p = \sum_{k=1}^{n} |x_k|^p$ define the usual ℓ^p norm, $p \geq 1$, with the usual modification when

$p = \infty$ and $p = 0$. As pointed out above, all these norms are equivalent on $\mathbb{C}^n$ and give rise to the same topology.

The goal of this chapter is to investigate optimal decompositions of a matrix $A \in S_+^n(\mathbb{C})$ as a sum of rank one matrices. In Sect. 2, we introduce some measures of optimality of the kinds of decompositions we seek and investigate the relationship between these measures. However, before doing so, we give an exact statement of the problems we shall address and review some results about the convex cone $S_+^n(\mathbb{C})$. In Sect. 3, we restrict our attention to some classes of matrices in $S_+^n(\mathbb{C})$, including diagonally dominant matrices. Finally, in Sect. 4, we report on some numerical experiments designed to find some of these optimal decompositions.

2 Preliminaries and Measures of Optimality

In the first part of this section, we collect some foundational facts on convex subsets of S^n. The second part will be devoted to introducing some quantities that will serve as measures of optimality of the decomposition results we seek.

2.1 Preliminaries

We denote the convex hull of a set S by co$S$. For the compact set $X = \{xx^* : x \in \mathbb{C}^n \text{ and } \|x\|_1 = 1\}$, we let $\Gamma = \text{co}X$ and $\Omega = \text{co}\,(X \cup \{0\})$. Observe that $\Omega \subset S_+^n(\mathbb{C})$. In fact, the following result holds.

Definition 2.1 An extreme point is a point such that it is not a convex combination of other points.

Lemma 2.2 *Ω is closed and compact convex subset of $S_+^n(\mathbb{C})$ with int $\Omega \neq \emptyset$. Furthermore, the set of extreme points of Ω is $X \cup \{0\}$.*

The proof is based on one of the versions of the Minkowski-Carathéodory Theorem, which, for completeness, we recall. We refer to [3, 4, 8] for more details and background.

Theorem 2.3 ([4, Proposition 3.1][8, Lemma 4.1] (Minkowski-Carathéodory Theorem)) *Let A be a compact convex subset of a normed vector space X of finite dimension n. Then any point in A is a convex combination of at most $n+1$ extreme points. Furthermore, we can fix one of these extreme points resulting in expressing any point in A is a convex combination of at most n extreme points in addition to the one we fixed.*

***Proof of Lemma* 2.2** Ω can be written as

$$\begin{aligned}\Omega &= \left\{\sum_{k=1}^{m} w_k x_k x_k^* : m \geq 1, \text{ an integer}, w_1, .., w_m \geq 0, \sum_{k=1}^{m} w_k \leq 1, \|x_k\|_1 = 1, 1 \leq k \leq m\right\} \\ &= \bigcup_{m\geq 1} \left\{\sum_{k=1}^{m} w_k x_k x_k^* : w_1, .., w_m \geq 0, \sum_{k=1}^{m} w_k \leq 1, \|x_k\|_1 = 1, 1 \leq k \leq m\right\} \\ &= \bigcup_{m\geq 1} \Omega_m,\end{aligned}$$

where $\Omega_m = \left\{ \sum_{k=1}^{m} w_k x_k x_k^* : w_1, .., w_m \geq 0, \sum_{k=1}^{m} w_k \leq 1, \|x_k\|_1 = 1, 1 \leq k \leq m\right\}$. Notice that $\Omega_1 \subset \Omega_2 \subset .. \subset \Omega_m \subset .. \subset \Omega$. By Minkowski-Carathéodory Theorem if $T \in \Omega$, then $T \in \Omega_{\dim S^n(\mathbb{C})+1}$. Therefore

$$\begin{aligned}\Omega &= \bigcup_{m\geq 1} \Omega_m = \Omega_1 \cup \ldots \cup \Omega_{n^2+1} = \Omega_{n^2+1} \\ &= \left\{\sum_{k=1}^{n^2+1} t_k x_k x_k^* : \sum_{k=1}^{n^2+1} t_k = 1, t_k \geq 0, \|x_k\|_1 = 1, \forall k, 1 \leq k \leq n^2 + 1\right\}\end{aligned}$$

We recall that the dimension of $S^n(\mathbb{C})$ as a real vector space over is n^2. As such, and since X is compact, we conclude that Ω as a convex hull of a compact set is compact.

To show that int $\Omega \neq \emptyset$, take $\frac{1}{2n^2} I \in \Omega$. We prove that for $0 < r < \frac{1}{2n^2}$, we have the ball

$$B_r\left(\frac{1}{2n^2} I\right) = \left\{\frac{1}{2n^2} I + T : T = T^*; \|T\|_{op} < r\right\} \subset \Omega.$$

Let $T = \sum_{k=1}^{n} \lambda_k v_k v_k^*$, $\|v_k\|_2 = 1$, and $|\lambda_k| \leq \|T\|_{op} < r$. Now

$$\begin{aligned}\frac{1}{2n^2} I + T &= \frac{1}{2n^2} \sum_{k=1}^{n} v_k v_k^* + \sum_{k=1}^{n} \lambda_k v_k v_k^* \\ &= \sum_{k=1}^{n} \left(\frac{1}{2n^2} + \lambda_k\right) \|v_k\|_1^2 \cdot \left(\frac{v_k}{\|v_k\|_1}\right) \cdot \left(\frac{v_k}{\|v_k\|_1}\right)^*.\end{aligned}$$

Also

$$\|v_k\|_1 = \sum_{j=1}^{n} |v_{k,j}| \leq \left(\sum_{j=1}^{n} |v_{k,j}|^2\right)^{\frac{1}{2}} \cdot \left(\sum_{j=1}^{n} 1\right)^{\frac{1}{2}} = \sqrt{n}\|v_k\|_2 = \sqrt{n}.$$

Hence

$$\|\frac{1}{2n^2}I + T\|_{1,1} \leq \sum_{k=1}^{n} \left(\frac{1}{2n^2} + \lambda_k\right) \|v_k\|_1^2 \leq n\left(\frac{1}{2n^2} + r\right) n = \frac{1}{2} + rn^2 < 1$$

In addition, because $r < \frac{1}{2n^2}$, we conclude that

$$\langle(\frac{1}{2n^2}I + T)x, x\rangle \geq \|x\|^2(\frac{1}{2n^2} - r) \geq 0$$

for all $x \in \mathbb{C}^n$. Consequently, $\frac{1}{2n^2}I + T \geq 0$. We conclude that $B_r\left(\frac{1}{2n^2}I\right) \subset \Omega$ where we use the norm $\|A\|_{1,1}$ for convenience. □

By a similar argument, Γ is also compact convex subset of $S_+^n(\mathbb{C})$.

2.2 *Measures of Optimality*

We next introduce and study the properties of some quantities defined on S^n and which will serve as measures of optimality of the rank one decompositions of matrices in S_+^n.

Definition 2.4 For $A \in S_+^n$, let

$$\gamma_+(A) := \inf_{A=\sum_{n\geq 1} g_n g_n^*} \sum_{n\geq 1} \|g_n\|_1^2. \tag{1}$$

If $A \in S^n$, we let

$$\gamma(A) := \inf_{A=\sum_{n\geq 1} g_n h_n^*} \sum_{n\geq 1} \|g_n\|_1 \|h_n\|_1, \tag{2}$$

and

$$\gamma_0(A) := \inf_{\substack{A=B-C,\\ B,C\in S_+^n}} (\gamma_+(B) + \gamma_+(C)) = \inf_{A=\sum_{n\geq 1} g_n g_n^* - \sum_{k\geq 1} h_k h_k^*} \left(\sum_{n\geq 1} \|g_n\|_1^2 + \sum_{k\geq 1} \|h_k\|_1^2\right). \tag{3}$$

We collect some of the properties of these functionals.

Proposition 2.5 *The functionals given in Definition 2.4 are sub-additive. In particular, the following statements hold.*

(a) *Given $A, B \in S_+^n$, we have $\gamma_+(A + B) \leq \gamma_+(A) + \gamma_+(B)$*
(b) *Given $A, B \in S^n$, we have $\gamma(A + B) \leq \gamma(A) + \gamma(B)$*

(c) *Given $A, B \in S^n$, we have $\gamma_0(A+B) \le \gamma_0(A) + \gamma_0(B)$*

In addition, if $a \ge 0$, we have $\gamma_+(aA) = a\gamma_+(A)$ when $A \in S^n_+$, and

$$\begin{cases} \gamma(aA) &= |a|\gamma(A) \\ \gamma_0(aA) &= |a|\gamma_0(A) \end{cases}$$

for $A \in S^n$ and $a \in \mathbb{R}$.

Proof Let $\epsilon > 0$ and choose $\{g_k\}_{k\ge1} \subset \mathbb{C}^n$ and $\{h_k\}_{k\ge1} \subset \mathbb{C}^n$ such that

$$\begin{cases} \sum_{k\ge1} \|g_k\|_1^2 &\le \gamma_+(A) + \epsilon/2 \\ \sum_{k\ge1} \|h_k\|_1^2 &\le \gamma_+(B) + \epsilon/2 \end{cases}$$

with $A = \sum_{k\ge1} g_k g_k^*$ and $B = \sum_{k\ge1} h_k h_k^*$. It follows that

$$A + B = \sum_{k\ge1} g_k g_k^* + \sum_{k\ge1} h_k h_k^* = \sum_{\ell\ge1} f_\ell f_\ell^*,$$

after reindexing. Furthermore,

$$\sum_{\ell\ge1} \|f_\ell\|_1^2 = \sum_{k\ge1} \|g_k\|_1^2 + \sum_{k\ge1} \|h_k\|_1^2 \le \gamma_+(A) + \gamma_+(B) + \epsilon.$$

The rest of the statements are proved in a similar manner, so we omit the details. □

The next result gives a comparison among the quantities defined above.

Proposition 2.6 *For any $A \in S^n$, the following statements hold.*

(a) $\gamma(A) \le \gamma_0(A) \le 2\gamma(A)$.
(b) $\|A\|_{\mathcal{I}_1} \le \|A\|_{1,1} \le \gamma_0(A) \le 2\gamma(A)$. *If, in addition, we assume that $A \in S^n_+$, then we have*

$$\|A\|_{\mathcal{I}_1} \le \|A\|_{1,1} \le \gamma_0(A) \le \gamma_+(A).$$

Proof

(a) Let $A \in S^n$ such that $A = A^* = \sum\limits_{k\ge1} g_k g_k^* - \sum\limits_{k\ge1} h_k h_k^*$. Then,

$$\gamma(A) \le \sum_{k\ge1} \|g_k\|_1^2 + \sum_{k\ge1} \|h_k\|_1^2.$$

Consequently, $\gamma(A) \le \gamma_0(A)$.

Fix $\varepsilon > 0$ and let $\{g_k\}_{k=1}^M$, $\{h_k\}_{k=1}^M$ be such that $A = \sum_{k=1}^M g_k h_k^*$ and

$$\sum_{k=1}^{M} \|g_k\|_1 \|h_k\|_1 \leq \gamma(A) + \varepsilon.$$

Furthermore, rescale g_k and h_k so that $\|g_k\|_1 = \|h_k\|_1$.

Let $x_k = \frac{1}{2}(g_k + h_k)$ and $y_k = \frac{1}{2}(g_k - h_k)$. Then

$$\sum_{k=1}^{M} x_k x_k^* - \sum_{k=1}^{M} y_k y_k^* = \frac{1}{2}\sum_{k=1}^{M} g_k h_k^* + \frac{1}{2}\sum_{k=1}^{M} h_k g_k^* = A$$

Note also $\|x_k\|_1 \leq \|g_k\|_1 = \|h_k\|_1$ and $\|y_k\|_1 \leq \|g_k\|_1 = \|h_k\|_1$. Thus

$$\gamma_0(A) \leq \sum_{k=1}^{M} \|x_k\|_1^2 + \sum_{k=1}^{M} \|y_k\|_1^2 \leq 2\sum_{k=1}^{M} \|g_k\|_1^2 \leq 2\gamma(A) + 2\varepsilon.$$

Since $\varepsilon > 0$ was arbitrary, the second inequality follows.

(b) Since $\|A\|_{\mathcal{I}_1} = \max_{U \in U(n)} \mathit{Real}\, \mathrm{tr}(AU)$, let $U_0 \in U(n)$ denote the unitary that achieves the maximum and makes the trace real. Then

$$\|A\|_{\mathcal{I}_1} = \mathrm{tr}(AU_0) = \sum_{k=1}^{n}\sum_{\ell=1}^{n} A_{k\ell}(U_0)_{\ell k} \leq \left(\sum_{k=1}^{n}\sum_{\ell=1}^{n} |A_{k\ell}|\right) \cdot \left(\max_{k}\max_{\ell} |(U_0)_{\ell k}|\right) \leq \sum_{k=1}^{n}\sum_{\ell=1}^{n} |A_{k\ell}| = \|A\|_{1,1}.$$

Suppose that $A \in S_+^n$ and let $\epsilon > 0$. Choose $\{g_k\}_{k\geq 1} \subset \mathbb{C}^n$ such that $A = \sum_{k\geq 1} g_k g_k^*$ and

$$\sum_{k\geq 1} \|g_k\|_1^2 < \gamma_+(A) + \epsilon.$$

It follows that

$$\gamma_0(A) \leq \sum_{k\geq 1} \|g_k\|_1^2 < \gamma_+(A) + \epsilon.$$

□

The upper bound $2\gamma(A)$ is tight as we show in Proposition 2.8. We next show that $\|\cdot\|_{1,1}$ and $\gamma(\cdot)$ are identical on S^n.

Lemma 2.7 *For any $A \in S^n$, we have $\|A\|_{1,1} = \gamma(A)$. Consequently, (S^n, γ) is a normed vector space.*

Proof Let $A \in S^n$ and $\epsilon > 0$. Choose $\{g_j\}_{j\geq 1}, \{h_j\}_{j\geq 1} \subset \mathbb{C}^n$ such that $A = \sum_j g_j h_j^*$ with $\sum_j \|g_j\|_1 \cdot \|h_j\|_1 \leq \gamma(A) + \epsilon$. It follows that

$$\|A\|_{1,1}=\sum_{i,j}|A_{i,j}|=\|\sum_j g_j h_j^*\|_{1,1}\le\sum_j \|g_j h_j^*\|_{1,1}\le\sum_j \|g_j\|_1\cdot\|h_j\|_1\le\gamma(A)+\epsilon.$$

Thus $\|A\|_{1,1} \leq \gamma(A)$.

On the other hand, for $A \in S^n$, we can write: $A = (A_{i,j})_{i,j} = (\sum_j (A_{i,j}))_i \cdot \delta_i^T$, then

$$\gamma(A) \leq \sum_j \|A_{i,j}\|_1 \cdot \|\delta_i\|_1 = \sum_{i,j} |A_{i,j}| = \|A\|_{1,1}.$$

Therefore $\|A\|_{1,1} = \gamma(A)$. □

In fact, γ_0 defines also a norm on S^n. More precisely, we have the following result.

Proposition 2.8 *(S^n, γ_0) is normed vector space. Furthermore, γ_0 is Lipschitz with constant 2 on S^n:*

$$\sup_{A,B\in S^n, A\neq B} \frac{|\gamma_0(A) - \gamma_0(B)|}{\|A - B\|_{1,1}} = 2. \tag{4}$$

Proof We have already established in Proposition 2.5 that γ_0 satisfies the triangle inequality and is homogenous. Furthermore, suppose that $\gamma_0(A) = 0$. It follows that $A = 0$.

For the last part, let $A, B \in S^n$. We have

$$\gamma_0(B) = \gamma_0(B - A + A) \leq \gamma_0(B - A) + \gamma_0(A)$$

$$\gamma_0(A) = \gamma_0(B - B + A) \leq \gamma_0(B) + \gamma_0(-B + A)$$

So $|\gamma_0(B) - \gamma_0(A)| \leq \gamma_0(B - A) \leq 2\gamma(B - A) \leq 2\|B - A\|_{1,1}$.

To show the Lipschitz constant is exactly 2 (and hence the upper bound 2 is tight in Proposition 2.6(a)), consider the matrix

$$A = \begin{bmatrix} 0 & 1 \\ 1 & 0 \end{bmatrix}.$$

Note $\|A\|_{1,1} = 2$. For any decomposition $A = B - C$ with $B, C \in S_+^2$, we have

$$B = \begin{bmatrix} a & b \\ b & c \end{bmatrix}, \quad C = \begin{bmatrix} a & e \\ e & c \end{bmatrix}$$

with $a, c \geq 0$ and $b - e = 1$. Then

$$\gamma_0(A) \geq \gamma_+(B) + \gamma_+(C) \geq \gamma(B) + \gamma(C) = 2a + 2|b| + 2|1 - b| + 2c \geq 4|b| + 4|1 - b| \geq 4,$$

thanks to $ac \geq b^2$ and $ac \geq e^2$. On the other hand,

$$A = \frac{1}{2}\begin{bmatrix} 1 \\ 1 \end{bmatrix}\begin{bmatrix} 1 & 1 \end{bmatrix} - \frac{1}{2}\begin{bmatrix} 1 \\ -1 \end{bmatrix}\begin{bmatrix} 1 & -1 \end{bmatrix}$$

which certifies $\gamma_0(A) = 4$. The proof is now complete. □

We have now established that γ_0, $\gamma = \|\cdot\|_{1,1}$ are equivalent norms on S^n. In addition, we proved in Proposition 2.6 that $\gamma(A) = \|A\|_{1,1} \leq \gamma_+(A)$ for $A \in S^n_+$. A natural question that arises is whether a converse estimate holds. More precisely, the rest of the chapter will be devoted to investigating the following questions.

Question 2.1 Fix $n \geq 2$.

(1) Does there exist a constant $C > 0$, independent of n such that for all $A \in S^n_+$, we have

$$\gamma_+(A) \leq C \cdot \|A\|_{1,1}.$$

(2) For a given $A \in S^n_+$, give an algorithm to find $\{h_1, h_2, .., h_M\}$ such that $A = \sum_{k=1}^{M} h_k h_k^*$ with

$$\gamma_+(A) = \sum_{k=1}^{M} \|h_k\|_1^2.$$

We begin by justifying why the second question makes sense. In particular, we prove that $\gamma_+(A)$ is achieved for a certain decomposition.

Theorem 2.9 *Given $T \in S^n_+$,*

$$\gamma_+(T) = \inf_{T=\sum\limits_{k\geq 1} g_k g_k^*} \sum_{k\geq 1} \|g_k\|_1^2 = \min_{T=\sum\limits_{k=1}^{n^2+1} g_k g_k^*} \sum_{k=1}^{n^2+1} \|g_k\|_1^2$$

for some $\{g_k\}_{k=1}^{n^2+1} \subset \mathbb{C}^n$.

Proof Let $T \in S^n_+(\mathbb{C})$,

$$\gamma_+(T) = \inf_{T=\sum\limits_{k\geq 1} g_k g_k^*} \sum_{k\geq 1} \|g_k\|_1^2.$$

Assume $T \neq 0$, then $\gamma_+(T) > 0$. Let $\tilde{T} = \frac{T}{\gamma_+(T)}$,

$$\tilde{T} = \frac{1}{\gamma_+(T)} \sum_{k\geq 1} g_k g_k^* = \sum_{k\geq 1} \frac{\|g_k\|_1^2}{\gamma_+(T)} \cdot \left(\frac{g_k}{\|g_k\|_1}\right) \cdot \left(\frac{g_k}{\|g_k\|_1}\right)^* = \sum_{k\geq 1} w_k \cdot e_k e_k^*,$$

where $w_k = \frac{\|g_k\|_1^2}{\gamma_+(T)}$ and $e_k = \frac{g_k}{\|g_k\|_1}$. Hence $\sum_{k\geq 1} w_k = \frac{1}{\gamma_+(T)} \sum_{k\geq 1} \|g_k\|_1^2 = 1$ and $\|e_k\|_1 = 1$. Therefore $\gamma_+(\tilde{T}) = 1$. It follows that $\tilde{T} \in \Gamma$.

By Minkowski-Carathéodory Theorem 2.3

$$\tilde{T} = \sum_{k=1}^{n^2+1} w_k \cdot e_k e_k^*, \; w_k \geq 0, \; \sum_{k=1}^{n^2+1} w_k = 1.$$

Therefore

$$\gamma_+(T) = \min_{\sum_{k=1}^{n^2+1} g_k g_k^*} \sum_{k=1}^{n^2+1} \|g_k\|_1^2.$$

□

The next question one could ask is how to find an optimal decomposition for $A \in S_+^n$ that achieves the value $\gamma_+(A)$. The following technical tool will be useful in addressing this question, at least for small size matrices.

Theorem 2.10 *Suppose that $A \in S_+^n(\mathbb{C})$ and $y \in \mathbb{C}^n$. Then $A - yy^* \in S_+^n(\mathbb{C})$ if and only if there exists $x \in \mathbb{C}^n$ such that $y = Ax$ and $\langle Ax, x\rangle \leq 1$. When equality holds, then $A - yy^*$ will have rank one less than that of A.*

Proof The case $y = 0$ is trivial, so we can assume without loss of generality that $y \neq 0$.

Suppose there exists a vector y such that $y = Ax$ and $\langle Ax, x\rangle \leq 1$. For any vector z and observe that $|\langle Ax, z\rangle|^2 \leq \langle Ax, x\rangle\langle Az, z\rangle$. Consequently,

$$\langle (A - yy^*)z, z\rangle = \langle Az, z\rangle - |\langle Ax, z\rangle|^2 \geq \langle Az, z\rangle - \langle Ax, x\rangle\langle Az, z\rangle = \langle Az, z\rangle(1 - \langle Ax, x\rangle) \geq 0.$$

When $\langle Ax, x\rangle = 1$, we $\langle (A - yy^*)x, x\rangle = \langle Ax, x\rangle - |\langle y, x\rangle|^2 = \langle Ax, x\rangle - |\langle Ax, x\rangle|^2 = 0$. It follows that $x \in \mathcal{N}(A - yy^*)$. Combining the fact that $x \notin \mathcal{N}(A)$, we have $\text{rank}(A - yy^*) < \text{rank}(A)$.

For the converse, suppose that $A - yy^*$ is positive semidefinite, where $y \in \mathbb{C}^n$. Write $y = Ax + z$ where $x \in \mathbb{C}^n$ and $Az = 0$. It follows that

$$\langle (A - yy^*)z, z\rangle = -|\langle y, z\rangle|^2 \leq 0$$

with equality only if $0 = \langle z, y\rangle = \langle z, Ax + z\rangle = \langle z, z\rangle$ which implies $z = 0$. In addition,

$$\langle (A - yy^*)x, x\rangle = \langle Ax, x\rangle - \langle Ax, x\rangle^2 \geq 0$$

implies $\langle Ax, x\rangle \leq 1$. □

The following result follows from Theorem 2.10

Corollary 2.11 *For any $A \in S_+^n(\mathbb{C})$, we have*

$$\begin{aligned}\gamma_+(A) &= \min_{\langle Ax,x\rangle \leq 1, x\neq 0} \gamma_+(A - Axx^*A) + \|Ax\|_1^2 \\ &\leq \min_{\langle Ax,x\rangle = 1} \gamma_+(A - Axx^*A) + \|Ax\|_1^2.\end{aligned}$$

Proof Let $A \in S_+^n$ and $0 \neq x \in \mathbb{C}^n$ such that $\langle Ax, x\rangle \leq 1$. Then by Theorem 2.10 and Proposition 2.5(a), we see that

$$\gamma_+(A) \leq \min_{\langle Ax,x\rangle \leq 1, x\neq 0} \gamma_+(A - Axx^*A) + \|Ax\|_1^2$$

On the other hand, let $A = \sum_{k=1}^N u_k u_k^*$ be an optimal decomposition, that is $\gamma_+(A) = \sum_{k=1}^N \|u_k\|_1^2$. Since $A - Axx^*A \in S_+^n$, we can write $A - Axx^*A = \sum_{k=1}^n v_k v_k^*$. Hence, $A = \sum_{k=1}^n v_k v_k^* + Axx^*A$, and by the optimality, we see that

$$\gamma_+(A - Axx^*A) + \|Ax\|_1^2 \leq \sum_{k=1} \|v_k\|_1^2 + \|Ax\|_1^2 \leq \gamma_+(A)$$

□

We recall that $\Omega = \text{co}\,(X \cup \{0\})$ where $X = \{xx^* \,: x \in \mathbb{C}^n \; \|x\|_1 = 1\}$. We now give a characterization of Ω in terms of γ_+ that is equivalent to the one proved in Lemma 2.2.

Lemma 2.12 *Using the notations of Lemma 2.2, the following result holds.* $\Omega = \{T \in S_+^n(\mathbb{C}) : \gamma_+(T) \leq 1\}$.

Proof Let $T \in \{T \in S_+^n(\mathbb{C}) : \gamma_+(T) \leq 1\}$. Then

$$T = \sum_{k=1}^{n^2+1} g_k g_k^* = \sum_{k=1}^{n^2+1} w_k X_k X_k^*,$$

where $w_k = \|g_k\|_1^2$ and $X_k = \frac{g_k}{\|g_k\|_1}$. Therefore $\gamma_+(T) = \sum_{k=1}^{n^2+1} w_k \leq 1$. Hence

$$T = \sum_{k=1}^{n^2+1} w_k X_k X_k^* + (1 - \gamma_+(T)) \cdot 0 \in \Omega.$$

Conversely, let $T \in \Omega$. Then $T = \sum_k w_k X_k X_k^*$, $w_k \geq 0$, and $\sum_k w_k \leq 1$. Hence

$$\gamma_+(T) \leq \sum_k w_k \cdot \gamma_+(X_k X_k^*) = \sum_k w_k \leq 1.$$

□

In fact, γ_+ can be identified with the following gauge-like function φ_Ω : $S_+^n(\mathbb{C}) \to \mathbb{R}$ defined as follows:

$$\varphi_\Omega(T) = \inf\{t > 0 : T \in t\Omega\}.$$

Let $\tau_T = \{t > 0 : T \in t\Omega\}$. Then τ_T is nonempty, since $\frac{T}{\gamma_+(T)} \in \Gamma \subset \Omega \Rightarrow T \in \gamma_+(T)\Omega \Rightarrow \gamma_+(T) \in \tau_T$. Therefore $\varphi_\Omega(T) \leq \gamma_+(T)$. In fact, the following stronger result holds.

Lemma 2.13 *For each $T \in S_+^n$, we have $\varphi_\Omega(T) = \gamma_+(T)$*

Proof We need to prove $\gamma_+(T) \leq \varphi_\Omega(T)$. If $t \in \tau_T$, then $\frac{T}{t} \in \Omega$,

$$\frac{T}{t} = \sum_{k=1}^{n^2+1} w_k x_k x_k^*, \; w_1, .., w_{n^2+1} \geq 0, \; \sum_{k=1}^{n^2+1} w_k \leq 1, \; \|x_k\|_1 = 1, \forall k.$$

$$T = \sum_{k=1}^{n^2+1} t w_k x_k x_k^* = \sum_{k=1}^{n^2+1} g_k g_k^*,$$

where $g_k = \sqrt{t w_k} x_k$. Now $\gamma_+(T) \leq \sum_{k=1}^{n^2+1} t w_k = t \sum_{k=1}^{n^2+1} w_k \leq t \Rightarrow \gamma_+(T) \leq \varphi_\Omega(T)$.

□

Remark It follows that φ_Ω is also positively homogeneous and sub-additive, hence convex. However, we point out that φ_Ω is not a Minkowski gauge function since Ω does not include a neighborhood of 0.

We close this section with a discussion of some regularity properties of γ_+.

Theorem 2.14 *Fix $\delta > 0$. Let $C_\delta = \{T \in S_+^n : T \geq \delta I, tr(T) \leq 1\}$, then $\gamma_+ : C_\delta \to \mathbb{R}$ is Lipschitz continuous on C_δ with Lipschitz constant $(n/\delta) + n^{3/2}$.*

Proof We show that $\forall\, T_1, T_2 \in C_\delta$,

$$|\gamma_+(T_1) - \gamma_+(T_2)| \leq \left(\frac{n}{\delta} + n^2\right) \|T_1 - T_2\|.$$

Define

$$\tilde{T} = T_2 + \frac{\delta}{\|T_2 - T_1\|}(T_2 - T_1).$$

Then

$$\lambda_{\min}(\tilde{T}) \geq \lambda_{\min}(T_2) - \left\| \frac{\delta}{\|T_2 - T_1\|}(T_2 - T_1) \right\| = \lambda_{\min}(T_2) - \delta \geq 0.$$

Consequently, $\tilde{T} \in S^n_+$.
Now

$$T_2 = \frac{\delta}{\delta + \|T_2 - T_1\|} T_1 + \frac{\|T_2 - T_1\|}{\delta + \|T_2 - T_1\|} \tilde{T}.$$

The convexity of γ_+ yields

$$\gamma_+(T_2) \leq \frac{\delta}{\delta + \|T_2 - T_1\|} \gamma_+(T_1) + \frac{\|T_2 - T_1\|}{\delta + \|T_2 - T_1\|} \gamma_+(\tilde{T}),$$

which implies

$$\gamma_+(T_2) - \gamma_+(T_1) \leq \frac{\|T_2 - T_1\| \left(\gamma_+(\tilde{T}) - \gamma_+(T_1)\right)}{\delta + \|T_2 - T_1\|}. \tag{5}$$

We have

$$\gamma_+(\tilde{T}) \leq n \cdot tr(\tilde{T}) = n \cdot \left[tr(T_2) + \delta \cdot tr\left(\frac{T_2 - T_1}{\|T_2 - T_1\|}\right)\right] \leq n \cdot tr(T_2) + \delta n^{3/2}. \tag{6}$$

$$\gamma_+(T_1) \geq \|T_1\|_{1,1} = \sum_{i,j} |(T_1)_{i,j}| \geq tr(T_1) \geq n\delta. \tag{7}$$

Using Equations (6) and (7), we get

$$\gamma_+(\tilde{T}) - \gamma_+(T_1) \leq n \cdot tr(T_2) + \delta n^{3/2} - n\delta \leq n \cdot tr(T_2) + \delta n^{3/2}. \tag{8}$$

Now

$$\gamma_+(T_2) - \gamma_+(T_1) \leq \frac{\|T_2 - T_1\|}{\delta} \left(\gamma_+(\tilde{T}) - \gamma_+(T_1)\right) \leq \|T_2 - T_1\| \left[\frac{n}{\delta} \cdot tr(T_2) + n^{3/2}\right]$$

$$\Rightarrow \frac{\gamma_+(T_2) - \gamma_+(T_1)}{\|T_2 - T_1\|} \leq \frac{n}{\delta} \cdot tr(T_2) + n^{3/2}. \tag{9}$$

Similarly

$$\frac{\gamma_+(T_1) - \gamma_+(T_2)}{\|T_1 - T_2\|} \leq \frac{n}{\delta} \cdot tr(T_1) + n^{3/2}. \tag{10}$$

Therefore

$$\frac{|\gamma_+(T_1) - \gamma_+(T_2)|}{\|T_1 - T_2\|} \leq \frac{n}{\delta} \cdot \max\left(tr(T_1), tr(T_2)\right) + n^{3/2} \leq \frac{n}{\delta} + n^{3/2}. \quad (11)$$

□

In fact, we can prove a stronger result if we restrict to S^n_{++}.

Corollary 2.15 *$\gamma_+ : S^n_{++}(\mathbb{C}) \to \mathbb{R}$ is continuous. Further, let $T \in S^n_{++}(\mathbb{C})$ and $\delta = \frac{1}{2}\lambda_{\min}(T) > 0$. Then for every $S \in S^n_{++}(\mathbb{C})$ with $\|T - S\| \leq \delta$,*

$$\frac{|\gamma_+(T) - \gamma_+(S)|}{\|T - S\|} \leq \frac{n}{\delta} \cdot tr(T) + 2n^{3/2}.$$

Proof Let $T \in S^n_{++}(\mathbb{C})$ and $\delta = \frac{1}{2}\lambda_{\min}(T) > 0$. For any $S \in S^n_{++}(\mathbb{C})$ with $\|T - S\| \leq \delta$, and every $x \in \mathbb{C}^n$, we have that

$$\langle Sx, x\rangle = \langle (S - T)x, x\rangle + \langle Tx, x\rangle \geq (-\delta + \lambda_{min}(T))\|x\|^2 = \delta\|x\|^2.$$

Using this (11) becomes

$$\frac{|\gamma_+(T) - \gamma_+(S)|}{\|T - S\|} \leq \frac{n}{\delta} \cdot \max\left(tr(T), tr(S)\right) + n^{3/2}.$$

However, $tr(S) \leq tr(T) + \sqrt{n}\delta$. Therefore,

$$\frac{|\gamma_+(T) - \gamma_+(S)|}{\|T - S\|} \leq \frac{n}{\delta} \cdot tr(T) + 2n^{3/2}.$$

□

3 Finding Optimal Rank One Decomposition for Some Special Classes of Matrices

In this section we consider several classes of matrices in S^n_+ for which the answer to Question 2.1 is affirmative.

3.1 Diagonally Dominant Matrices

Recall that a matrix $A \in S^n_+(\mathbb{C})$ is said to be diagonally dominant if $A_{ii} \geq \sum_{j=1}^n |A_{ij}|$ for each $i = 1, 2, \ldots, n$. If the inequality is strict for each i, we say that the matrix is strictly diagonally dominant. The following result can be proved for any diagonally dominant matrix in S_+^n.

Theorem 3.1 *Let $A \in S_+^n$ be a diagonally dominant matrix. Then $\gamma(A) = \gamma_0(A) = \gamma_+(A)$.*

Proof Let $e_i = (0, \ldots, 0, 1, 0, \ldots, 0)$ and $u_{ij}(x) = (0, \ldots, \sqrt{x}, \ldots, \overline{\sqrt{x}}, \ldots, 0)$. Given a diagonally dominant matrix A, we consider the following decomposition of A ([2])

$$A = \sum_{i<j} u_{ij}(A_{ij})u_{ij}(A_{ij})^* + \sum_i (A_{ii} - \sum_{j\in\{1,\ldots,n\}\setminus\{i\}} |A_{ij}|)e_i e_i^*.$$

It follows that

$$\begin{aligned}
\gamma_+(A) &\leq \sum_{i<j} 4|A_{ij}| + \sum_i (A_{ii} - \sum_{j\in\{1,\ldots,n\}\setminus\{i\}} |A_{ij}|) \\
&= \sum_{i<j} 4|A_{ij}| + \sum_i A_{ii} - \sum_i \sum_{j\in\{1,\ldots,n\}\setminus\{i\}} |A_{ij}| \\
&= \sum_{i<j} 4|A_{ij}| + \sum_i A_{ii} - \sum_{i<j} 2|A_{ij}| \\
&= \|A\|_{1,1}.
\end{aligned}$$

□

The case of diagonally dominant matrices is a particular case of the following more general decomposition result:

Theorem 3.2 *Assume $A \in S_+^n$ admits a decomposition*

$$A = \sum_{1\leq i<j\leq n} u_{ij}u_{ij}^* + \sum_{i=1}^n v_i v_i^* \tag{12}$$

where each $u_{i,j}$ has non-zero entries at most on positions i and j and each v_i has non-zero entries at most on position i. Then $\gamma_+(A) = \|A\|_{1,1}$.

Proof The hypothesis implies

$$u_{ij} = \left[0 \cdots 0\; c_{ij;i}\; 0 \cdots 0\; c_{ij;j}\; 0 \cdots 0\right]^T$$

and

$$v_i = \left[0 \cdots 0\; d_i\; 0 \cdots 0\right]^T$$

where $c_{ij;i}$ is on position i, $c_{ij;j}$ is on position j, and d_i is on position i. Without loss of generality, we can assume $d_i \in \mathbb{R}$ and $c_{ij;i}, c_{ij;j} \in \mathbb{C}$. We write $A = (a_{ij})_{i,j=1}^n$ where for $1 \le i < j \le n$, $a_{ij} = c_{ij;i}\overline{c_{ij;j}}$, whereas for $1 \le i \le n$,

$$a_{ii} = d_i^2 + \sum_{j=1}^{i-1} |c_{ji;i}|^2 + \sum_{j=i+1}^{n} |c_{ij;i}|^2.$$

These imply

$$\sum_{1 \le i < j \le n} \|u_{ij}\|_1^2 + \sum_{i=1}^{n} \|v_i\|_1^2 = \sum_{1 \le i < j \le n} \left(|u_{ij;i}| + |u_{ij;j}|\right)^2 + \sum_{i=1}^{n} d_i^2 = \sum_{1 \le i,j \le n} |a_{i,j}| = \|A\|_{1,1}.$$

Now the proof is complete. □

3.2 The Cases for Matrices in $S_+^n(\mathbb{C})$ for $n \in \{2, 3\}$

Proposition 3.3 *Suppose that $A \in S_+^2$, then*

$$\gamma_+(A) = \|A\|_{1,1}.$$

Proof If $A = uu^*$ is a rank 1 matrix in S_+^2, the proof is straightforward. Suppose $A \in S_+^2$ is rank 2. $A = \begin{bmatrix} a & c \\ \bar{c} & b \end{bmatrix}$ with $ab - |c|^2 > 0$. Using the Lagrange decomposition [10], we can write

$$A = \begin{bmatrix} \sqrt{a} \\ \frac{\bar{c}}{\sqrt{a}} \end{bmatrix} \begin{bmatrix} \sqrt{a} & \frac{c}{\sqrt{a}} \end{bmatrix} + \begin{bmatrix} 0 \\ \sqrt{b - \frac{|c|^2}{a}} \end{bmatrix} \begin{bmatrix} 0 & \sqrt{b - \frac{|c|^2}{a}} \end{bmatrix}$$

The result then follows. □

For certain 3×3 matrices, the Lagrange decomposition [10] is optimal. In particular, we have the following result.

Proposition 3.4 *Let $A \in S_+^3$ be of rank 2 or 3. If*

$$A = \begin{bmatrix} a & b & c \\ \bar{b} & d & e \\ \bar{c} & \bar{e} & f \end{bmatrix}$$

then

$$\gamma_+(A) \leq \|A\|_{1,1} + \frac{2(|ae-\overline{b}c|+|b||c|-a|e|)}{a}.$$

In particular, if $|ae - \overline{b}c| + |b||c| = a|e|$, *then* $\gamma_+(A) = \|A\|_{1,1}$ *and the Lagrange decomposition (which in this case is the LDL factorization) is optimal.*

Proof We first assume that A has rank 3. In this case, A must be positive definite and $adf \neq 0$. Indeed, if one of the diagonal term, say $f = 0$, then using the fact that $A \in S^3_+$ would imply that $df - |e|^2 = -|e|^2 > 0$ which is impossible.

Let

$$u_1 = \frac{1}{\sqrt{a}} A\delta_1 = \begin{bmatrix} \sqrt{a} \\ \frac{\overline{b}}{\sqrt{a}} \\ \frac{\overline{c}}{\sqrt{a}} \end{bmatrix},$$

where $\{\delta_i\}_{i=1}^3$ is the standard ONB for $\mathbb{C}^3$. By Theorem 2.10, the matrix $A - u_1u_1^*$. In fact, in this case, this is a rank 2 matrix given by

$$A - u_1u_1^* = \begin{bmatrix} 0 & 0 & 0 \\ 0 & d - \frac{|b|^2}{a} & e - \frac{\overline{b}c}{a} \\ 0 & \overline{e} - \frac{\overline{c}b}{a} & f - \frac{|c|^2}{a} \end{bmatrix}$$

Let

$$u_2 = \frac{1}{\sqrt{d-\frac{|b|^2}{a}}}(A - u_1u_1^*)\delta_2 = \begin{bmatrix} 0 \\ \sqrt{d - \frac{|b|^2}{a}} \\ \frac{\overline{e}-\frac{\overline{c}b}{a}}{\sqrt{d-\frac{|b|^2}{a}}} \end{bmatrix}.$$

It follows that $A - u_1u_1^* - u_2u_2^* = u_3u_3^*$ where

$$u_3 = \begin{bmatrix} 0 \\ 0 \\ \sqrt{\frac{\det A}{ad-|b|^2}} \end{bmatrix}.$$

Consequently, the Lagrange decomposition of A is $A = u_1u_1^* + u_2u_2^* + u_3u_3^*$ which implies that

$$\gamma_+(A) \leq \sum_{k=1}^{3} \|u_k\|_1^2 = \|A\|_{1,1} + \frac{2(|ae-\overline{b}c|+|b||c|-a|e|)}{a}.$$

Now suppose that the rank of A is 2. In this case, it is possible for $adf = 0$. However, only one of the diagonal element can be 0. So assume that $f = 0$, then we also get that $e = c = 0$. In this case

$$A \begin{bmatrix} a & b & 0 \\ \overline{b} & d & 0 \\ 0 & 0 & 0 \end{bmatrix}$$

which reduces to Proposition 3.3. Thus, we may assume without loss of generality that $adf \neq 0$. In this case, we can proceed as above. However, because the rank of the matrix A is now 2, we see that $A = u_1u_1^* + u_2u_2^*$ and

$$\gamma_+(A) \leq \|u_1\|_1^2 + \|u_2\|_1^2 = \|A\|_{1,1} + \frac{2(|ae-\overline{b}c|+|b||c|-a|e|)}{a}.$$

□

Remark

(1) If one of the off diagonal elements b, or c is 0, then Proposition 3.4 shows that the Lagrange decomposition is optimal for $\gamma_+(A)$.

(2) Suppose $n = 4$ and let $V = \frac{1}{\sqrt{14}} \begin{bmatrix} 1 & 0 \\ 0 & 1 \\ 1 & -1 \\ 1 & 1 \end{bmatrix}$, and consider

$$A = VV^T = \frac{1}{14} \begin{bmatrix} 1 & 0 & 1 & 1 \\ 0 & 1 & -1 & 1 \\ 1 & -1 & 2 & 0 \\ 1 & 1 & 0 & 2 \end{bmatrix}$$

Then A has rank 2, and the $\|A\|_{1,1} = 1$. However, $\gamma_+(A) \neq \gamma(A)$.

4 Numerics

Here we inspect upper bounds of $\gamma_+(A)/\|A\|_{1,1}$ for A an N x N matrix with simulated data. We randomly generate symmetric positive definite matrices and compute upper bounds on $\gamma_+(A)/\|A\|_{1,1}$ with different decompositions of A. The first step is generating Gaussian distributed realizations in a matrix size N by N. Then by multiplying by its transpose, the result is symmetric positive semidefinite, denoted A. Let $\mathcal{A}_N$ denote a collection of 30 independent realizations of this random matrix.

We consider two factorizations of the matrix A: the LDL and the Eigen matrix decomposition. Specifically:

$$LDL: \quad A = \sum_{k=1}^{N} v_k v_k^*$$

with v_k vectors that have the top $k-1$ entries 0, and

$$\text{Eigen}: \quad A = \sum_{k=1}^{N} g_k g_k^*$$

where $\{g_1, \ldots, g_n\}$ are the eigenvectors, each scaled by the corresponding eigenvalue's square-root. For each decomposition, denote:

$$J_{LDL}(A) = \sum_{k=1}^{N} \|v_k\|_1^2 \text{ and } J_{\text{Eigen}}(A) = \sum_{k=1}^{N} \|g_k\|_1^2$$

Let F_{LDL} and F_{Eigen} denote the worst upper bounds over the N realization ensemble:

$$F_{LDL}(N) = \max_{A \in \mathcal{A}_N} \frac{J_{LDL}(A)}{\|A\|_{1,1}}$$

$$F_{\text{Eigen}}(N) = \max_{A \in \mathcal{A}_N} \frac{J_{\text{Eigen}}(A)}{\|A\|_{1,1}}$$

We plot these worst upper bounds after 30 realizations for various N in Figure 1.

In the same figure, we plot the analytic approximations of these two curves using a square root functions and a logarithmic function. The square root function was scaled as $c\sqrt{N}$ to closely fit the Eigen decomposition bound, $F_{\text{Eigen}}(N)$. Numerically we obtained $c = 4/5$.

From these plots, we notice a clearly strictly increasing trend. Furthermore, the LDL factorization produces a smaller (tighter) upper bound than the Eigen decomposition. On the other hand, as we show in Theorem 2.9, any optimal decomposition may take $N^2 + 1$ vectors. By limiting the number of vector to N, one should not expect to achieve the optimal bound $\gamma_+(A)$ with any decomposition.

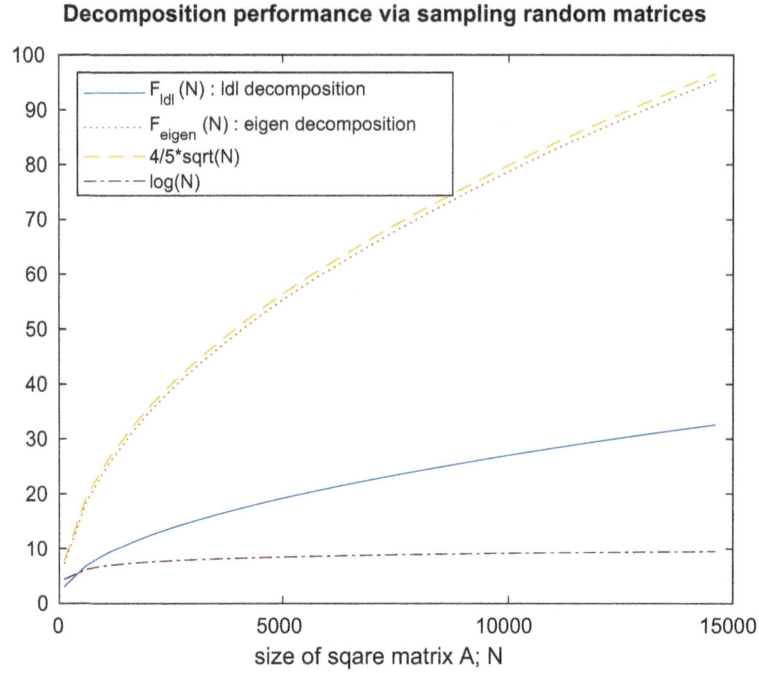

Fig. 1 For each size N, 30 random matrices are sampled and decomposed in different ways. The worst upper bound of $\gamma_+(A)$ is plotted for various N. Reference curves are also plotted to indicate trend

Acknowledgments R. Balan was partially supported by the National Science Foundation grant DMS-1816608 and Laboratory for Telecommunication Sciences under grant H9823031D00560049. K. A. Okoudjou was partially supported by the U S Army Research Office grant W911NF1610008, the National Science Foundation grant DMS 1814253, and an MLK visiting professorship.

References

1. R. Balan, K.A. Okoudjou, A. Poria, On a Feichtinger problem. Oper. Matrices **12**(3), 881–891 (2018)
2. G.P. Barker, D.H. Carlson, Cones of diagonally dominant matrices. Pac. J. Math. **57**(1), 15–32 (1975)
3. F. Clarke, *Functional Analysis, Calculus of Variations and Optimal Control*. Graduate Texts in Mathematics, vol. 264 (Springer, London, 2013)
4. J.A. De Loera, X. Goaoc, F. Meunier, N.H. Mustafa, The discrete yet ubiquitous theorems of Carathéodory, Helly, Sperner, Tucker, and Tverberg. Bull. Am. Math. Soc. **56**(3), 415–511 (2019)
5. N. Dunford, J.T. Schwartz, *Linear Operators, Part II* (Wiley, New York, 1988)

6. H. Feichtinger, P. Jorgensen, D. Larson, G. Ólafsson, *Mini-Workshop: Wavelets and Frames*, Abstracts from the mini-workshop held 15–21 Feb 2004. Oberwolfach Rep. **1**(1), 479–543 (2004)
7. C. Heil, D. Larson, Operator theory and modulation spaces. Contemp. Math. **451**, 137–150 (2008)
8. J. Reay, Generalizations of a theorem of Carathéodory. Mem. Am. Math. Soc. **54** (1965)
9. B. Simon, *Trace Ideals and Their Applications* (Cambridge University Press, Cambridge, 1979)
10. B. Ycart, Extreme points in convex sets of symmetric matrices. Proc. Am. Math. Soc. **95**(4), 607–612 (1985)

An Arithmetical Function Related to Báez-Duarte's Criterion for the Riemann Hypothesis

Michel Balazard

To the memory of my friend, Luis Báez-Duarte.

Abstract In this mainly expository article, we revisit some formal aspects of Báez-Duarte's criterion for the Riemann hypothesis. In particular, starting from Weingartner's formulation of the criterion, we define an arithmetical function ν, which is equal to the Möbius function if and only if the Riemann hypothesis is true. We record the basic properties of the Dirichlet series of ν and state a few questions.

1 The Spaces D and D_0

We will denote by $\mathbb{N}$ (resp. $\mathbb{N}^*$) the set of non-negative (resp. positive) integers, by H the Hilbert space $L^2(0, \infty; t^{-2}dt)$, with inner product

$$\langle f, g\rangle = \int_0^\infty f(t)\overline{g(t)}\,\frac{dt}{t^2},$$

and by $\mathrm{Vect}(\mathscr{F})$ the set of finite linear combinations of elements of a family $\mathscr{F}$ of elements of H.

For $k \in \mathbb{N}^*$, we define

$$e_k(t) = \{t/k\} \quad (t > 0),$$

where $\{u\} = u - \lfloor u\rfloor$ denotes the fractional part of the real number u and $\lfloor u\rfloor$ its integer part. The functions e_k belong to H, as do the functions χ and κ defined by

$$\chi(t) = [t \geq 1] \quad ; \quad \kappa(t) = t[0 < t < 1]$$

M. Balazard (✉)
Aix-Marseille University, CNRS, Centrale Marseille, Marseille, France
e-mail: balazard@math.cnrs.fr

M. Th. Rassias (ed.), *Harmonic Analysis and Applications*, Springer Optimization and Its Applications 168, https://doi.org/10.1007/978-3-030-61887-2_3

(here, and in the following, we use Iverson's notation: $[P] = 1$ if the assertion P is true, $[P] = 0$ if it is false).

Let D be the closed subspace of functions $f \in H$ of the type

$$f(t) = \lambda t + \varphi(t), \tag{1}$$

where φ is constant on each interval $[j, j+1[$, $j \in \mathbb{N}$ (for $j = 0$, the constant must be 0). The functions e_k belong to D.

Let D_0 be the subspace of D defined by taking $\lambda = 0$ in (1), that is, the subspace of functions $\varphi \in H$ which are constant on each interval $[j, j+1[$, $j \in \mathbb{N}$. The functions χ and $e_k - e_1/k$ belong to D_0.

A Hilbertian basis for D_0 is given by the family of step functions ε_k defined by

$$\varepsilon_k(t) = \sqrt{k(k+1)} \cdot [k \le t < k+1] \quad (k \in \mathbb{N}^*,\ t > 0).$$

The mapping $h \mapsto \big(h(j)\big)_{j \ge 1}$ is a Hilbert space isomorphism of D_0 onto the sequence space $\mathfrak{h}$ of complex sequences $(x_j)_{j \ge 1}$ such that

$$\sum_{j \ge 1} \frac{|x_j|^2}{j(j+1)} < \infty.$$

Observe that, for $f \in D$, written as (1), one has

$$\lambda = \langle f, \kappa \rangle \tag{2}$$

$$f = \lambda e_1 + h, \text{ where } h \in D_0. \tag{3}$$

Thus, the subspace D is the (non-orthogonal) direct sum of $\mathrm{Vect}(e_1)$ and D_0.

In formula (2), the function κ could be replaced by its orthogonal projection κ' on D. The definition of the families (ψ_n) of Proposition 2 and (g_n) of Proposition 4 could be modified accordingly. We compute κ' in the appendix.

To every function in D, one can associate certain arithmetical functions. Let $f \in D$, with λ and h as in (2), (3). We first define the arithmetical function

$$u(n) = u(n; f) = -\lambda + h(n) - h(n-1) \quad (n \in \mathbb{N}^*). \tag{4}$$

With this definition, we see that the function φ of (1) is given by

$$\varphi(t) = -\lambda t + f(t) = -\lambda t + \lambda\{t\} + h(t) = \sum_{n \le t} u(n).$$

Thus, $f(t)$ is the remainder term in the approximation of the sum function $\varphi(t)$ of the arithmetical function u by the linear function $-\lambda t$. The fact that f belongs to H implies, and is stronger than, the asymptotic relation $f(t) = o(t)$.

For $f \in D$, we will also consider the arithmetical function $w = \mu * u$, where μ denotes the Möbius function:

$$w(n) = w(n; f) = \sum_{d|n} \mu(n/d)u(d; f) \quad (n \in \mathbb{N}^*).$$

For instance,

$$u(n; \chi) = [n = 1] \quad ; \quad w(n; \chi) = \mu(n) \quad (n \in \mathbb{N}^*).$$

The arithmetical functions u and w depend linearly on f and the correspondences are one-to-one.

Proposition 1 *For $f \in D$,*

$$f = 0 \Leftrightarrow u = 0 \Leftrightarrow w = 0.$$

Proof The second equivalence follows from $w = u * \mu$ and $u = w * 1$ (Möbius inversion). It remains to prove that $u = 0 \Rightarrow f = 0$. By (4), $u = 0$ implies $h(n) = \lambda n$ for all n, hence $\lambda = 0$ since $h \in D_0$, and $h = 0$. □

Since $u = w * 1$, one has

$$f(t) = \lambda t + \sum_{n \le t} u(n) = \lambda t + \sum_{n \ge 1} w(n) \lfloor t/n \rfloor.$$

In Proposition 7, we will prove the identity

$$\sum_{n \ge 1} \frac{w(n)}{n} = -\lambda, \tag{5}$$

so that, for every f in D and every $t > 0$, one has

$$f(t) = -\sum_{n \ge 1} w(n) e_n(t). \tag{6}$$

Of course, it does not mean that the series $\sum_{n \ge 1} w(n; f) e_n$ converges in H (in fact, it diverges if $f = \chi$, cf. [1], Theorem 2.2, p. 6), but, if it does, its sum is $-f$.

2 Vasyunin's Biorthogonal System

In Theorem 7 of his paper [7], Vasyunin defined a family $(f_k)_{k \ge 2}$, which, together with the family $(e_k - e_1/k)_{k \ge 2}$, yields a biorthogonal system in D_0, which means that

$$\langle e_j - e_1/j, f_k \rangle = [j = k] \quad (j \geq 2,\ k \geq 2). \tag{7}$$

We will recall Vasyunin's construction, which can be thought of as a Hilbert space formulation of Möbius inversion, and add several comments.

2.1 The Sequence (φ_k)

First one defines, for $k \in \mathbb{N}^*$, a step function $\varphi_k \in D_0$ by

$$\varphi_k(t) = k(k-1)[k-1 \leq t < k] - k(k+1)[k \leq t < k+1]$$

(Vasyunin's φ_k have the opposite sign, according to his definition for e_k). Thus

$$\varphi_k = \sqrt{k(k-1)} \cdot \varepsilon_{k-1} - \sqrt{k(k+1)} \cdot \varepsilon_k \quad (k \in \mathbb{N}^*),$$

with $\varepsilon_0 = 0$ by convention. One sees that the family $(\varphi_k)_{k \geq 1}$ is total in D_0.

One checks that

$$\langle h, \varphi_k \rangle = h(k-1) - h(k) \quad (k \in \mathbb{N}^*), \tag{8}$$

for $h \in D_0$ with constant value $h(k)$ on $[k, k+1[$ $(h(0) = 0)$. In particular,

$$\langle e_j - e_1/j, \varphi_k \rangle = [j \mid k] - 1/j \quad (j \geq 1,\ k \geq 1).$$

Using the family (φ_k), one can write the values $u(n; f)$, for $f \in D$, as scalar products.

Proposition 2 *For $f \in D$, with λ and h as in* (2) *and* (3), *one has*

$$u(n; f) = \langle f, \psi_n \rangle,$$

where

$$\psi_n = (\langle e_1, \varphi_n \rangle - 1)\kappa - \varphi_n \quad (n \in \mathbb{N}^*).$$

In particular, $f \mapsto u(n; f)$ is a continuous linear form on D, for every $n \in \mathbb{N}^*$.

Proof By (2), (4) and (8), one has

$$\begin{aligned} u(n; f) &= -\langle f, \kappa \rangle - \langle h, \varphi_n \rangle \\ &= -\langle f, \kappa \rangle - \langle f - \langle f, \kappa \rangle e_1, \varphi_n \rangle \\ &= -\langle f, \kappa \rangle - \langle f, \varphi_n \rangle + \langle e_1, \varphi_n \rangle \langle f, \kappa \rangle \end{aligned}$$

$$= \langle f, \psi_n \rangle \quad (n \in \mathbb{N}^*).$$ □

We compute the scalar product $\langle e_1, \varphi_n \rangle$ in the appendix.

The next proposition describes the behavior of the series $\sum_k \varphi_k/k$.

Proposition 3 *The series*

$$\sum_{k \geq 1} \frac{\varphi_k}{k}$$

is weakly convergent in D_0, *with weak sum* $-\chi$.

Proof The partial sum

$$\sum_{k \leq K} \frac{\varphi_k}{k}$$

is the step function with values

$$\begin{aligned} 0 \text{ on } (0, 1) \text{ and } (K+1, \infty) \\ -1 \text{ on } (1, K) \\ -(K+1) \text{ on } (K, K+1) \end{aligned}$$

This partial sum is thus equal to $-\chi$ on every fixed bounded segment of $(0, \infty)$, if K is large enough, and the norm of this partial sum in H is the constant $\sqrt{2}$. The result follows. □

2.2 *The Sequence* (f_k)

Vasyunin defined

$$f_k = \sum_{d|k} \mu(k/d)\varphi_d \quad (k \in \mathbb{N}^*).$$

Equivalently,

$$\varphi_k = \sum_{d|k} f_d \quad (k \in \mathbb{N}^*),$$

by Möbius inversion; this implies that the family $(f_k)_{k \geq 1}$ is also total in D_0.

A slightly more general form of (7), namely,

$$\langle e_j - e_1/j, f_k\rangle = [j = k] - [k = 1]/j \quad (j, k \in \mathbb{N}^*), \tag{9}$$

is proved by means of the identity

$$\sum_{j|d|k} \mu(k/d) = [j = k].$$

Using the family (f_k), one can write the values $w(n; f)$, for $f \in D$, as scalar products.

Proposition 4 *For $f \in D$, with λ and h as in* (2) *and* (3), *one has*

$$w(n; f) = \langle f, g_n\rangle,$$

where

$$g_n = (\langle e_1, f_n\rangle - [n = 1])\kappa - f_n \quad (n \in \mathbb{N}^*).$$

In particular, $f \mapsto w(n; f)$ is a continuous linear form on D, for every $n \in \mathbb{N}^*$.

Proof By Proposition 2, one has

$$\begin{aligned} w(n; f) &= \sum_{d|n} \mu(n/d) u(d; f) \\ &= \langle f, \sum_{d|n} \mu(n/d)\psi_d\rangle \quad (n \in \mathbb{N}^*). \end{aligned}$$

Now,

$$\begin{aligned} \sum_{d|n} \mu(n/d)\psi_d &= \sum_{d|n} \mu(n/d)\big((\langle e_1, \varphi_d\rangle - 1)\kappa - \varphi_d\big) \\ &= (\langle e_1, f_n\rangle - [n = 1])\kappa - f_n. \end{aligned}$$ □

We compute the scalar product $\langle e_1, f_n\rangle$ in the appendix.

In order to study the series $\sum_k f_k/k$, we will need the following auxiliary proposition.

Proposition 5 *Let*

$$f(x) = \sum_{k \le x} \eta(k) \quad (x > 0),$$

where η is a complex arithmetical function such that $\eta(k) = O(1/k)$, for $k \ge 1$.

Then, for every fixed $\alpha > 1$,

$$\sum_{k\geq 1} \left|f(x/k) - f\big(x/(k+1)\big)\right|^{\alpha} = O(1) \quad (x > 0).$$

Proof The series is in fact a finite sum, as

$$f(x/k) = f\big(x/(k+1)\big) = 0 \quad (k > x).$$

We will use the estimate

$$f(y) - f(x) \ll \sum_{x<k\leq y} \frac{1}{k} \ll \frac{1}{x} + \ln(y/x) \quad (y > x \geq 1).$$

Thus,

$$f(x/k) - f\big(x/(k+1)\big) \ll \frac{k}{x} + \frac{1}{k} \ll \frac{1}{k} \quad (k \leq \sqrt{x}),$$

and

$$\sum_{k\leq\sqrt{x}} \left|f(x/k) - f\big(x/(k+1)\big)\right|^{\alpha} \ll \sum_{k\geq 1} \frac{1}{k^{\alpha}} \ll 1 \quad (x > 0).$$

If $k > \sqrt{x}$, then

$$\frac{x}{k} - \frac{x}{k+1} < 1,$$

so that the interval $]x/(k+1), x/k]$ contains at most one integer, say n, and, if n exists, one has $k = \lfloor x/n \rfloor$ and

$$f(x/k) - f\big(x/(k+1)\big) = \eta(n) \ll \frac{1}{n}\cdot$$

Hence

$$\sum_{k>\sqrt{x}} \left|f(x/k) - f\big(x/(k+1)\big)\right|^{\alpha} \ll \sum_{n\geq 1} \frac{1}{n^{\alpha}} \ll 1 \quad (x > 0).$$

The result follows. □

Proposition 6 *The series*

$$\sum_{k\geq 1} \frac{f_k}{k}$$

is weakly convergent in D_0 (hence in H), with weak sum 0.

Proof Let $K \in \mathbb{N}^*$. One has

$$S_K = \sum_{k\le K} \frac{f_k}{k} = \sum_{d\le K} \frac{m(K/d)}{d}\varphi_d,$$

where

$$m(x) = \sum_{n\le x} \frac{\mu(n)}{n} \quad (x > 0).$$

Hence,

$$\begin{aligned} S_K &= \sum_{d\le K} \frac{m(K/d)}{d}\big(\sqrt{d(d-1)}\cdot\varepsilon_{d-1} - \sqrt{d(d+1)}\cdot\varepsilon_d\big) \\ &= \sum_{d\le K-1}\Big(\frac{m\big(K/(d+1)\big)}{d+1} - \frac{m(K/d)}{d}\Big)\sqrt{d(d+1)}\cdot\varepsilon_d - \sqrt{1+1/K}\cdot\varepsilon_K \end{aligned}$$

For every fixed $d \in \mathbb{N}^*$, the fact that $\langle S_K, \varepsilon_d\rangle$ tends to 0 when K tends to infinity follows from this formula and the classical result of von Mangoldt, asserting that $m(x)$ tends to 0 when x tends to infinity.

It remains to show that $\|S_K\|$ is bounded. One has

$$\begin{aligned} \|S_K\|^2 &= \sum_{d\le K-1} d(d+1)\Big(\frac{m(K/d)}{d} - \frac{m\big(K/(d+1)\big)}{d+1}\Big)^2 + 1 + 1/K \\ &\le 2\sum_{d\le K-1} d(d+1)\Big(\frac{m(K/d) - m\big(K/(d+1)\big)}{d}\Big)^2 \\ &\quad + 2\sum_{d\le K-1} d(d+1)\Big(\frac{m\big(K/(d+1)\big)}{d(d+1)}\Big)^2 + 1 + 1/K \\ &\ll 1 + \sum_{d\le K-1}\Big(m(K/d) - m\big(K/(d+1)\big)\Big)^2 \end{aligned}$$

The boundedness of $\|S_K\|$ then follows from Proposition 5. □

We are now able to prove (5).

Proposition 7 *Let $f \in D$, with λ and h as in* (2), (3). *The series*

$$\sum_{n\ge 1} \frac{w(n; f)}{n}$$

is convergent and has sum $-\lambda$.

Proof Putting $\beta_N = \sum_{n \le N} f_n/n$ for $N \in \mathbb{N}^*$, one has

$$\sum_{n \le N} \frac{g_n}{n} = \sum_{n \le N} \frac{(\langle e_1, f_n \rangle - [n = 1])\kappa - f_n}{n}$$
$$= \big(\langle e_1, \beta_N \rangle - 1\big)\kappa - \beta_N,$$

which tends weakly to $-\kappa$, as N tends to infinity, by Proposition 6.

Hence,

$$\sum_{n \le N} \frac{w(n; f)}{n} = \sum_{n \le N} \frac{\langle f, g_n \rangle}{n} = \langle f, \sum_{n \le N} g_n/n \rangle \to -\langle f, \kappa \rangle = -\lambda \quad (N \to \infty). \;\square$$

3 Dirichlet Series

For $f \in D$ we define

$$F(s) = \sum_{n \ge 1} \frac{u(n; f)}{n^s},$$

and we will say that F is the Dirichlet series of f.

We will denote by σ the real part of the complex variable s. The following proposition summarizes the basic facts about the correspondence between elements f of D and their Dirichlet series F. We keep the notations of (2) and (3).

Proposition 8 *For $f \in D$, the Dirichlet series $F(s)$ is absolutely convergent in the half-plane $\sigma > 3/2$ and convergent in the half-plane $\sigma > 1$. It has a meromorphic continuation to the half-plane $\sigma > 1/2$ (we will denote it also by $F(s)$), with a unique pole in $s = 1$, simple and with residue $-\lambda$. In the strip $1/2 < \sigma < 1$, one has*

$$F(s)/s = \int_0^\infty f(t) t^{-s-1} dt. \tag{10}$$

If $f \in D_0$, that is $\lambda = 0$, there is no pole at $s = 1$, and the Mellin transform (10) *represents the analytic continuation of $F(s)/s$ to the half-plane $\sigma > 1/2$. Moreover, the Dirichlet series $F(s)$ converges on the line $\sigma = 1$.*

Proof If $h = 0$ in (3), the arithmetical function u is the constant $-\lambda$, and $F = -\lambda\zeta$. In this case, the assertion about (10) follows from (2.1.5), p. 14 of [6].

If $\lambda = 0$, then $f = h \in D_0$ and $u(n) = h(n) - h(n-1)$ by (4). Therefore,

$$\begin{aligned}\sum_{n\geq 1}\frac{|u(n)|}{n^\sigma} &\leq 2\sum_{n\geq 1}\frac{|h(n)|}{n^\sigma}\\ &\leq 2\Big(\sum_{n\geq 1}\frac{|h(n)|^2}{n^2}\Big)^{1/2}\Big(\sum_{n\geq 1}\frac{1}{n^{2\sigma-2}}\Big)^{1/2}\\ &\leq 2\sqrt{2}\,\zeta(2\sigma-2)^{1/2}\|h\| < \infty,\end{aligned}$$

if $\sigma > 3/2$, where we used Cauchy's inequality for sums.

The convergence of the series $F(1)$ follows from the formula $u(n) = -\langle h, \varphi_n\rangle$ and Proposition 3. It implies the convergence of $F(s)$ in the half-plane $\sigma > 1$.

Using the Bunyakovsky-Schwarz inequality for integrals, and the fact that $h = 0$ on $(0, 1)$, one sees that the integral (10) now converges absolutely and uniformly in every half-plane $\sigma \geq 1/2 + \varepsilon$ (with $\varepsilon > 0$), thus defining a holomorphic function in the half-plane $\sigma > 1/2$. It is the analytic continuation of $F(s)/s$ since one has, for $\sigma > 3/2$,

$$\begin{aligned}\int_0^\infty h(t)t^{-s-1}dt &= \frac{1}{s}\sum_{n\geq 1}h(n)\big(n^{-s}-(n+1)^{-s}\big)\\ &= \frac{1}{s}\sum_{n\geq 1}\frac{h(n)-h(n-1)}{n^s} = \frac{F(s)}{s}.\end{aligned}$$

Finally, the convergence of the Dirichlet series $F(s)$ on the line $\sigma = 1$ follows from the convergence at $s = 1$ and the holomorphy of F on the line, by a theorem of Marcel Riesz (cf. [5], Satz I, p. 350).

One combines the two cases, $h = 0$ and $\lambda = 0$, to obtain the statement of the proposition. □

The Dirichlet series $F(s)$ of functions in D_0 are precisely those which converge in some half-plane and have an analytic continuation to $\sigma > 1/2$ such that $F(s)/s$ belongs to the Hardy space H^2 of this last half-plane. As we will not use this fact in the present paper, we omit its proof.

We now investigate the Dirichlet series

$$\frac{F(s)}{\zeta(s)} = \sum_{n\geq 1}\frac{w(n; f)}{n^s}.$$

Proposition 9 *Let $f \in D$, and let $F(s)$ be the Dirichlet series of f. The Dirichlet series $F(s)/\zeta(s)$ is absolutely convergent if $\sigma > 3/2$ and convergent if $\sigma \geq 1$.*

Proof The Dirichlet series $F(s)$ converges for $\sigma > 1$, and converges absolutely for $\sigma > 3/2$ (Proposition 8). The Dirichlet series $1/\zeta(s)$ converges absolutely for $\sigma > 1$. The Dirichlet product $F(s)/\zeta(s)$ thus converges absolutely for $\sigma > 3/2$ and converges for $\sigma > 1$.

If $s = 1$, the series is convergent by Proposition 7. Since the function $F(s)/\zeta(s)$ is holomorphic in the closed half-plane $\sigma \geq 1$, Riesz' convergence theorem applies again to ensure convergence on the line $\sigma = 1$. □

4 Báez-Duarte's Criterion for the Riemann Hypothesis

We now define

$$\mathscr{B} = \mathrm{Vect}(e_n,\ n \in \mathbb{N}^*) \quad ; \quad \mathscr{B}_0 = \mathrm{Vect}(e_n - e_1/n,\ n \in \mathbb{N}^*,\ n \geq 2).$$

Since $e_n \in D$ and $e_n - e_1/n \in D_0$ for all $n \in \mathbb{N}^*$, one sees that

$$\overline{\mathscr{B}} \subset D \quad ; \quad \overline{\mathscr{B}_0} \subset D_0 \quad ; \quad \overline{\mathscr{B}_0} = \overline{\mathscr{B}} \cap D_0.$$

The subspace $\overline{\mathscr{B}}$ is the (non-orthogonal) direct sum of $\mathrm{Vect}(e_1)$ and $\overline{\mathscr{B}_0}$.

We will consider the orthogonal projection $\widetilde{\chi}$ (resp. $\widetilde{\chi}_0$) of χ on $\overline{\mathscr{B}}$ (resp. $\overline{\mathscr{B}_0}$). In 2003, Báez-Duarte gave the following criterion for the Riemann hypothesis.

Proposition 10 *The following seven assertions are equivalent.*

(i) $\overline{\mathscr{B}} = D$; (i)$_0$ $\overline{\mathscr{B}_0} = D_0$

(ii) $\chi \in \overline{\mathscr{B}}$; (ii)$_0$ $\chi \in \overline{\mathscr{B}_0}$

(iii) $\widetilde{\chi} = \chi$; (iii)$_0$ $\widetilde{\chi}_0 = \chi$

(iv) *the Riemann hypothesis is true.*

In fact, Báez-Duarte's paper [2] contains the proof of the equivalence of (ii) and (iv); the other equivalences are mere variations. The statements (i)$_0$, (ii)$_0$, and (iii)$_0$ allow one to work in the sequence space $\mathfrak{h}$ instead of the function space H; see [3] for an exposition in this setting.

The main property of Dirichlet series of elements of $\overline{\mathscr{B}}$ is given in the following proposition.

Proposition 11 *If $f \in \overline{\mathscr{B}}$, the Dirichlet series $F(s)/\zeta(s)$ has a holomorphic continuation to the half-plane $\sigma > 1/2$.*

Proof Write $f = \lambda e_1 + h$, with $\lambda \in \mathbb{R}$ and $h \in D_0$. If $h = 0$, one has $F = -\lambda\zeta$, and the result is true.

Now suppose $\lambda = 0$. The function h is the limit in H of finite linear combinations, say h_j ($j \geq 1$), of the $e_k - e_1/k$ ($k \geq 2$), when $j \to \infty$. The Dirichlet series of $e_k - e_1/k$ is

$$(k^{-1} - k^{-s})\zeta(s),$$

so that the result is true for each h_j. It remains to see what happens when one passes to the limit.

By the relation between the Dirichlet series of h_j and the Mellin transform of h_j, one sees that the Mellin transform of h_j must vanish at each zero ρ of ζ in the half-plane $\sigma > 1/2$, with a multiplicity no less than the corresponding multiplicity of ρ as a zero of ζ. Thus

$$\int_1^\infty h_j(t) t^{-\rho-1} \ln^k t \, dt = 0 \tag{11}$$

for every zero ρ of the Riemann zeta function, such that $\Re\rho > 1/2$, and for every non-negative integer k smaller than the multiplicity of ρ as a zero of ζ. When $j \to \infty$, one gets (11) with h_j replaced by h, which proves the result for h.

One combines the two cases, $h = 0$ and $\lambda = 0$, to obtain the statement of the proposition. □

5 The ν Function

5.1 Weingartner's Form of Báez-Duarte's Criterion

For $N \in \mathbb{N}^*$, we will consider the orthogonal projections of χ on $\mathrm{Vect}(e_1, \dots, e_N)$ and on $\mathrm{Vect}(e_2 - e_1/2, \dots, e_N - e_1/N)$:

$$\chi_N = \sum_{k=1}^{N} c(k, N) e_k \tag{12}$$

$$\chi_{0,N} = \sum_{k=2}^{N} c_0(k, N)(e_k - e_1/k), \tag{13}$$

thus defining the coefficients $c(k, N)$ and $c_0(k, N)$. In [8], Weingartner gave a formulation of Báez-Duarte's criterion in terms of the coefficients $c_0(k, N)$ of (13). The same can be done with the $c(k, N)$ of (12). First, we state a basic property of these coefficients.

Proposition 12 *For every $k \in \mathbb{N}^*$, the coefficients $c(k, N)$ in* (12) *and $c_0(k, N)$ in* (13) *(here, with $k \geq 2$) converge when N tends to infinity.*

Proof With the notations of Section 4,

$$\widetilde{\chi} = \lim_{N \to \infty} \chi_N$$

$$\widetilde{\chi}_0 = \lim_{N \to \infty} \chi_{0,N},$$

where the limits are taken in H.

Using the identity (6), we observe that, for every $N \in \mathbb{N}^*$,

$$c(k, N) = -w(k; \chi_N) \qquad (k \geq 1)$$
$$c_0(k, N) = -w(k; \chi_{0,N}) \quad (k \geq 2),$$

Therefore, Proposition 4 yields, for every k,

$$c(k, N) \to -w(k; \widetilde{\chi}) \qquad (N \to \infty)$$
$$c_0(k, N) \to -w(k; \widetilde{\chi_0}) \quad (N \to \infty).$$

□

Definition 1 The arithmetical functions ν and ν_0 are defined by

$$\nu(n) = w(n; \widetilde{\chi})$$
$$\nu_0(n) = w(n; \widetilde{\chi_0}).$$

Note that

$$\nu_0(1) = \lim_{N\to\infty} \sum_{2\leq k\leq N} \frac{c_0(k, N)}{k} = -\sum_{k\geq 2} \frac{\nu_0(k)}{k},$$

by Proposition 7.

We can now state Báez-Duarte's criterion in Weingartner's formulation.

Proposition 13 *The following assertions are equivalent.*

(i) $\nu = \mu$

(ii) $\nu_0 = \mu$ *on* $\mathbb{N}^* \setminus \{1\}$

(iii) *the Riemann hypothesis is true.*

Proof By Báez-Duarte's criterion, (iii) is equivalent to $\chi = \widetilde{\chi}$. By Proposition 1, this is equivalent to $w(n; \chi) = w(n; \widetilde{\chi})$ for all $n \geq 1$, that is, $\mu = \nu$.

Similarly, (iii) implies $\mu = \nu_0$. Conversely, if $\mu(n) = \nu_0(n)$ for all $n \geq 2$, then one has $w(n; \chi - \widetilde{\chi_0}) = 0$ for $n \geq 2$, which means that $\chi - \widetilde{\chi_0}$ is a scalar multiple of e_1. This implies $\chi = \widetilde{\chi_0}$ since χ and $\widetilde{\chi_0}$ belong to D_0. □

5.2 *The Dirichlet Series* $\sum_n \nu(n)n^{-s}$

Since $\nu(n) = w(n; \widetilde{\chi})$, the following proposition is a corollary of Propositions 9 and 11.

Proposition 14 *The Dirichlet series*

$$\sum_{n\geq 1} \frac{\nu(n)}{n^s}$$

is absolutely convergent for $\sigma > 3/2$, *convergent for* $\sigma \geq 1$, *and has a holomorphic continuation to the half-plane* $\sigma > 1/2$.

6 Questions

Here are three questions related to the preceding exposition.

Question 1 Is it true that $\widetilde{\chi} = \widetilde{\chi}_0$?

Question 2 Let $f \in D$ such that the Dirichlet series $F(s)/\zeta(s)$ has a holomorphic continuation to the half-plane $\sigma > 1/2$. Is it true that $f \in \overline{\mathscr{B}}$?

A positive answer would be a discrete analogue of Bercovici's and Foias' Corollary 2.2, p. 63 of [4].

Question 3 Is the Dirichlet series

$$\sum_{n\geq 1} \frac{\nu(n)}{n^s}$$

convergent in the half-plane $\sigma > 1/2$?

Another open problem is to obtain any quantitative estimate beyond the tautologies $\|\widetilde{\chi} - \widetilde{\chi}_N\| = o(1)$ and $\|\widetilde{\chi}_0 - \widetilde{\chi}_{0,N}\| = o(1)$ $(N \to \infty)$.

Appendix: Some Computations

Scalar Products

1. One has

$$\langle e_1, \varepsilon_k \rangle = \sqrt{k(k+1)} \int_k^{k+1} (t-k)\frac{dt}{t^2} = \sqrt{k(k+1)}\big(\ln(1+1/k) - 1/(k+1)\big). \tag{14}$$

2. For $k \in \mathbb{N}^*$, one has

$$\langle e_1, \varphi_k \rangle = \int_{k-1}^{k} k(k-1)(t-k+1)\frac{dt}{t^2} - \int_k^{k+1} k(k+1)(t-k)\frac{dt}{t^2}$$

$$= 2k^2 \ln k - k(k-1)\ln(k-1) - k(k+1)\ln(k+1) + 1$$
$$= -\omega(1/k),$$

where

$$\omega(z) = z^{-2}\big((1-z)\ln(1-z) + (1+z)\ln(1+z)\big) - 1$$
$$= \sum_{j\geq 1} \frac{z^{2j}}{(j+1)(2j+1)} \qquad (|z| \leq 1).$$

3. For $n \in \mathbb{N}^*$, one has

$$\langle e_1, f_n\rangle = \sum_{k|n} \mu(n/k)\langle e_1, \varphi_k\rangle = -\sum_{k|n} \mu(n/k)\omega(1/k)$$
$$= -\sum_{j\geq 1} \frac{\sum_{k|n} \mu(n/k)k^{-2j}}{(j+1)(2j+1)} = -\sum_{j\geq 1} \frac{n^{-2j}\prod_{p|n}(1-p^{2j})}{(j+1)(2j+1)}.$$

In particular,

$$\sup_{n\in\mathbb{N}^*} |\langle e_1, f_n\rangle| = \sum_{j\geq 1} \frac{1}{(j+1)(2j+1)} = \ln 4 - 1.$$

Projections

By (14), the orthogonal projection e_1' of e_1 on D_0 is

$$e_1' = \sum_{k\geq 1} \langle e_1, \varepsilon_k\rangle\varepsilon_k = \sum_{k\geq 1} \sqrt{k(k+1)}\big(\ln(1+1/k) - 1/(k+1)\big)\varepsilon_k.$$

Since $e_1'(k)$ has limit $1/2$ when k tends to infinity, one sees that $e_1 - e_1'$ "interpolates" between the fractional part (on $[0, 1[$) and the first Bernoulli function (at infinity). One has the Hilbertian decomposition

$$D = D_0 \oplus \mathrm{Vect}(e_1 - e_1').$$

Since $\kappa \perp D_0$ and $\langle \kappa, e_1\rangle = 1$, the orthogonal projection of κ on D is

$$\kappa' = \frac{e_1 - e_1'}{\|e_1 - e_1'\|^2}.$$

Acknowledgments I thank Andreas Weingartner for useful remarks on the manuscript.

References

1. L. Báez-Duarte, A class of invariant unitary operators. Adv. Math. **144**, 1–12 (1999)
2. L. Báez-Duarte, A strengthening of the Nyman-Beurling criterion for the Riemann hypothesis. Atti Accad. Naz. Lincei Cl. Sci. Fis. Mat. Natur. Rend. Lincei (9) Mat. Appl. **14**, 5–11 (2003)
3. B. Bagchi, On Nyman, Beurling and Baez-Duarte's Hilbert space reformulation of the Riemann hypothesis. Proc. Indian Acad. Sci. Math. Sci. **116**, 139–146 (2006)
4. H. Bercovici et C. Foias, A real variable restatement of Riemann's hypothesis. Isr. J. Math. **48**(1), 57–68 (1984)
5. M. Riesz, Ein Konvergenzsatz für *Dirichlet*sche Reihen. Acta Math. **40**, 349–361 (1916)
6. E.C. Titchmarsh, *The Theory of the Riemann Zeta Function*, 2nd edn. (Clarendon Press, Oxford, 1986), revised by D. R. Heath-Brown
7. V. Vasyunin, On a biorthogonal system associated with the Riemann hypothesis. St. Petersbg. Math. J. **7**, 405–419 (1996)
8. A. Weingartner, On a question of Balazard and Saias related to the Riemann hypothesis. Adv. Math. **208**, 905–908 (2007)

Large Sets Avoiding Rough Patterns

Jacob Denson, Malabika Pramanik, and Joshua Zahl

Abstract The pattern avoidance problem seeks to construct a set $X \subset \mathbf{R}^d$ with large dimension that avoids a prescribed pattern. Examples of such patterns include three-term arithmetic progressions (solutions to $x_1 - 2x_2 + x_3 = 0$), geometric structures such as simplices, or more general patterns of the form $f(x_1, \dots, x_n) = 0$. Previous work on the subject has considered patterns described by polynomials or by functions f satisfying certain regularity conditions. We consider the case of "rough" patterns, not prescribed by functional zeros.

There are several problems that fit into the framework of rough pattern avoidance. As a first application, if $Y \subset \mathbf{R}^d$ is a set with Minkowski dimension α, we construct a set X with Hausdorff dimension $d - \alpha$ such that $X + X$ is disjoint from Y. As a second application, if C is a Lipschitz curve with Lipschitz constant less than one, we construct a set $X \subset C$ of dimension $1/2$ that does not contain the vertices of an isosceles triangle.

A major question in modern geometric measure theory is whether sufficiently large sets are forced to contain copies of certain patterns. Intuitively, one expects the answer to be yes, and many results in the literature support this intuition. For example, the Lebesgue density theorem implies that a set of positive Lebesgue measure contains an affine copy of every finite set. Similarly, any set $X \subset \mathbf{R}^2$ with Hausdorff dimension exceeding one must contain three collinear points. On the other hand, there is a distinct genre of results that challenges this intuition. Keleti [1] constructs a full-dimensional set $X \subset \mathbf{R}$ that avoids all solutions of the equation $x_2 - x_1 = x_4 - x_3$ with $x_1 < x_2 \leq x_3 < x_4$ and which consequently does not contain any nontrivial arithmetic progression. Given any triangle, Falconer [2] constructs a full-dimensional planar set that does not contain any similar copy of the vertex set of the triangle. Maga [3] provides a set $X \subset \mathbf{R}^2$ of full Hausdorff dimension such that no four points in X form the vertices of a parallelogram. The pattern avoidance

J. Denson · M. Pramanik · J. Zahl (✉)
University of British Columbia, Vancouver, BC, Canada
e-mail: denson@math.ubc.ca; malabika@math.ubc.ca; jzahl@math.ubc.ca

M. Th. Rassias (ed.), *Harmonic Analysis and Applications*, Springer Optimization and Its Applications 168, https://doi.org/10.1007/978-3-030-61887-2_4

problem (informally stated) asks: for a given pattern, how large can the dimension of a set $X \subset \mathbf{R}^d$ be before it is forced to contain a copy of this pattern?

One way to formalize the notion of a pattern is as follows. If $d \geq 1$ and $n \geq 2$ are integers, we define a pattern to be a set $Z \subset \mathbf{R}^{dn}$. We say that a set $X \subset \mathbf{R}^d$ avoids the pattern Z if for every n-tuple of distinct points $x_1, \ldots, x_n \in X$, we have $(x_1, \ldots, x_n) \notin Z$. For example, a set $X \subset \mathbf{R}^2$ does not contain three collinear points if and only if it avoids the pattern

$$Z = \{(x_1, x_2, x_3) \in \mathbf{R}^6 : |(x_1 - x_2) \wedge (x_1 - x_3)| = 0\}.$$

Similarly, a set $X \subset \mathbf{R}^2$ avoids the pattern

$$Z = \{(x_1, x_2, x_3, x_4) \in \mathbf{R}^8 : x_1 + x_4 = x_2 + x_3\}$$

if and only if no four points in X form the vertices of a (possibly degenerate) parallelogram.

A number of recent articles have established pattern avoidance results for increasingly general patterns. In [4], Máthé constructs a set $X \subset \mathbf{R}^d$ that avoids a pattern specified by a countable union of algebraic varieties of controlled degree. In [5], Fraser and the second author consider the pattern avoidance problem for countable unions of C^1 manifolds. In this paper, we consider the pattern avoidance problem for an even more general class of "rough" patterns $Z \subset \mathbf{R}^{dn}$, which are the countable union of sets with controlled lower Minkowski dimension.

Theorem 1 *Let $\alpha \geq d$, and let $Z \subset \mathbf{R}^{dn}$ be a countable union of compact sets, each with lower Minkowski dimension at most α. Then there exists a compact set $X \subset [0, 1]^d$ with Hausdorff dimension at least $(nd - \alpha)/(n - 1)$ such that whenever $x_1, \ldots, x_n \in X$ are distinct, we have $(x_1, \ldots, x_n) \notin Z$.*

Remarks

1. When $\alpha < d$, the pattern avoidance problem is trivial, since $X = [0, 1]^d - \pi(Z)$ is full dimensional and solves the pattern avoidance problem, where $\pi(x_1, \ldots, x_n) = x_1$ is a projection map from $\mathbf{R}^{dn}$ to $\mathbf{R}^d$. We will therefore assume that $\alpha \geq d$ in our proof of the theorem. Note that obtaining a full dimensional avoiding set in the case $\alpha = d$, however, is still interesting.
2. Theorem 1 is trivial when $\alpha = dn$, since we can set $X = \emptyset$. We will therefore assume that $\alpha < dn$ in our proof of the theorem.
3. When Z is a countable union of smooth manifolds in $\mathbf{R}^{nd}$ of co-dimension m, we have $\alpha = nd - m$. In this case Theorem 1 yields a set in $\mathbf{R}^d$ with Hausdorff dimension at least $(nd - \alpha)/(n - 1) = m/(n - 1)$. This recovers Theorem 1.1 and 1.2 from [5], making Theorem 1 a generalization of these results.
4. Since Theorem 1 does not require any regularity assumptions on the set Z, it can be applied in contexts that cannot be addressed using previous methods. Two such applications, new to the best of our knowledge, have been recorded in Section 5; see Theorems 2 and 3 there.

The set X in Theorem 1 is obtained by constructing a sequence of approximations to X, each of which avoids the pattern Z at different scales. For a sequence of lengths $l_k \searrow 0$, we construct a nested family of compact sets $\{X_k\}$, where X_k is a union of closed cubes of sidelength l_k that avoids Z at scales close to l_k. The set $X = \bigcap X_k$ avoids Z at all scales. While this proof strategy is not new, our method for constructing the sets $\{X_k\}$ has several innovations that simplify the analysis of the resulting set $X = \bigcap X_k$. In particular, through a probabilistic selection process, we are able to avoid the complicated queuing techniques used in [1] and [5], which required storage of data from each step of the iterated construction, to be retrieved at a much later stage of the construction process.

At the same time, our construction continues to share certain features with [5]. For example, between each pair of scales l_{k-1} and l_k, we carefully select an intermediate scale r_k. The set $X_k \subset X_{k-1}$ avoids Z at scale l_k, and it is "evenly distributed" at scale r_k; the set X_k is a union of cubes of length l_k whose midpoints resemble (a large subset of) generalized arithmetic progression of step size r_k. The details of a single step of this construction are described in Section 2. In Section 3, we explain how the length scales l_k and r_k for X are chosen and prove its avoidance property. In Section 4, we analyze the size of X and show that it satisfies the conclusions of Theorem 1.

1 Frequently Used Notation and Terminology

1. A *dyadic length* is a number l of the form 2^{-k} for some non-negative integer k.
2. Given a length $l > 0$, we let $\mathcal{B}_l^d$ denote the set of all closed cubes in $\mathbf{R}^d$ with sidelength l and corners on the lattice $(l \cdot \mathbf{Z})^d$, i.e.,
$$\mathcal{B}_l^d = \{[a_1, a_1 + l] \times \cdots \times [a_d, a_d + l] : a_k \in l \cdot \mathbf{Z}\}.$$
3. A set $E \subset \mathbf{R}^d$ is *l discretized* if it is a union of cubes in $\mathcal{B}_l^d$. For any set $E \subset \mathbf{R}^d$, and any length $l > 0$, we let
$$E(l) = \bigcup \{I \in \mathcal{B}_l^d : I \cap E \neq \emptyset\}.$$
Then $E(l)$ is the smallest l discretized set with $E \subset E(l)^\circ$. Here and throughout the paper A° will denote the interior of the set A. Given an l discretized set E, we let
$$\mathcal{B}_l^d(E) = \{I \in \mathcal{B}_l^d : I \subset E\}.$$
Then $E = \bigcup \mathcal{B}_l^d(E)$.
4. The *lower Minkowski dimension* of a bounded set $Z \subset \mathbf{R}^d$ is defined as

$$\underline{\dim}_{\mathbf{M}}(Z) = \liminf_{l \to 0} \frac{\log\left[\# \mathcal{B}_l^d(Z(l))\right]}{\log[1/l]}.$$

5. If $\alpha \geq 0$ and $\delta > 0$, we define the dyadic Hausdorff content of a set $E \subset \mathbf{R}^d$ as

$$H_\delta^\alpha(E) = \inf\left\{\sum_{k=1}^\infty l_k^\alpha : E \subset \bigcup_{k=1}^\infty I_k\right\},$$

where the infinum is taken over all families of cubes $\{I_k\}$ such that for each k, there exists a dyadic length $l_k \leq \delta$ such that $I_k \in \mathcal{B}_{l_k}^d$. The α-dimensional dyadic Hausdorff measure H^α on $\mathbf{R}^d$ is $H^\alpha(E) = \lim_{\delta \to 0} H_\delta^\alpha(E)$, and the *Hausdorff dimension* of a set E is $\dim_{\mathbf{H}}(E) = \inf\{\alpha \geq 0 : H^\alpha(E) = 0\}$.
6. Given $K \in \mathcal{B}_l^{dn}$, we can decompose K as $K_1 \times \cdots \times K_n$ for unique cubes $K_1, \ldots, K_n \in \mathcal{B}_l^d$. We say K is *strongly non-diagonal* if the cubes $K_1, \ldots, K_n$ are distinct. Strongly non-diagonal cubes will play an important role in Section 2, when we solve a discrete version of Theorem 1.
7. Adopting the terminology of [6], we say a collection of sets $\{U_k\}$ is a *strong cover* of a set E if $E \subset \limsup U_k$, which means every element of E is contained in infinitely many of the sets U_k. This idea will be useful in Section 3.
8. A *Frostman measure* of dimension α is a non-zero compactly supported probability measure μ on $\mathbf{R}^d$ such that for every dyadic cube I of sidelength l, $\mu(I) \lesssim l^\alpha$. Note that a measure μ satisfies this inequality for every dyadic cube I if and only if it satisfies a similar inequality, possibly with a different implicit constant, for all cubes I. *Frostman's lemma* says that

$$\dim_{\mathbf{H}}(E) = \sup\left\{\alpha : \begin{array}{c}\text{there is a Frostman measure of} \\ \text{dimension } \alpha \text{ supported on } E\end{array}\right\}.$$

2 Avoidance at Discrete Scales

In this section we describe a method for avoiding Z at a single scale. We apply this technique in Section 3 at many scales to construct a set X avoiding Z at all scales. This single scale avoidance technique is the core building block of our construction, and the efficiency with which we can avoid Z at a single scale has direct consequences on the Hausdorff dimension of the set X obtained in Theorem 1.

At a single scale, we solve a discretized version of the problem, where all sets are unions of cubes at two dyadic lengths $l > s$. In this discrete setting, Z is replaced by a discretized version of itself, namely, a union of cubes in $\mathcal{B}_s^{dn}$ denoted by G. Given a set E, which is a union of cubes in $\mathcal{B}_l^d$, our goal is to construct a set $F \subset E$ that is a union of cubes in $\mathcal{B}_s^d$, such that F^n is disjoint from strongly non-diagonal cubes (see Definition 6) in $\mathcal{B}_s^{dn}(G)$. Using the setup mentioned at the end of the introduction, we will later choose $l = l_k$, $s = l_{k+1}$, and $E = X_k$. The set X_{k+1} will be defined as the set F constructed.

In order to ensure the final set X obtained in Theorem 1 has large Hausdorff dimension regardless of the rapid decay of scales used in the construction of X, it is crucial that F is uniformly distributed at intermediate scales between l and s. We achieve this by decomposing E into sub-cubes in $\mathcal{B}_r^d$ for some intermediate scale $r \in [s, l]$ and distributing F as evenly among these intermediate sub-cubes as possible. We achieve this by assuming a mild regularity condition on the number of cubes in G; see equation (2.1).

Lemma 1 *Fix two distinct dyadic lengths l and s, with $l > 4ds$. Let $E \subseteq [0, 1]^d$ be a nonempty and l discretized set, and let $G \subset \mathbf{R}^{dn}$ be a nonempty s discretized set such that*

$$(l/s)^d \leq \#\mathcal{B}_s^{dn}(G) \leq \frac{(l/s)^{dn}}{(4d)^{d(n-1)}}. \tag{2.1}$$

Then there exists a dyadic length $r \in [4ds, l]$ of size

$$r \sim_d \left(l^{-d} s^{dn} \# \mathcal{B}_s^{dn}(G)\right)^{\frac{1}{d(n-1)}}, \tag{2.2}$$

and an s discretized set $F \subset E$ satisfying the following four properties:

1. Disjointness*: The cubes in $\mathcal{B}_s^{dn}(F)$ are disjoint from one another.*
2. Avoidance*: For any n distinct cubes $J_1, \ldots, J_n \in \mathcal{B}_s^d(F)$, $J_1 \times \cdots \times J_n \notin \mathcal{B}_s^d(G)$.*
3. Non-concentration*: For any $I \in \mathcal{B}_r^d(E)$, $\#(\mathcal{B}_s^d(F \cap I)) \leq 1$.*
4. Large Size*: For every $I \in \mathcal{B}_l^d(E)$, $\#(\mathcal{B}_s^d(F \cap I)) \geq \#(\mathcal{B}_r^d(E \cap I))/2$.*

Remark Property 2 says that F avoids strongly non-diagonal cubes in $\mathcal{B}_s^{dn}(G)$. Properties 3 and 4 together imply that for every $I \in \mathcal{B}_l^d(E)$, at least half of the cubes in $\mathcal{B}_r^d(I)$ contribute a single sub-cube of sidelength s to F; the rest contribute none. One of the many consequences of Property 4 is that every sidelength l cube in E contains a sidelength s cube in F; in other words, no sidelength l cube in E "dies out."

Proof Let r be the smallest dyadic length at least as large as R, where

$$R = (4d)\left(l^{-d} s^{dn} \# \mathcal{B}_s^{dn}(G)\right)^{\frac{1}{d(n-1)}}. \tag{2.3}$$

This choice of r satisfies (2.2). The inequalities in (2.1) ensure that $r \in [4ds, l]$; more precisely, the left inequality in (2.1) implies R is bounded from below by $4ds$, and the right inequality implies R is bounded from above by l. The minimality of r ensures $4ds \leq r \leq l$.

For each $I \in \mathcal{B}_r^d(E)$, pick J_I uniformly at random from $\mathcal{J}_I = \{J \in \mathcal{B}_s^d(I) : J \subset I^\circ\}$; these choices are independent as I ranges over the elements of $\mathcal{B}_r^d(E)$. Since two distinct dyadic cubes I and I' in $\mathcal{B}_r^d(E)$ have disjoint interiors, we have that

$J_I \cap J_{I'} = \emptyset$. Define

$$U = \bigcup \left\{ J_I : I \in \mathcal{B}_r^d(E) \right\},$$

and

$$\mathcal{K}(U) = \{K \in \mathcal{B}_s^{dn}(G) : K \in U^n, K \text{ strongly non-diagonal}\}.$$

Note that the sets U and $\mathcal{K}(U)$ are random, in the sense that they depend on the random variables $\{J_I\}$. Define

$$F(U) = U - \bigcup \{\pi(K) : K \in \mathcal{K}(U)\} = \bigcup \left\{ \mathcal{B}_s^d(U) - \{\pi(K) : K \in \mathcal{K}(U)\} \right\}, \tag{2.4}$$

where $\pi : \mathbf{R}^{dn} \to \mathbf{R}^d$ is the projection map $(x_1, \dots, x_n) \mapsto x_1$, for $x_i \in \mathbf{R}^d$. We will verify that $F = F(U)$ always obeys the first three properties claimed in Lemma 1 and satisfies Property 4 with non-zero probability.

Our construction ensures that U is s discretized and that the cubes of $\mathcal{B}_s^d(U)$ are disjoint from one another. Since $F(U) \subset U$, it follows that $F(U)$ satisfies Property 1.

Given any strongly non-diagonal cube $K = J_1 \times \cdots \times J_n \in \mathcal{B}_s^{dn}(G)$, either $K \notin \mathcal{B}_s^{dn}(U^n)$ or $K \in \mathcal{B}_s^{dn}(U^n)$. If the former occurs then $K \notin \mathcal{B}_s^{dn}(F(U)^n)$ since $F(U) \subset U$, so $\mathcal{B}_s^{dn}(F(U)^n) \subset \mathcal{B}_s^{dn}(U^n)$. If the latter occurs then $K \in \mathcal{K}(U)$, and since $\pi(K) = J_1$, $J_1 \notin \mathcal{B}_s^d(F(U))$. In either case, $K \notin \mathcal{B}_s^{dn}(F(U)^n)$, so $F(U)$ satisfies Property 2.

By construction, U contains at most one sub-cube $J \in \mathcal{B}_s^{dn}$ for each $I \in \mathcal{B}_l^{dn}(E)$. Since $F(U) \subset U$, $F(U)$ satisfies Property 3.

It remains to verify that with non-zero probability, the set $F(U)$ satisfies Property 4. For each cube $J \in \mathcal{B}_s^d(E)$, there is a unique "parent" cube $I \in \mathcal{B}_r^d(E)$ such that $J \subset I$. Since each $(d-1)$-dimensional face of I intersects $(r/s)^{d-1}$ cubes in $\mathcal{B}_s^d(I)$, and I has $2d$ faces, the relation $r \geq 4ds$ implies

$$\#(\mathcal{J}_I) \geq (r/s)^d - 2d(r/s)^{d-1} = (r/s)^d[1 - 2d(s/r)] \geq (1/2) \cdot (r/s)^d. \tag{2.5}$$

Since J_I is chosen uniformly at random from $\mathcal{J}_I$, (2.5) shows

$$\mathbf{P}(J \subset U) = \mathbf{P}(J_I = J) \leq 2(s/r)^d.$$

The cubes $\{J_I\}$ are chosen independently, so if $J_1, \dots, J_n$ are distinct cubes in $\mathcal{B}_s^d(E)$, then either the cubes $J_1, \dots, J_n$ have distinct parents, in which case we apply the independence of J_I to conclude that $\mathbf{P}(J_1, \dots, J_n \subset U) \leq 2^n (s/r)^{dn}$. If the cubes $J_1, \dots, J_n$ do not have distinct parents, Property 3 implies $\mathbf{P}(J_1, \dots, J_n \subset U) = 0$. In either case, we conclude that

$$\mathbf{P}(J_1, \dots, J_n \subset U) \leq 2^n (s/r)^{dn}. \tag{2.6}$$

Let $K = J_1 \times \cdots \times J_n$ be a strongly non-diagonal cube in $\mathcal{B}_s^{dn}(G)$. We deduce from (2.6) that

$$\mathbf{P}(K \subset U^n) = \mathbf{P}(J_1, \ldots, J_n \subset U) \leq 2^n (s/r)^{dn}. \tag{2.7}$$

By (2.7) and the linearity of expectation,

$$\mathbf{E}(\#\mathcal{K}(U)) = \sum_{K \in \mathcal{B}_s^{dn}(G)} \mathbf{P}(K \subset U^n) \leq \# \mathcal{B}_s^{dn}(G) \cdot 2^n (s/r)^{dn} \leq (1/2) \cdot (l/r)^d.$$

The last inequality can be deduced from (2.3) and the condition $r \geq 4ds$. In particular, there exists at least one (non-random) set U_0 such that

$$\#\mathcal{K}(U_0) \leq \mathbf{E}(\#\mathcal{K}(U)) \leq (1/2) \cdot (l/r)^d. \tag{2.8}$$

In other words, $F(U_0) \subset U_0$ is obtained by removing at most $(1/2) \cdot (l/r)^d$ cubes in $\mathcal{B}_s^d$ from U_0. For each $I \in \mathcal{B}_l^d(E)$, we know that $\# \mathcal{B}_s^d(I \cap U_0) = (l/r)^d$. Combining this with (2.8), we arrive at the estimate

$$\begin{aligned} \# \mathcal{B}_s^d(I \cap F(U_0)) &= \mathcal{B}_s^d(I \cap U_0) - \#\{\pi(K) : \ K \in \mathcal{K}(U_0), \pi(K) \in U_0\} \\ &\geq \mathcal{B}_s^d(I \cap U_0) - \#(\mathcal{K}(U_0)) \\ &\geq (l/r)^d - (1/2) \cdot (l/r)^d \geq (1/2) \cdot (l/r)^d. \end{aligned}$$

In other words, $F(U_0)$ satisfies Property 4.

The set $F(U)$ satisfies Properties 1, 2, and 3 regardless of which values are assumed by the random variables $\{J_I\}$. Furthermore, there is at least one set U_0 such that $F(U_0)$ satisfies Property 4. Setting $F = F(U_0)$ completes the proof. □

Remarks

1. While Lemma 1 uses probabilistic arguments, the conclusion of the lemma is not a probabilistic statement. In particular, one can find a suitable F constructively by checking every possible choice of U (there are finitely many) to find one particular choice U_0 which satisfies (2.8) and then defining F by (2.4). Thus the set we obtain in Theorem 1 exists by purely constructive means.
2. At this point, it is possible to motivate the numerology behind the dimension bound $\dim(X) \geq (dn - \alpha)/(n - 1)$ from Theorem 1, albeit in the context of Minkowski dimension. We will pause to do so here before returning to the proof of Theorem 1. For simplicity, let $\alpha > d$, and suppose that $Z \subset \mathbf{R}^{dn}$ satisfies

$$\# \mathcal{B}_s^{dn}(Z(s)) \sim s^{-\alpha} \quad \text{for every } s \in (0, 1]. \tag{2.9}$$

Let $l = 1$ and $E = [0, 1]^d$, and let $s > 0$ be a small parameter. If s is chosen sufficiently small compared to d, n, and α, then (2.1) is satisfied with $G = \bigcup \mathcal{B}_s^{dn}(Z(s))$. We can then apply Lemma 1 to find a dyadic scale $r \sim s^{(dn-\alpha)/d(n-1)}$ and a set F that avoids the strongly non-diagonal cubes of $\mathcal{B}_s^{dn}(Z(s))$. The set F is a union of approximately $r^{-d} \sim s^{-(dn-\alpha)/(n-1)}$ cubes of sidelength s. Thus informally, the set F resembles a set with Minkowski dimension α when viewed at scale s.

The set X constructed in Theorem 1 will be obtained by applying Lemma 1 iteratively at many scales. At each of these scales, X will resemble a set of Minkowski dimension $(dn - \alpha)/(n - 1)$. A careful analysis of the construction (performed in Section 4) shows that X actually has Hausdorff dimension at least $(dn - \alpha)/(n - 1)$.

3. Lemma 1 is the core method in our avoidance technique. The remaining argument is fairly modular. If, for a special case of Z, one can improve the result of Lemma 1 so that r is chosen on the order of $s^{\beta/d}$, then the remaining parts of our paper can be applied near verbatim to yield a set X with Hausdorff dimension β, as in Theorem 1.

3 Fractal Discretization

In this section, we construct the set X from Theorem 1 by applying Lemma 1 at many scales. Let us start by fixing a strong cover Z that we will work with in the sequel.

Lemma 2 *Let $Z \subset \mathbf{R}^{dn}$ be a countable union of bounded sets with Minkowski dimension at most α, and let $\epsilon_k \searrow 0$ with $2\epsilon_k < dn - \alpha$ for all k. Then there exists a sequence of dyadic lengths $\{l_k\}$ and a sequence of sets $\{Z_k\}$, such that*

1. Strong Cover*: The interiors $\{Z_k^\circ\}$ of the sets $\{Z_k\}$ form a strong cover of Z.*
2. Discreteness*: For all $k \geq 0$, Z_k is an l_k discretized subset of $\mathbf{R}^{dn}$.*
3. Sparsity*: For all $k \geq 0$, $l_k^{-d} \leq \#\mathcal{B}_{l_k}^{dn}(Z_k) \leq l_k^{-\alpha-\epsilon_k}$.*
4. Rapid Decay*: For all $k > 1$,*

$$l_k^{dn-\alpha-\varepsilon_k} \leq (1/4d)^{d(n-1)} \cdot l_{k-1}^{dn}, \tag{3.1}$$

$$l_k^{\epsilon_k} \leq l_{k-1}^{2d}. \tag{3.2}$$

Proof By hypothesis, there exists a sequence of sets $\{Y_i\}$ so that $Z \subset \bigcup_{i=1}^{\infty} Y_i$ and $\underline{\dim}_{\mathbf{M}}(Y_i) \leq \alpha$ for each index i. Without loss of generality, we may assume that for each length l,

$$\#\mathcal{B}_l^{nd}(Y_i(l)) \geq l^{-d}. \tag{3.3}$$

If (3.3) fails to be satisfied for some set Y_i, we consider the d dimensional hyperplane

$$H = \{(x_1, \dots, x_1) : x_1 \in [0, 1]^d\}.$$

and replace Y_i with $Y_i \cup H$. Let $\{i_k\}$ be a sequence of integers that repeats each integer infinitely often.

The lengths $\{l_k\}$ and sets $\{Z_k\}$ are defined inductively. As a base case, set $l_0 = 1$ and $Z_0 = [0, 1]^{nd}$. Suppose that the lengths $l_0, \dots, l_{k-1}$ have been chosen. Since $\underline{\dim}_{\mathbf{M}}(Y_{i_k}) \leq \alpha$, and $\varepsilon_k > 0$, Definition 4 implies that there exist arbitrarily small dyadic lengths l that satisfy

$$\# \mathcal{B}_l^{dn}(Y_{i_k}(l)) \leq l^{-\alpha-\varepsilon_k}. \tag{3.4}$$

In particular, we can choose a dyadic length $l = l_k$ small enough to satisfy (3.1), (3.2), and (3.4). With this choice of l_k, Property 4 is satisfied. Define $Z_k = Y_{i_k}(l_k)$. This choice of Z_k clearly satisfies Property 2, and Property 3 is implied by (3.3) and (3.4).

It remains to verify that the sets $\{Z_k^\circ\}$ strongly cover Z. Fix a point $z \in Z$. Then there exists an index i such that $z \in Y_i$, and there is a subsequence $k_1, k_2, \dots$ such that $i_{k_j} = i$ for each j. But then $z \in Y_i \subset Z_{i_{k_j}}^\circ$, so z is contained in each of the sets $Z_{i_{k_j}}^\circ$, and thus $z \in \limsup Z_i^\circ$. □

To construct X, we consider a nested, decreasing family of compact sets $\{X_k\}$, where each X_k is an l_k discretized subset of $\mathbf{R}^d$. We then set $X = \bigcap X_k$. Then X is a nonempty compact set. The goal is to choose X_k such that X_k^n does not contain any *strongly non diagonal* cubes in Z_k.

Lemma 3 *Let $\{l_k\}$ be a sequence of positive numbers converging to zero, and let $Z \subset \mathbf{R}^{dn}$. Let $\{Z_k\}$ be a sequence of sets, with each Z_k an l_k discretized subset of $\mathbf{R}^{dn}$, such that the interiors $\{Z_k^\circ\}$ strongly cover Z. For each index k, let X_k be an l_k discretized subset of $\mathbf{R}^d$. Suppose that for each k, $\mathcal{B}_{l_k}^{dn}(X_k^n) \cap \mathcal{B}_{l_k}^{dn}(Z_k)$ contains no strongly non diagonal cubes. If $X = \bigcap X_k$, then for any distinct $x_1, \dots, x_n \in X$, we have $(x_1, \dots, x_n) \notin Z$.*

Proof Let $z \in Z$ be a point with distinct coordinates $z_1, \dots, z_n$. Define

$$\Delta = \{(w_1, \dots, w_n) \in \mathbf{R}^{dn} : \text{there exists } i \neq j \text{ such that } w_i = w_j\}.$$

Then $d(\Delta, z) > 0$, where d is the Hausdorff distance between Δ and z. Since $\{Z_k^\circ\}$ strongly covers Z, there is a subsequence $\{k_m\}$ such that $z \in Z_{k_m}^\circ$ for every index m. Since l_k converges to 0 and thus l_{k_m} converges to 0, if m is sufficiently large, then $\sqrt{dn} \cdot l_{k_m} < d(\Delta, z)$. Note that $\sqrt{dn} \cdot l_{km}$ is the diameter of a cube in $\mathcal{B}_{l_{km}}^{dn}$. For such a choice of m, any cube $I \in \mathcal{B}_{l_{km}}^{d}$ which contains z is strongly non-diagonal. Furthermore, $z \in Z_{k_m}^\circ$. Since X_{k_m} and Z_{k_m} share no cube which contains z, this implies $z \notin X_{k_m}$. In particular, this means $z \notin X^n$. □

All that remains is to apply the discrete lemma to choose the sets X_k.

Lemma 4 *Given a sequence of dyadic length scales* $\{l_k\}$ *obeying,* (3.1), (3.2), *and* (3.4) *as above, there exists a sequence of sets* $\{X_k\}$ *and a sequence of dyadic intermediate scales* $\{r_k\}$ *with* $l_k \le r_k \le l_{k-1}$ *for each* $k \ge 1$*, such that each set* X_k *is an* l_k *discretized subset of* $[0,1]^d$ *and such that* $\mathcal{B}^{dn}_{l_k}(X^d_k) \cap \mathcal{B}^{dn}_{l_k}(Z_k)$ *contains no strongly non diagonal cubes. Furthermore, for each index* $k \ge 1$*, we have*

$$r_k \lesssim l_k^{(dn-\alpha-\epsilon_k)/d(n-1)}, \tag{3.5}$$

$$\#\mathcal{B}^d_{l_k}(X_k \cap I) \ge (1/2)\cdot(l_{k-1}/r_k)^d \quad \textit{for each } I \in \mathcal{B}^d_{l_{k-1}}(X_{k-1}), \tag{3.6}$$

$$\#\mathcal{B}^d_{l_k}(X_k \cap I) \le 1 \quad \textit{for each } I \in \mathcal{B}^d_{r_k}(X_{k-1}). \tag{3.7}$$

Proof We construct X_k by induction, using Lemma 1 at each step. Set $X_0 = [0,1]^d$. Next, suppose that the sets $X_0, \ldots, X_{k-1}$ have been defined. Our goal is to apply Lemma 1 to $E = X_{k-1}$ and $G = Z_k$ with $l = l_{k-1}$ and $s = l_k$. This will be possible once we verify the hypothesis (2.1), which in this case takes the form

$$(l_{k-1}/l_k)^d \le \#\mathcal{B}^{dn}_{l_k}(Z_k) \le (1/4d)^{d(n-1)} \cdot (l_{k-1}/l_k)^{dn}. \tag{3.8}$$

The right-hand side follows from Property 3 of Lemma 2 and (3.1). On the other hand, Property 3 of Lemma 2 and the fact that $l_{k-1} \le 1$ implies that

$$(l_{k-1}/l_k)^d \le l_k^{-d} \le \#\mathcal{B}^{dn}_{l_k}(Z_k),$$

establishing the left inequality in (3.8). Applying Lemma 1 as described above now produces a dyadic length

$$r \sim_d \left(l_{k-1}^{-d} l_k^{dn} \#\mathcal{B}^{dn}_{l_k}(Z_k)\right)^{\frac{1}{d(n-1)}} \tag{3.9}$$

and an l_k discretized set $F \subset X_{k-1}$. The set F satisfies Properties 2, 3, and 4 from the statement of Lemma 1. Define $r_k = r$ and $X_k = F$. The estimate (3.5) on r_k follows from (3.9) using the known bounds (3.2) and (3.4):

$$\begin{aligned} r_k &\lesssim \left(l_{k-1}^{-d} l_k^{dn-\alpha-0.5\epsilon_k}\right)^{\frac{1}{d(n-1)}} = \left(l_{k-1}^{-d} l_k^{0.5\epsilon_k} l_k^{dn-\alpha-\epsilon_k}\right)^{\frac{1}{d(n-1)}} \\ &= \left(l_{k-1}^{-2d} l_k^{\epsilon_k}\right)^{\frac{1}{2d(n-1)}} l_k^{\frac{dn-\alpha-\epsilon_k}{d(n-1)}} \lesssim l_k^{\frac{dn-\alpha-\epsilon_k}{d(n-1)}}. \end{aligned}$$

The requirements (3.6) and (3.7) follow from Properties 3 and 4 of Lemma 1, respectively. □

Now we have defined the sets $\{X_k\}$, we set $X = \bigcap X_k$. Since X_k avoids strongly non-diagonal cubes in Z_k, Lemma 3 implies that if $x_1, \ldots, x_n \in X$ are distinct, then $(x_1, \ldots, x_n) \notin Z$. To finish the proof of Theorem 1, we must show that $\dim_{\mathbf{H}}(X) \ge (dn-\alpha)/(n-1)$. This will be done in the next section.

4 Dimension Bounds

To complete the proof of Theorem 1, we must show that $\dim_{\mathbf{H}}(X) \geq (dn - \alpha)/(n - 1)$. In view of Item 8 in Section 1, this will follow from the existence of a Frostman measure of appropriate dimension supported on X.

We start by recursively defining a positive function $w : \bigcup_{k=0}^{\infty} \mathcal{B}^d_{l_k}(X_k) \to [0, 1]$, following the well-known mass distribution principle. Set $w([0, 1]^d) = 1$. Suppose now that $w(I)$ has been defined for all cubes $I \in \mathcal{B}^d_{l_{k-1}}(X_{k-1})$. Let $I \in \mathcal{B}^d_{l_k}(X_k)$. Consider the unique "parent cube" $I^* \in \mathcal{B}^d_{l_{k-1}}(X_{k-1})$ for which $I \subset I^*$. Define

$$w(I) = \frac{w(I^*)}{\# \mathcal{B}^d_{l_k}(X_k \cap I^*)}. \tag{4.1}$$

In other words, the mass $w(I^*)$ of I^* is divided equally among its descendants. The construction ensures that for every $I^* \in \mathcal{B}^d_{l_{k-1}}(X_{k-1})$, $\# \mathcal{B}^d_{l_k}(X_k \cap I^*) > 0$, i.e., I^* has a non-zero number of descendants; hence no mass is allowed to escape. Observe that for each index $k \geq 1$, if $I^* \in \mathcal{B}^d_{l_{k-1}}(X_{k-1})$,

$$\sum_{I \in \mathcal{B}^d_{l_k}(I^* \cap X_k)} w(I) = w(I^*) \neq 0. \tag{4.2}$$

In particular, for each index k we have

$$\sum_{I \in \mathcal{B}^d_{l_k}(X_k)} w(I) = 1.$$

Let $\mathcal{B}$ denote the collection of all dyadic cubes in $\mathcal{B}^d_{l_k}(X_k)$ for all $k \geq 0$.

We now apply a standard procedure due to Caratheodory, modelled after the approach in [7, Theorem 4.2], to obtain a Borel measure μ supported on X. For any $k \geq 1$ and any set $E \subset \mathbf{R}^d$, we define an exterior measure

$$\begin{aligned}\mu_k(E) &= \inf\left\{ \sum_{i=1}^{\infty} w(I_i) : E \cap X \subset \bigcup_{i=1}^{\infty} I_i,\, I_i \in \bigcup_{j=k}^{\infty} \mathcal{B}^d_{l_j}(X_j) \right\} \\ &= \inf\left\{ \sum_{i=1}^{\infty} w(I_i) : E \cap X \subset \bigcup_{i=1}^{\infty} I_i,\ \operatorname{diam}(I_i) \leq l_k,\ I_i \in \mathcal{B} \right\}.\end{aligned}$$

Then $\mu_k(E)$ is monotone in k for each set E. We set $\mu(E) = \lim_{k \to \infty} \mu_k(E)$. It follows from [7, Theorem 4.2] that μ is Borel measure supported on X. Property 1 of Lemma 1 implies $\mathcal{B}_{l_k}(X_k)$ is a family of disjoint closed cubes for each k. In particular, this means that $I \cap X$ is relatively open in X for each $I \in \mathcal{B}$. Since X is a compact set, this means that for any $I \in \mathcal{B}$, $\mu_k(I \cap X)$ is equal to the infinum of $\sum_{i=1}^{N} w(I_i)$ for covers of $I \cap X$ by *finitely* many sets $I_1, \ldots, I_N$. Combined with the relation (4.2), this gives that $\mu_{k'}(I) = w(I)$ for all $I \in \mathcal{B}^d_{l_k}$ and all $k' \geq k$. In particular, $\mu_k(\mathbf{R}^d) = \mu_k([0,1]^d) = 1$ for all $k \geq 1$; hence μ is a probability measure.

To complete the proof of Theorem 1, we will show that μ is a Frostman measure of dimension $(dn - \alpha)/(n-1) - \epsilon$ for every $\epsilon > 0$.

Lemma 5 *For each $k \geq 1$, if $I \in \mathcal{B}^d_{l_k}$,*

$$\mu(I) \lesssim l_k^{\frac{dn-\alpha}{n-1} - \eta_k}, \quad \textit{where} \quad \eta_k = \frac{n+1}{2(n-1)} \cdot \epsilon_k \searrow 0 \textit{ as } k \to \infty.$$

Proof Let $I \in \mathcal{B}^d_{l_k}$ and let $I^* \in \mathcal{B}^d_{l_{k-1}}$ be the parent cube of I. If $\mu(I) > 0$, then $I \subset X_k$, and combining (4.1), (3.6), and (3.5) with the fact that $\mu(I^*) \leq 1$, we obtain

$$\mu(I) = \frac{\mu(I^*)}{\# \mathcal{B}^d_{l_k}(X_k \cap I^*)} \leq 2\left(\frac{r_k}{l_{k-1}}\right)^d \lesssim \frac{l_k^{\frac{dn-\alpha-\epsilon_k}{n-1}}}{l_{k-1}^d} \leq l_k^{\frac{dn-\alpha}{n-1} - \eta_k} \left(\frac{l_k^{\epsilon_k/2}}{l_{k-1}^d}\right) \leq l_k^{\frac{dn-\alpha}{n-1} - \eta_k}.$$

□

Corollary 1 *For each $k \geq 1$ and each $I \in \mathcal{B}^d_{r_k}$, $\mu(I) \lesssim (r_k/l_{k-1})^d l_{k-1}^{\frac{dn-\alpha}{n-1} - \eta_{k-1}}$.*

Proof We begin by noting that $\mathcal{B}^d_{r_k}(I(r_k)) \lesssim 1$. If we combine this with (3.7), we find that

$$\#(\mathcal{B}^d_{l_k}(I(r_k) \cap X_k)) \lesssim 1. \tag{4.3}$$

For each $I \in \mathcal{B}^d_{l_k}(I(r_k) \cap X_k)$, let $I^* \in \mathcal{B}^d_{l_{k-1}}$ denote the parent cube of I. Working as in Lemma 5, but using its conclusion combined with (4.1) and (3.6), we find

$$\mu(I) = \frac{\mu(I^*)}{\# \mathcal{B}^d_{l_k}(X_k \cap I^*)} \lesssim (r_k/l_{k-1})^d l_{k-1}^{\frac{dn-\alpha}{n-1} - \eta_k}.$$ □

Lemma 5 and Corollary 1 allow us to control the behavior of μ at all scales.

Lemma 6 *For every $\alpha \in [d, dn)$, and for each $\epsilon > 0$, there is a constant C_ϵ so that for all dyadic lengths $l \in (0, 1]$ and all $I \in \mathcal{B}^d_l$, we have*

$$\mu(I) \leq C_\epsilon l^{\frac{dn-\alpha}{n-1} - \epsilon}. \tag{4.4}$$

Proof Fix $\epsilon > 0$. Since $\eta_k \searrow 0$ as $k \to \infty$, there is a constant C_ϵ so that $l_k^{-\eta_k} \leq C_\epsilon l_k^{-\epsilon}$ for each $k \geq 1$. For instance, if ε_k is decreasing, we could choose $C_\epsilon = l_{k_0}^{-\eta_{k_0}}$, where k_0 is the largest integer for which $\eta_{k_0} \geq \epsilon$. Next, let k be the (unique) index so that $l_{k+1} \leq l < l_k$. We will split the proof of (4.4) into two cases, depending on the position of l within $[l_{k+1}, l_k]$.

Case 1: If $r_{k+1} \leq l \leq l_k$, we can cover I by $(l/r_{k+1})^d$ cubes in $\mathcal{B}^d_{r_{k+1}}$. A union bound combined with Corollary 1 gives

$$\begin{aligned} \mu(I) &\lesssim (l/r_{k+1})^d (r_{k+1}/l_k)^d l_k^{\frac{dn-\alpha}{n-1}-\eta_k} \\ &= (l/l_k)^d l_k^{\frac{dn-\alpha}{n-1}-\eta_k} \\ &= l^{\frac{dn-\alpha}{n-1}} (l/l_k)^{\frac{\alpha-d}{n-1}} l_k^{-\eta_k} \\ &\leq l^{\frac{dn-\alpha}{n-1}-\eta_k} \\ &\leq C_\epsilon l^{\frac{dn-\alpha}{n-1}-\epsilon}. \end{aligned} \tag{4.5}$$

The penultimate inequality is a consequence of our assumption $\alpha \geq d$.

Case 2: If $l_{k+1} \leq l \leq r_{k+1}$, we can cover I by a single cube in $\mathcal{B}^d_{r_{k+1}}$. By (3.7), each cube in $\mathcal{B}^d_{r_{k+1}}$ contains at most one cube $I_0 \in \mathcal{B}^d_{l_{k+1}}(X_{k+1})$, so by Lemma 5,

$$\mu(I) \leq \mu(I_0) \lesssim l_{k+1}^{\frac{dn-\alpha}{n-1}-\eta_{k+1}} \leq C_\epsilon l_{k+1}^{\frac{dn-\alpha}{n-1}-\epsilon} \leq C_\epsilon l^{\frac{dn-\alpha}{n-1}-\epsilon}. \qquad \square$$

Applying Frostman's lemma to Lemma 6 gives $\dim_{\mathbf{H}}(X) \geq \frac{dn-\alpha}{n-1} - \epsilon$ for every $\epsilon > 0$, which concludes the proof of Theorem 1.

5 Applications

As discussed in the introduction, Theorem 1 generalizes Theorems 1.1 and 1.2 from [5]. In this section, we present two applications of Theorem 1 in settings where previous methods do not yield any results.

5.1 Sum-Sets Avoiding Specified Sets

Theorem 2 *Let $Y \subset \mathbf{R}^d$ be a countable union of sets of Minkowski dimension at most $\beta < d$. Then there exists a set $X \subset \mathbf{R}^d$ with Hausdorff dimension at least $d-\beta$ such that $X + X$ is disjoint from Y.*

Proof Define $Z = Z_1 \cup Z_2$, where

$$Z_1 = \{(x, y) : x + y \in Y\} \quad \text{and} \quad Z_2 = \{(x, y) : y \in Y/2\}.$$

Since Y is a countable union of sets of Minkowski dimension at most β, Z is a countable union of sets with lower Minkowski dimension at most $d + \beta$. Applying Theorem 1 with $n = 2$ and $\alpha = d + \beta$ produces a set $X \subset \mathbf{R}^d$ with Hausdorff dimension $2d - (d + \beta) = d - \beta$ such that $(x, y) \notin Z$ for all $x, y \in X$ with $x \neq y$. We claim that $X + X$ is disjoint from Y. To see this, first suppose $x, y \in X, x \neq y$. Since X avoids Z_1, we conclude that $x + y \notin Y$. Suppose now that $x = y \in X$. Since X avoids Z_2, we deduce that $X \cap (Y/2) = \emptyset$, and thus for any $x \in X$, $x + x = 2x \notin Y$. This completes the proof. □

5.2 Subsets of Lipschitz Curves Avoiding Isosceles Triangles

In [5], Fraser and the second author prove that there exists a set $S \subset [0, 1]$ with dimension $\log_3 2$ such that for any simple C^2 curve $\gamma : [0, 1] \to \mathbf{R}^n$ with bounded non-vanishing curvature, $\gamma(S)$ does not contain the vertices of an isosceles triangle. Our method enables us to obtain a result that works for Lipschitz curves with small Lipschitz constants. The dimensional bound that we provide is slightly worse than [5] ($1/2$ instead of $\log_3 2$), and the set we obtain only works for a single Lipschitz curve, not for many curves simultaneously.

Theorem 3 *Let $f : [0, 1] \to \mathbf{R}^{n-1}$ be Lipschitz with*

$$\|f\|_{Lip} := \sup \left\{ |f(x) - f(y)|/|x - y| : x, y \in [0, 1], x \neq y \right\} < 1.$$

Then there is a set $X \subset [0, 1]$ of Hausdorff dimension $1/2$ so that the set $\{(t, f(t)) : t \in X\}$ does not contain the vertices of an isosceles triangle.

Corollary 2 *Let $f : [0, 1] \to \mathbf{R}^{n-1}$ be C^1. Then there is a set $X \subset [0, 1]$ of Hausdorff dimension $1/2$ so that the set $\{(t, f(t)) : t \in X\}$ does not contain the vertices of an isosceles triangle.*

Proof of Corollary 2 The graph of any C^1 function can be locally expressed, after possibly a translation and rotation, as the graph of a Lipschitz function with small Lipschitz constant. In particular, there exists an interval $I \subset [0, 1]$ of positive length so that the graph of f restricted to I, after being suitably translated and rotated, is the graph of a Lipschitz function $g : [0, 1] \to \mathbf{R}^{n-1}$ with Lipschitz constant at most $1/2$. Since isosceles triangles remain invariant under translations and rotations, the corollary is a consequence of Theorem 3. □

Proof of Theorem 3 Set

$$Z = \left\{ (x_1, x_2, x_3) \in [0, 1]^3 : \begin{array}{l} \text{The three points } p_j = (x_j, f(x_j)), 1 \leq j \leq 3 \\ \text{form the vertices of an isosceles triangle} \end{array} \right\}. \tag{5.1}$$

In the next lemma, we show Z has lower Minkowski dimension at most two. By Theorem 1, there is a set $X \subset [0, 1]$ of Hausdorff dimension $1/2$ so that for each distinct $x_1, x_2, x_3 \in X$, we have $(x_1, x_2, x_3) \notin Z$. This is precisely the statement that for each $x_1, x_2, x_3 \in X$, the points $(x_1, f(x_1))$, $(x_2, f(x_2))$, and $(x_3, f(x_3))$ do not form the vertices of an isosceles triangle. □

Lemma 7 *Let $f : [0, 1] \to \mathbf{R}^{n-1}$ be Lipschitz with $\|f\|_{Lip} < 1$. Then the set Z given by* (5.1) *has Minkowski dimension at most two.*

Proof First, notice that three points $p_1, p_2, p_3 \in \mathbf{R}^n$ form an isosceles triangle, with p_3 as the apex, if and only if $p_3 \in H_{p_1,p_2}$, where

$$H_{p_1,p_2} = \left\{ x \in \mathbf{R}^n : \left(x - \frac{p_1 + p_2}{2} \right) \cdot (p_2 - p_1) = 0 \right\}. \tag{5.2}$$

To prove Z has Minkowski dimension at most two, it suffices to show that the set

$$W = \left\{ x \in [0, 1]^3 : p_3 = (x_3, f(x_3)) \in H_{p_1,p_2} \right\}$$

has upper Minkowski dimension at most 2. This is because Z is the union of three copies of W, obtained by permuting coordinates. To bound the upper Minkowski dimension of W, we prove the estimate

$$\#\big(\mathcal{B}^3_\delta(W(\delta))\big) \leq C\delta^{-2} \log(1/\delta) \quad \text{for all dyadic } 0 < \delta < 1, \tag{5.3}$$

where C is a constant independent of δ.

Fix $0 < \delta < 1$. Note that

$$\#\big(\mathcal{B}^3_\delta(W)\big) = \sum_{k=0}^{1/\delta} \sum_{\substack{I_1, I_2 \in \mathcal{B}^1_\delta[0,1] \\ d(I_1,I_2)=k\delta}} \#\mathcal{B}^3_\delta(W(\delta) \cap I_1 \times I_2 \times [0, 1]). \tag{5.4}$$

Our next task is to bound each of the summands in (5.4). Let $I_1, I_2 \in \mathcal{B}^1_\delta[0, 1]$, and let $k = \delta^{-1} d(I_1, I_2)$. Let x_1 be the midpoint of I_1, and x_2 the midpoint of I_2. Let $(y_1, y_2, y_3) \in W \cap I_1 \times I_2 \times [0, 1]$. Then it follows from (5.2) that

$$\left(y_3 - \frac{y_1 + y_2}{2} \right) \cdot (y_2 - y_1) + \left(f(y_3) - \frac{f(y_2) + f(y_1)}{2} \right) \cdot (f(y_2) - f(y_1)) = 0.$$

We know $|x_1 - y_1|, |x_2 - y_2| \leq \delta/2$, so

$$\left| \left(y_3 - \frac{y_1 + y_2}{2} \right) (y_2 - y_1) - \left(y_3 - \frac{x_1 + x_2}{2} \right) (x_2 - x_1) \right|$$

$$\leq \frac{|y_1 - x_1| + |y_2 - x_2|}{2} |y_2 - y_1| + \Big(|y_1 - x_1| + |y_2 - x_2|\Big) \left| y_3 - \frac{x_1 + x_2}{2} \right| \tag{5.5}$$

$$\leq (\delta/2) \cdot 1 + \delta \cdot 1 \leq 3\delta/2.$$

Conversely, we know $|f(x_1) - f(y_1)|, |f(x_2) - f(y_2)| \leq \delta/2$ because $\|f\|_{\text{Lip}} < 1$, and a similar calculation yields

$$\left| \left(f(y_3) - \frac{f(y_1) + f(y_2)}{2} \right) \cdot (f(y_2) - f(y_1)) \right.$$

$$\left. - \left(f(y_3) - \frac{f(x_1) + f(x_2)}{2} \right) \cdot (f(x_2) - f(x_1)) \right| \leq 3\delta/2. \tag{5.6}$$

Putting (5.5) and (5.6) together, we conclude that

$$\left| \left(y_3 - \frac{x_1 + x_2}{2} \right) (x_2 - x_1) + \left(f(y_3) - \frac{f(x_2) + f(x_1)}{2} \right) \cdot (f(x_2) - f(x_1)) \right| \leq 3\delta. \tag{5.7}$$

Since $|(x_2 - x_1, f(x_2) - f(x_1))| \geq |x_2 - x_1| \geq k\delta$, we can interpret (5.7) as saying the point $(y_3, f(y_3))$ is contained in a $3/k$ thickening of the hyperplane $H_{(x_1, f(x_1)),(x_2, f(x_2))}$. Given another value $y' \in W \cap I_1 \cap I_2 \cap [0, 1]$, it satisfies a variant of the inequality (5.7), and we can subtract the difference between the two inequalities to conclude

$$\left| (y_3 - y_3') (x_2 - x_1) + (f(y_3) - f(y_3')) \cdot (f(x_2) - f(x_1)) \right| \leq 6\delta. \tag{5.8}$$

The triangle difference inequality applied with (5.8) implies

$$\begin{aligned}(f(y_3) - f(y_3')) \cdot (f(x_2) - f(x_1)) &\geq |y_3 - y_3'||x_2 - x_1| - 6\delta \\ &= (k + 1)\delta \cdot |y_3 - y_3'| - 6\delta.\end{aligned} \tag{5.9}$$

Conversely,

$$\begin{aligned}(f(y_3) - f(y_3')) \cdot (f(x_2) - f(x_1)) &\leq \|f\|_{\text{Lip}}^2 |y_3 - y_3'||x_2 - x_1| \\ &= \|f\|_{\text{Lip}}^2 \cdot (k + 1)\delta \cdot |y_3 - y_3'|.\end{aligned} \tag{5.10}$$

Combining (5.9) and (5.10) and rearranging, we see that

$$|y_3 - y_3'| \leq \frac{6}{(k + 1)(1 - \|f\|_{\text{Lip}}^2)} \lesssim \frac{1}{k + 1}, \tag{5.11}$$

where the implicit constant depends only on $\|f\|_{\mathrm{Lip}}$ (and blows up as $\|f\|_{\mathrm{Lip}}$ approaches 1). We conclude that

$$\#\mathcal{B}_\delta^3(W(\delta) \cap I_1 \times I_2 \times [0,1]) \lesssim \frac{\delta^{-1}}{k+1}, \tag{5.12}$$

which holds uniformly over any value of k and δ.

We are now ready to bound the sum from (5.4). Note that for each value of k, there are at most $2/\delta$ pairs (I_1, I_2) with $d(I_1, I_2) = k\delta$. Indeed, there are $1/\delta$ choices for I_1 and then at most two choices for I_2. Equation (5.12) shows

$$\#\mathcal{B}_\delta^3(W(\delta))$$
$$= \sum_{k=0}^{1/\delta} \sum_{\substack{I_1, I_2 \in \mathcal{B}_\delta^1[0,1] \\ d(I_1, I_2) = k\delta}} \#\mathcal{B}_\delta^3(W(\delta) \cap I_1 \times I_2 \times [0,1]) \lesssim \delta^{-2} \sum_{k=0}^{1/\delta} \frac{1}{k+1} \lesssim \delta^{-2} \log(1/\delta).$$

In the above inequalities, the implicit constants depend on $\|f\|_{\mathrm{Lip}}$, but they are independent of δ. This establishes (5.3) and completes the proof. □

References

1. T. Keleti, A 1-dimensional subset of the reals that intersects each of its translates in at most a single point. Real Anal. Exchange **24**, 843–845 (1999)
2. K. Falconer, On a problem of Erdös on Fractal Combinatorial Geometry. J. Combin. Theory Ser. A. **59**, 142–148 (1992)
3. P. Maga, Full dimensional sets without given patterns. Real Anal. Exchange **36**, 79–90 (2010)
4. A. Máthé, Sets of large dimension not containing polynomial configurations. Adv. Math. **316**, 691–709 (2017)
5. R. Fraser, M. Pramanik, Large sets avoiding patterns. Anal. PDE **11**, 1083–1111 (2018)
6. N.H. Katz, T. Tao, Some connections between Falconer's distance set conjecture, and sets of Furstenburg type. New York J. Math. **7**, 149–187 (2001)
7. P. Mattila, *Geometric of Sets and Measures in Euclidean Spaces* (Cambridge University Press, Cambridge, 1995)

PDE Methods in Random Matrix Theory

Brian C. Hall

Abstract This article begins with a brief review of random matrix theory, followed by a discussion of how the large-N limit of random matrix models can be realized using operator algebras. I then explain the notion of "Brown measure," which play the role of the eigenvalue distribution for operators in an operator algebra.

I then show how methods of partial differential equations can be used to compute Brown measures. I consider in detail the case of the circular law and then discuss more briefly the case of the free multiplicative Brownian motion, which was worked out recently by the author with Driver and Kemp.

1 Random Matrices

Random matrix theory consists of choosing an $N \times N$ matrix at random and looking at natural properties of that matrix, notably its eigenvalues. Typically, interesting results are obtained only for *large* random matrices, that is, in the limit as N tends to infinity. The subject began with the work of Wigner [43], who was studying energy levels in large atomic nuclei. The subject took on new life with the discovery that the eigenvalues of certain types of large random matrices resemble the energy levels of quantum chaotic systems—that is, quantum mechanical systems for which the underlying classical system is chaotic. (See, e.g., [20] or [39].) There is also a fascinating conjectural agreement, due to Montgomery [35], between the statistical behavior of zeros of the Riemann zeta function and the eigenvalues of random matrices. See also [30] or [6].

We will review briefly some standard results in the subject, which may be found in textbooks such as those by Tao [40] or Mehta [33].

Supported in part by a Simons Foundation Collaboration Grant for Mathematicians

B. C. Hall (✉)
Department of Mathematics, University of Notre Dame, Notre Dame, IN, USA
e-mail: bhall@nd.edu

M. Th. Rassias (ed.), *Harmonic Analysis and Applications*, Springer Optimization and Its Applications 168, https://doi.org/10.1007/978-3-030-61887-2_5

1.1 The Gaussian Unitary Ensemble

The first example of a random matrix is the **Gaussian unitary ensemble** (GUE) introduced by Wigner [43]. Let H_N denote the real vector space of $N \times N$ Hermitian matrices, that is, those with $X^* = X$, where X^* is the conjugate transpose of X. We then consider a Gaussian measure on H_N given by

$$d_N e^{-N\operatorname{trace}(X^2)/2}\, dX, \quad X \in H_N, \tag{1}$$

where dX denotes the Lebesgue measure on H_N and where d_N is a normalizing constant. If X^N is a random matrix having this measure as its distribution, then the diagonal entries are normally distributed real random variables with mean zero and variance $1/N$. The off-diagonal entries are normally distributed complex random variables, again with mean zero and variance $1/N$. Finally, the entries are as independent as possible given that they are constrained to be Hermitian, meaning that the entries on and above the diagonal are independent (and then the entries below the diagonal are determined by those above the diagonal). The factor of N in the exponent in (1) is responsible for making the variance of the entries of order $1/N$. This scaling of the variances, in turn, guarantees that the eigenvalues of the random matrix X^N do not blow up as N tends to infinity.

In order to state the first main result of random matrix theory, we introduce the following notation.

Definition 1 For any $N \times N$ matrix X, the **empirical eigenvalue distribution** of X is the probability measure on $\mathbb{C}$ given by

$$\frac{1}{N}\sum_{j=1}^{N} \lambda_j,$$

where $\{\lambda_1, \ldots, \lambda_N\}$ are the eigenvalues of X, listed with their algebraic multiplicity.

We now state **Wigner's semicircle law**.

Theorem 2 *Let X^N be a sequence of independently chosen $N \times N$ random matrices, each chosen according to the probability distribution in* (1). *Then as $N \to \infty$, the empirical eigenvalue distribution of X^N converges almost surely in the weak topology to Wigner's semicircle law, namely, the measure supported on* $[-2, 2]$ *and given there by*

$$\frac{1}{2\pi}\sqrt{4 - x^2}\, dx, \quad -2 \le x \le 2. \tag{2}$$

Figure 1 shows a simulation of the Gaussian unitary ensemble for $N = 2{,}000$, plotted against the semicircular density in (2). One notable aspect of Theorem 2 is that the limiting eigenvalue distribution (i.e., the semicircular measure in (2))

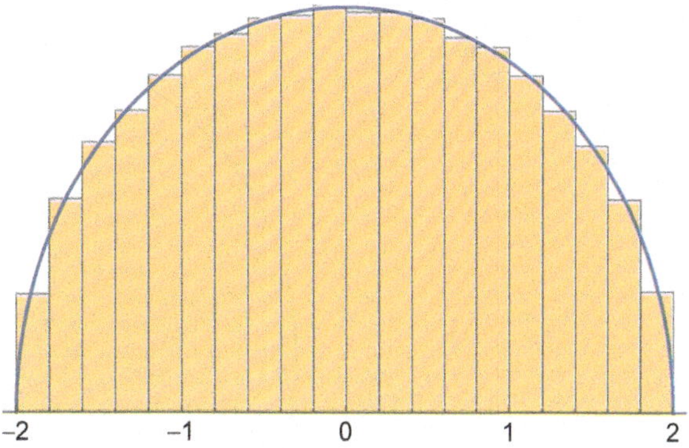

Fig. 1 A histogram of the eigenvalues of a GUE random variable with $N = 2{,}000$, plotted against a semicircular density

is *nonrandom*. That is to say, we are choosing a matrix at random, so that its eigenvalues are random, but in the large-N limit, the randomness in the bulk eigenvalue distribution disappears—it is always semicircular. Thus, if we were to select another GUE matrix with $N = 2{,}000$ and plot its eigenvalues, the histogram would (with high probability) look very much like the one in Figure 1.

It is important to note, however, that if one zooms in with a magnifying glass so that one can see the individual eigenvalues of a large GUE matrix, the randomness in the eigenvalues will persist. The behavior of these individual eigenvalues is of considerable interest, because they are supposed to resemble the energy levels of a "quantum chaotic system" (i.e., a quantum mechanical system whose classical counterpart is chaotic). Nevertheless, in this article, I will deal only with the bulk properties of the eigenvalues.

1.2 The Ginibre Ensemble

We now discuss the non-Hermitian counterpart to the Gaussian unitary ensemble, known as the **Ginibre ensemble** [15].We let $M_N(\mathbb{C})$ denote the space of *all* $N \times N$ matrices, not necessarily Hermitian. We then make a measure on $M_N(\mathbb{C})$ using a formula similar to the Hermitian case:

$$f_N\, e^{-N\,\mathrm{trace}(Z^*Z)}\, dZ, \quad Z \in M_N(\mathbb{C}), \tag{3}$$

where dZ denotes the Lebesgue measure on H_N and where f_N is a normalizing constant. In this case, the eigenvalues need not be real, and they follow the **circular law**.

Theorem 3 *Let Z^N be a sequence of independently chosen $N \times N$ random matrices, each chosen according to the probability distribution in* (3). *Then as $N \to \infty$, the empirical eigenvalue distribution of Z^N converges almost surely in the weak topology to the uniform measure on the unit disk.*

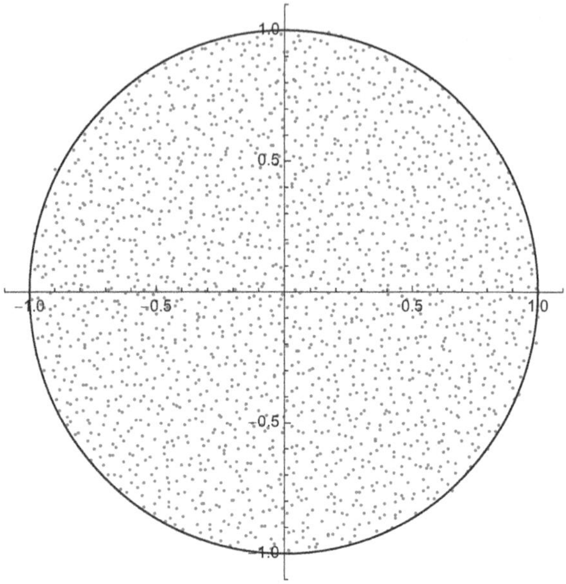

Fig. 2 A plot of the eigenvalues of a Ginibre matrix with $N = 2{,}000$

Figure 2 shows the eigenvalues of a random matrix chosen from the Ginibre ensemble with $N = 2{,}000$. As in the GUE case, the bulk eigenvalue distribution becomes deterministic in the large-N limit. As in the GUE case, one can also zoom in with a magnifying glass on the eigenvalues of a Ginibre matrix until the individual eigenvalues become visible, and the local behavior of these eigenvalues is an interesting problem—which will not be discussed in this article.

1.3 The Ginibre Brownian Motion

In this article, I will discuss a certain approach to analyzing the behavior of the eigenvalues in the Ginibre ensemble. The main purpose of this analysis is not so much to obtain the circular law, which can be proved by various other methods. The main purpose is rather to develop tools that can be used to study a more complex random matrix model in the group of *invertible* $N \times N$ matrices. The Ginibre case then represents a useful prototype for this more complicated problem.

It is then useful to introduce a time parameter into the description of the Ginibre ensemble, which we can do by studying the **Ginibre Brownian motion**. Specifically, in any finite-dimensional real inner product space V, there is a natural notion of Brownian motion. The Ginibre Brownian motion is obtained by taking V to be $M_N(\mathbb{C})$, viewed as a real vector space of dimension $2N^2$, and using the (real) inner product $\langle\cdot,\cdot\rangle_N$ given by

$$\langle X, Y\rangle_N := N\,\mathrm{Re}(\mathrm{trace}(X^*Y)).$$

We let C_t^N denote this Brownian motion, assumed to start at the origin.

At any one fixed time, the distribution of C_t^N is just the same as $\sqrt{t}Z^N$, where Z^N is distributed as the Ginibre ensemble. The *joint* distribution of the process C_t^N for various values of t is determined by the following property: For any collection of times $0 = t_0 < t_1 < t_2 < \cdots < t_k$, the "increments"

$$C_{t_1}^N - C_{t_0}^N, C_{t_2}^N - C_{t_1}^N, \ldots, C_{t_k}^N - C_{t_{k-1}}^N \tag{4}$$

are independent and distributed as $\sqrt{t_j - t_{j-1}}Z^N$.

2 Large-N Limits in Random Matrix Theory

Results in random matrix theory are typically expressed by first computing some quantity (e.g., the empirical eigenvalue distribution) associated to an $N \times N$ random matrix and then letting N tend to infinity. It is nevertheless interesting to ask whether there is some sort of limiting object that captures the large-N limit of the entire random matrix model. In this section, we discuss one common approach constructing such a limiting object.

2.1 Limit in $*$-Distribution

Suppose we have a matrix-valued random variable X, not necessarily normal. Then we can then speak about the $*$-moments of X, which are expressions like

$$\mathbb{E}\left\{\frac{1}{N}\mathrm{trace}(X^2(X^*)^3X^4X^*)\right\}.$$

Generally, suppose $p(a, b)$ is a polynomial in two noncommuting variables, that is, a linear combination of words involving products of a's and b's in all possible orders. We may then consider

$$\mathbb{E}\left\{\frac{1}{N}\mathrm{trace}[p(X, X^*)]\right\}.$$

If, as usual, we have a *family* X^N of $N \times N$ random matrices, we may consider the limits of such $*$-moments (if the limits exist):

$$\lim_{N\to\infty} \mathbb{E}\left\{\frac{1}{N}\mathrm{trace}[p(X^N, (X^N)^*)]\right\}. \tag{5}$$

2.2 Tracial von Neumann Algebras

Our goal is now to find some sort of limiting object that can encode *all* of the limits in (5). Specifically, we will try to find the following objects: (1) an operator algebra $\mathcal{A}$, (2) a "trace" $\tau : \mathcal{A} \to \mathbb{C}$, and (3) an element x of $\mathcal{A}$, such that for each polynomial p in two noncommuting variables, we have

$$\lim_{N\to\infty} \mathbb{E}\left\{\frac{1}{N}\mathrm{trace}[p(X^N, (X^N)^*)]\right\} = \tau[p(x, x^*)]. \tag{6}$$

We now explain in more detail what these objects should be. First, we generally take $\mathcal{A}$ to be a von Neumann algebra, that is, an algebra of operators that contains the identity, is closed under taking adjoints, and is closed under taking weak operator limits. Second, the "trace" τ is not actually computed by taking the trace of elements of $\mathcal{A}$, which are typically not of trace class. Rather, τ is a linear functional that has properties similar to the properties of the *normalized* trace $\frac{1}{N}\mathrm{trace}(\cdot)$ for matrices. Specifically, we require the following properties:

- $\tau(1) = 1$, where on the left-hand side, 1 denotes the identity operator,
- $\tau(a^*a) \geq 0$ with equality only if $a = 0$, and
- $\tau(ab) = \tau(ba)$, and
- τ should be continuous with respect to the weak-$*$ topology on $\mathcal{A}$.

Last, x is a single element of $\mathcal{A}$.

We will refer to the pair $(\mathcal{A}, \tau)$ as a **tracial von Neumann algebra**. We will not discuss here the methods used for actually constructing interesting examples of tracial von Neumann algebras. Instead, we will simply accept as a known result that certain random matrix models admit large-N limits as operators in a tracial von Neumann algebra. (The interested reader may consult the work of Biane and Speicher [5], who use a Fock space construction to find tracial von Neumann algebras of the sort we will be using in this article.)

Let me emphasize that although X^N is a matrix-valued random variable, x is *not* an operator-valued random variable. Rather, x is a *single* operator in the operator algebra $\mathcal{A}$. This situation reflects a typical property of random matrix models, which we have already seen an example of in Sections 1.1 and 1.2, that certain random quantities become nonrandom in the large-N limit. In the present context, it is often the case that we have a stronger statement than (6), as follows: If we sample the X^N's independently for different N's, then with probability one, we will have

$$\lim_{N\to\infty} \frac{1}{N}\mathrm{trace}[p(X^N, (X^N)^*)] = \tau[p(x, x^*)].$$

That is to say, in many cases, the *random* quantity $\frac{1}{N}\mathrm{trace}[p(X^N, (X^N)^*)]$ converges almost surely to the *single, deterministic* number $\tau[p(x, x^*)]$ as N tends to infinity.

2.3 Free Independence

In random matrix theory, it is often convenient to construct random matrices as sums or products of other random matrices, which are frequently assumed to be independent of one another. The appropriate notion of independence in the large-N limit—that is, in a tracial von Neumann algebra—is the notion of "freeness" or "free independence." This concept was introduced by Voiculescu [41, 42] and has become a powerful tool in random matrix theory. (See also the monographs [36] by Nica and Speicher and [34] by Mingo and Speicher.) Given an element a in a tracial von Neumann algebra $(\mathcal{A}, \tau)$ and a polynomial p, we may form the element $p(a)$. We also let $\dot{p}(a)$ denote the corresponding "centered" element, given by

$$\dot{p}(a) = p(a) - \tau(p(a))$$

We then say that elements $a_1, \ldots, a_k$ are **freely independent** (or, more concisely, **free**) if the following condition holds. Let $j_1, \ldots, j_n$ be any sequence of indices taken from $\{1, \ldots, k\}$, with the property that j_l is distinct from j_{l+1}. Let $p_{j_1}, \ldots, p_{j_n}$ be any sequence $p_{j_1}, \ldots, p_{j_n}$ of polynomials. Then we should have

$$\tau(\dot{p}_{j_1}(a_{j_1})\dot{p}_{j_2}(a_{j_2}) \cdots \dot{p}_{j_n}(a_{j_n})) = 0.$$

Thus, for example, if a and b are freely independent, then

$$\tau[(a^2 - \tau(a^2))(b^2 - \tau(b^2))(a - \tau(a))] = 0.$$

The concept of freeness allows us, in principle, to disentangle traces of arbitrary words in freely independent elements, thereby reducing the computation to the traces of powers of individual elements. As an example, let us do a few computations with two freely independent elements a and b. We form the corresponding centered elements $a - \tau(a)$ and $b - \tau(b)$ and start applying the definition:

$$\begin{aligned} 0 &= \tau[(a - \tau(a))(b - \tau(b))] \\ &= \tau[ab] - \tau[\tau(a)b] - \tau[a\tau(b)] + \tau[\tau(a)\tau(b)] \\ &= \tau[ab] - \tau(a)\tau(b) - \tau(a)\tau(b) + \tau(a)\tau(b) \\ &= \tau[ab] - \tau(a)\tau(b), \end{aligned}$$

where we have used that scalars can be pulled outside the trace and that $\tau(1) = 1$. We conclude, then, that

$$\tau(ab) = \tau(a)\tau(b).$$

A similar computation shows that $\tau(a^2 b) = \tau(a^2)\tau(b)$ and that $\tau(ab^2) = \tau(a)\tau(b^2)$.

The first really interesting case comes when we compute $\tau(abab)$. We start with

$$0 = \tau[(a - \tau(a))(b - \tau(b))(a - \tau(a))(b - \tau(b))]$$

and expand out the right-hand side as $\tau(abab)$ plus a sum of fifteen terms, all of which reduce to previously computed quantities. Sparing the reader the details of this computation, we find that

$$\tau(abab) = \tau(a^2)\tau(b)^2 + \tau(a)^2\tau(b^2) - \tau(a)^2\tau(b)^2.$$

Although the notion of free independence will not explicitly be used in the rest of this article, it is certainly a key concept that is always lurking in the background.

2.4 The Circular Brownian Motion

If Z^N is a Ginibre random matrix (Section 1.2), then the $*$-moments of Z^N converge to those of a "circular element" c in a certain tracial von Neumann algebra $(\mathcal{A}, \tau)$. The $*$-moments of c can be computed in an efficient combinatorial way (e.g., Example 11.23 in [36]). We have, for example, $\tau(c^*c) = 1$ and $\tau(c^k) = 0$ for all positive integers k.

More generally, we can realize the large-N limit of the entire Ginibre Brownian motion C_t^N, for all $t > 0$, as a family of elements c_t in a tracial von Neumann algebra $(\mathcal{A}, \tau)$. In the limit, the ordinary independence conditions for the increments of C_t^N (Section 1.3) is replaced by the free independence of the increments of c_t. That is, for all $0 = t_0 < t_1 < \cdots < t_k$, the elements

$$c_{t_1} - c_{t_0}, c_{t_2} - c_{t_1}, \ldots, c_{t_k} - c_{t_{k-1}}$$

are freely independent, in the sense described in the previous subsection. For any $t > 0$, the $*$-distribution of c_t is the same as the $*$-distribution of $\sqrt{t}c_1$.

3 Brown Measure

3.1 The Goal

Recall that if A is an $N \times N$ matrix with eigenvalues $\lambda_1, \ldots, \lambda_N$, the empirical eigenvalue distribution μ_A of A is the probability measure on $\mathbb{C}$ assigning mass $1/N$ to each eigenvalue:

$$\mu_A = \frac{1}{N}\sum_{j=1}^{N} \delta_{\lambda_j}.$$

Goal 4 *Given an arbitrary element x in a tracial von Neumann algebra $(\mathcal{A}, \tau)$, construct a probability measure μ_x on $\mathbb{C}$ analogous to the empirical eigenvalue distribution of a matrix.*

If $x \in \mathcal{A}$ is normal, then there is a standard way to construct such a measure. The spectral theorem allows us to construct a projection-valued measure γ_x [23, Section 10.3] associated to x. For each Borel set E, the projection $\gamma_x(E)$ will, again, belong to the von Neumann algebra $\mathcal{A}$, and we may therefore define

$$\mu_x(E) = \tau[\gamma_x(E)]. \tag{7}$$

We refer to μ_x as the **distribution** of x (relative to the trace τ). If x is not normal, we need a different construction—but one that we hope will agree with the above construction in the normal case.

3.2 A Motivating Computation

If A is an $N \times N$ matrix, define a function $s : \mathbb{C} \to \mathbb{R} \cup \{-\infty\}$ by

$$s(\lambda) = \log(|\det(A - \lambda)|^{2/N}),$$

where the logarithm takes the value $-\infty$ when $\det(A - \lambda) = 0$. Note that s is computed from the characteristic polynomial $\det(A - \lambda)$ of A. We can compute s in terms of its eigenvalues $\lambda_1, \ldots, \lambda_N$ (taken with their algebraic multiplicity) as

$$s(\lambda) = \frac{2}{N}\sum_{j=1}^{N} \log\left|\lambda - \lambda_j\right|. \tag{8}$$

See Figure 3 for a plot of (the negative of) $s(\lambda)$.

We then recall that the function $\log|\lambda|$ is a multiple of the Green's function for the Laplacian on the plane, meaning that the function is harmonic away from the origin and that

$$\Delta \log|\lambda| = 2\pi\,\delta_0(\lambda),$$

where δ_0 is a δ-measure at the origin. Thus, if we take the Laplacian of $s(\lambda)$, with an appropriate normalizing factor, we get the following nice result.

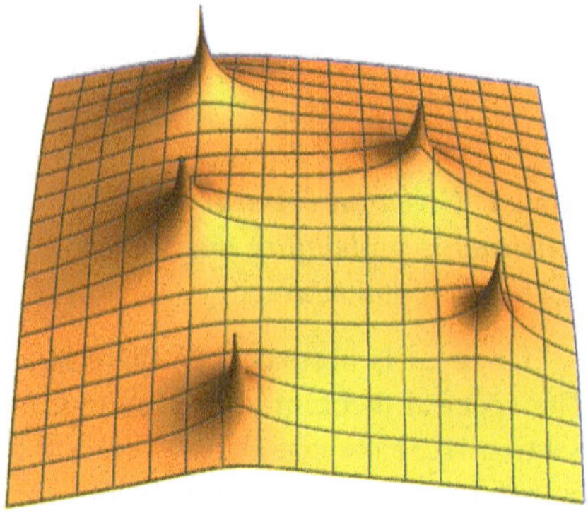

Fig. 3 A plot of the function $-s(\lambda)$ for a matrix with five eigenvalues. The function is harmonic except at the singularities

Proposition 5 *The Laplacian, in the distribution sense, of the function* $s(\lambda)$ *in* (8) *satisfies*

$$\frac{1}{4\pi}\Delta s(\lambda) = \frac{1}{N}\sum_{j=1}^{N}\delta_{\lambda_j}(\lambda),$$

where δ_{λ_j} *is a* δ*-measure at* λ_j*. That is to say,* $\frac{1}{4\pi}\Delta s$ *is the empirical eigenvalue distribution of* A *(Definition* 1*).*

Recall that if B is a strictly positive self-adjoint matrix, then we can take the logarithm of B, which is the self-adjoint matrix obtained by keeping the eigenvectors of B fixed and taking the logarithm of the eigenvalues.

Proposition 6 *The function* s *in* (8) *can also be computed as*

$$s(\lambda) = \frac{1}{N}\text{trace}[\log((A-\lambda)^*(A-\lambda))] \tag{9}$$

or as

$$s(\lambda) = \lim_{\varepsilon\to 0^+}\frac{1}{N}\text{trace}[\log((A-\lambda)^*(A-\lambda)+\varepsilon)]. \tag{10}$$

Here the logarithm is the self-adjoint logarithm of a positive self-adjoint matrix.

Note that in (9), the logarithm is undefined when λ is an eigenvalue of A. In (10), inserting $\varepsilon > 0$ guarantees that the logarithm is well defined for all λ, but a singularity of $s(\lambda)$ at each eigenvalue still arises in the limit as ε approaches zero.

Proof An elementary result [24, Theorem 2.12] says that for any matrix X, we have $\det(e^X) = e^{\text{trace}(X)}$. If P is a strictly positive matrix, we may apply this result with $X = \log P$ (so that $e^X = P$) to get

$$\det(P) = e^{\text{trace}(X)}$$

or

$$\operatorname{trace}(\log P) = \log[\det P].$$

Let us now apply this identity with $P = (A - \lambda)^*(A - \lambda)$, whenever λ is not an eigenvalue of A, to obtain

$$\begin{aligned}\frac{1}{N}\operatorname{trace}[\log((A-\lambda)^*(A-\lambda))] &= \frac{1}{N}\log[\det((A-\lambda)^*(A-\lambda))] \\ &= \frac{1}{N}\log[\det(A-\lambda)^*\det(A-\lambda)] \\ &= \log(|\det(A-\lambda)|^{2/N}),\end{aligned}$$

where this last expression is the definition of $s(\lambda)$.

Continuity of the matrix logarithm then establishes (10). □

3.3 *Definition and Basic Properties*

To define the Brown measure of a general element x in a tracial von Neumann algebra $(\mathcal{A}, \tau)$, we use the obvious generalization of (10). We refer to Brown's original paper [7] along with Chapter 11 of [34] for general references on the material in this section.

Theorem 7 *Let $(\mathcal{A}, \tau)$ be a tracial von Neumann algebra and let x be an arbitrary element of $\mathcal{A}$. Define*

$$S(\lambda, \varepsilon) = \tau[\log((x-\lambda)^*(x-\lambda)+\varepsilon)] \tag{11}$$

for all $\lambda \in \mathbb{C}$ and $\varepsilon > 0$. Then

$$s(\lambda) := \lim_{\varepsilon \to 0^+} S(\lambda, \varepsilon) \tag{12}$$

exists as an almost-everywhere-defined subharmonic function. Furthermore, the quantity

$$\frac{1}{4\pi}\Delta s, \tag{13}$$

where the Laplacian is computed in the distribution sense, is represented by a probability measure on the plane. We call this measure the Brown measure of x and denote it by μ_x.

The Brown measure of x is supported on the spectrum $\sigma(x)$ of x and has the property that

$$\int_{\sigma(x)} \lambda^k \, d\mu_x(\lambda) = \tau(x^k) \tag{14}$$

for all non-negative integers k.

See the original article [7] or Chapter 11 of the monograph [34] of Mingo and Speicher. We also note that the quantity $s(\lambda)$ is the logarithm of the **Fuglede–Kadison determinant** of $x - \lambda$; see [13, 14]. It is important to emphasize that, in general, the moment condition (14) *does not uniquely determine the measure* μ_x. After all, $\sigma(x)$ is an arbitrary nonempty compact subset of $\mathbb{C}$, which could, for example, be a closed disk. To uniquely determine the measure, we would need to know the value of $\int_{\sigma(x)} \lambda^k \bar{\lambda}^l \, d\mu_x(\lambda)$ for all non-negative integers k and l. There is not, however, any simple way to compute the value of $\int_{\sigma(x)} \lambda^k \bar{\lambda}^l \, d\mu_x(\lambda)$ in terms of the operator x. In particular, unless x is normal, this integral need not be equal to $\tau[x^k (x^*)^l]$. Thus, to compute the Brown measure of a general operator $x \in \mathcal{A}$, we actually have to work with the rather complicated definition in (11), (12), and (13).

We note two important special cases.

- Suppose $\mathcal{A}$ is the space of all $N \times N$ matrices and τ is the normalized trace, $\tau[x] = \frac{1}{N}\text{trace}(x)$. Then the Brown measure of any $x \in \mathcal{A}$ is simply the empirical eigenvalue distribution of x, which puts mass $1/N$ at each eigenvalue of x.
- If x is normal, then the Brown measure μ_x of x agrees with the measure defined in (7) using the spectral theorem.

3.4 Brown Measure in Random Matrix Theory

Suppose one has a family of $N \times N$ random matrix models X^N and one wishes to determine the large-N limit of the empirical eigenvalue distribution of X^N. (Recall Definition 1.) One may naturally use the following three-step process.

Step 1. Construct a large-N limit of X^N as an operator x in a tracial von Neumann algebra $(\mathcal{A}, \tau)$.

Step 2. Determine the Brown measure μ_x of x.

Step 3. Prove that the empirical eigenvalue distribution of X^N converges almost surely to μ_x as N tends to infinity.

It is important to emphasize that Step 3 in this process is not automatic. Indeed, this can be a difficult technical problem. Nevertheless, this article is concerned with exclusively with Step 2 in the process (in situations where Step 1 has been carried out). For Step 3, the main tool is the Hermitization method developed in Girko's pioneering paper [16] and further refined by Bai [1]. (Although neither of these authors explicitly uses the terminology of Brown measure, the idea is lurking there.)

There exist certain pathological examples where the limiting eigenvalue distribution does not coincide with the Brown measure. In light of a result of Śniady [38], we can say that such examples are associated with spectral instability, that is, matrices where a small change in the matrix produces a large change in the eigenvalues. Śniady shows that if we add to X^N a small amount of random Gaussian noise, then eigenvalues distribution of the perturbed matrices will converge to the Brown measure of the limiting object. (See also the papers [19] and [12], which obtain similar results by very different methods.) Thus, if the original random matrices X^N are somehow "stable," adding this noise should not change the eigenvalues of X^N by much, and the eigenvalues of the original and perturbed matrices should be almost the same. In such a case, we should get convergence of the eigenvalues of X^N to the Brown measure of the limiting object.

The canonical example in which instability occurs is the case in which $X^N = \mathrm{nil}_N$, the deterministic $N \times N$ matrix having 1s just above the diagonal and 0s elsewhere. Then of course nil_N is nilpotent, so all of its eigenvalues are zero. We note however that both $\mathrm{nil}_N^*\mathrm{nil}_N$ and $\mathrm{nil}_N\mathrm{nil}_N^*$ are diagonal matrices whose diagonal entries have $N-1$ values of 1 and only a single value of 0. Thus, when N is large, nil_N is "almost unitary," in the sense that $\mathrm{nil}_N^*\mathrm{nil}_N$ and $\mathrm{nil}_N\mathrm{nil}_N^*$ are close to the identity. Furthermore, for any positive integer k, we have that nil_N^k is again nilpotent, so that $\mathrm{trace}[\mathrm{nil}_N^k] = 0$. Using these observations, it is not hard to show that the limiting object is a "Haar unitary," that is, a unitary element u of a tracial von Neumann algebra satisfying $\tau(u^k) = 0$ for all positive integers k. The Brown measure of a Haar unitary is the uniform probability measure on the unit circle, while of course the eigenvalue distribution X^N is entirely concentrated at the origin.

In Figure 4 we see that even under a quite small perturbation (adding 10^{-6} times a Ginibre matrix), the spectrum of the nilpotent matrix X^N changes quite a lot. After the perturbation, the spectrum clearly resembles a uniform distribution over the unit circle. In Figure 5, by contrast, we see that even under a much larger perturbation (adding 10^{-1} times a Ginibre matrix), the spectrum of a GUE matrix changes only slightly. (Note the vertical scale in Figure 5.)

3.5 The Case of the Circular Brownian Motion

We now record the Brown measure of the circular Brownian motion.

Proposition 8 *For any $t > 0$, the Brown measure of c_t is the uniform probability measure on the disk of radius $\sqrt{t}$ centered at the origin.*

Now, as we noted in Section 2.4, the $*$-distribution of the circular Brownian motion at any time $t > 0$ is the same as the $*$-distribution of $\sqrt{t}c_1$. Thus, the proposition will follow if we know that the Brown measure of a circular element c is the uniform probability measure on the unit disk. This result, in turn, is well known; see, for example, Section 11.6.3 of [34].

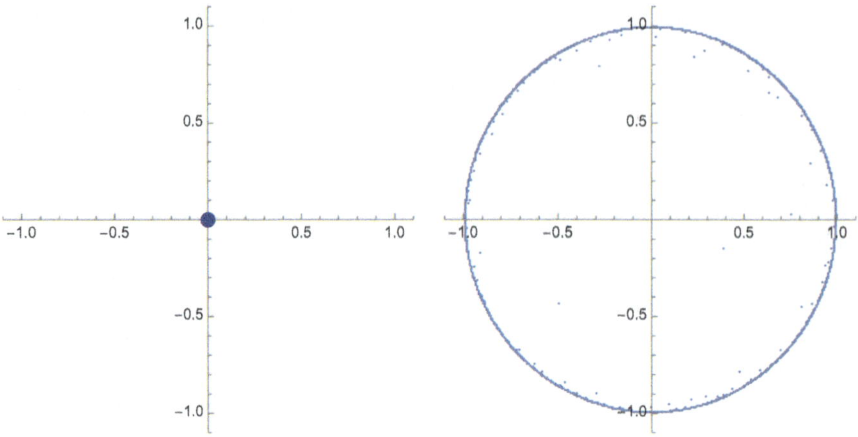

Fig. 4 Spectra of the nilpotent matrix nil_N (left) and of $\text{nil}_N + \varepsilon(\text{Ginibre})$ with $\varepsilon = 10^{-5}$ (right), with $N = 2{,}000$

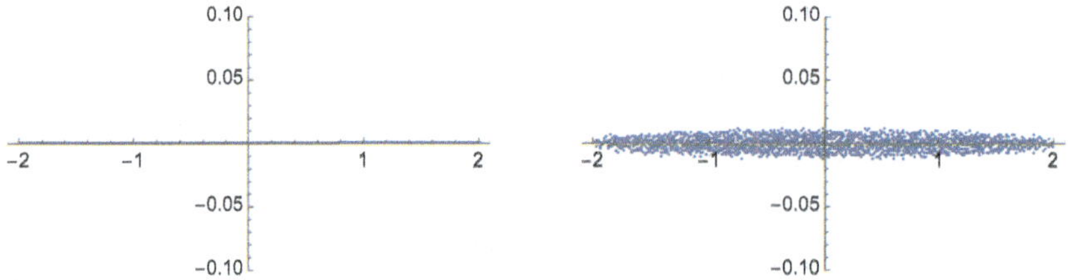

Fig. 5 Spectrum of a GUE matrix X (left) and $X + \varepsilon(\text{Ginibre})$ with $\varepsilon = 10^{-1}$ (right), with $N = 2{,}000$

4 PDE for the Circular Law

In this article, I present a different proof of Proposition 8 using the PDE method developed in [10]. The significance of this method is not so much that it gives another computation of the Brown measure of a circular element. Rather, it is a helpful warm-up case on the path to tackling the much more complicated problem in [10], namely, the computation of the Brown measure of the free multiplicative Brownian motion. In this section and the two that follow, I will show how the PDE method applies in the case of the circular Brownian motion. Then in the last section, I will describe the case of the free multiplicative Brownian motion.

The reader may also consult the recent preprint [29], which extends the results of [10] to case of the free multiplicative Brownian motion with arbitrary unitary initial distribution. Section 3 of this paper also analyzes the case of the free circular Brownian motion (with an arbitrary Hermitian initial distribution) using PDE methods.

We let c_t be the circular Brownian motion (Section 2.4). Then, following the construction of the Brown measure in Theorem 7, we define, for each $\lambda \in \mathbb{C}$, a function S^λ given by

$$S^\lambda(t,\varepsilon) = \tau[\log((c_t - \lambda)^*(c_t - \lambda) + \varepsilon)] \tag{15}$$

for all $t > 0$ and $\varepsilon > 0$. The Brown measure of c_t will then be obtained by letting ε tend to zero, taking the Laplacian with respect to λ, and dividing by 4π. Our first main result is that, for each λ, $S^\lambda(t,\varepsilon)$ satisfies a PDE in t and ε.

Theorem 9 *For each $\lambda \in \mathbb{C}$, the function S^λ satisfies the first-order, nonlinear differential equation*

$$\frac{\partial S^\lambda}{\partial t} = \varepsilon \left(\frac{\partial S^\lambda}{\partial \varepsilon}\right)^2 \tag{16}$$

subject to the initial condition

$$S^\lambda(0,\varepsilon) = \log(|\lambda|^2 + \varepsilon).$$

We now see the motivation for making λ a parameter rather than a variable for S: since λ does not appear in the PDE (16), we can think of solving the same equation for each different value of λ, with the dependence on λ entering only through the initial conditions.

On the other hand, we see that the regularization parameter ε plays a crucial role here as one of the variables in our PDE. Of course, we are ultimately interested in letting ε tend to zero, but since derivatives with respect to ε appear, we cannot merely set $\varepsilon = 0$ in the PDE.

Of course, the reader will point out that, formally, setting $\varepsilon = 0$ in (16) gives $\partial S^\lambda(t,0)/\partial t = 0$, because of the leading factor of ε on the right-hand side. This conclusion, however, is not actually correct, because $\partial S^\lambda/\partial\varepsilon$ can blow up as ε approaches zero. Actually, it will turn out that $S^\lambda(t,0)$ *is* independent of t when $|\lambda| > \sqrt{t}$, but not in general.

4.1 The Finite-N Equation

In this subsection, we give a heuristic argument for the PDE in Theorem 9. Although the argument is not rigorous as written, it should help explain what is going on. In particular, the computations that follow should make it clear why the PDE is only valid after taking the large-N limit.

4.1.1 The Result

We introduce a finite-N analog of the function S^λ in Theorem 9 and compute its time derivative. Let C_t^N denote the Ginibre Brownian motion introduced in Section 1.3.

Proposition 10 *For each N, let*

$$S^{\lambda,N}(t,\varepsilon) = \mathbb{E}\{\mathrm{tr}[\log((C_t^N - \lambda)^*(C_t^N - \lambda) + \varepsilon)]\}.$$

Then we have the following results.

(1) *The time derivative of $S^{\lambda,N}$ may be computed as*

$$\frac{\partial S^{\lambda,N}}{\partial t} = \varepsilon \mathbb{E}\{(\mathrm{tr}[((C_t^N - \lambda)^*(C_t^N - \lambda) + \varepsilon)^{-1}])^2\}. \tag{17}$$

(2) *We also have*

$$\frac{\partial}{\partial \varepsilon}\mathrm{tr}[\log((C_t^N - \lambda)^*(C_t - \lambda) + \varepsilon)] = \mathrm{tr}[((C_t^N - \lambda)^*(C_t^N - \lambda) + \varepsilon)^{-1}]. \tag{18}$$

(3) *Therefore, if we set*

$$T^{\lambda,N} = \mathrm{tr}[((C_t^N - \lambda)^*(C_t^N - \lambda) + \varepsilon)^{-1}],$$

we may rewrite the formula for $\partial S^{\lambda,N}/\partial t$ as

$$\frac{\partial S^{\lambda,N}}{\partial t} = \varepsilon \left(\frac{\partial S^{\lambda,N}}{\partial \varepsilon}\right)^2 + \mathrm{Cov}, \tag{19}$$

where Cov *is a "covariance term" given by*

$$\mathrm{Cov} = \mathbb{E}\{(T^{\lambda,N})^2\} - (\mathbb{E}\{T^{\lambda,N}\})^2.$$

The key point to observe here is that in the formula (17) for $\partial S^{\lambda,N}/\partial t$, we have the *expectation value of the square* of a trace. On the other hand, if we computed $(\partial S^{\lambda,N}/\partial \varepsilon)^2$ by taking the expectation value of both sides of (18) and squaring, we would have the *square of the expectation value* of a trace. Thus, there is no PDE for $S^{\lambda,N}$—we get an unavoidable covariance term on the right-hand side of (19).

On the other hand, the Ginibre Brownian motion C_t^N exhibits a **concentration phenomenon** for large N. Specifically, let us consider a family $\{Y^N\}$ of random variables of the form

$$Y^N = \mathrm{tr}[\text{word in } C_t^N \text{ and } (C_t^N)^*].$$

(Thus, e.g., we might have $Y^N = \mathrm{tr}[C_t^N(C_t^N)^* C_t^N(C_t^N)^*]$.) Then it is known that (1) the large-N limit of $\mathbb{E}\{Y^N\}$ exists, and (2) the variance of Y^N goes to zero. That is to say, when N is large, Y^N will be, with high probability, close to its expectation value. It then follows that $\mathbb{E}\{(Y^N)^2\}$ will be close to $(\mathbb{E}\{Y^N\})^2$. (This concentration phenomenon was established by Voiculescu in [42] for the analogous case of the "GUE Brownian motion." The case of the Ginibre Brownian motion is similar.)

Now, although the quantity

$$((C_t^N - \lambda)^*(C_t^N - \lambda) + \varepsilon)^{-1}$$

is not a word in C_t^N and $(C_t^N)^*$, it is expressible—at least for large ε—as a power series in such words. It is therefore reasonable to expect—this is not a proof!—that the variance of X^N will go to zero as N goes to infinity and the covariance term in (19) will vanish in the limit.

4.1.2 Setting Up the Computation

We view $M_N(\mathbb{C})$ as a real vector space of dimension $2N^2$ and we use the following real-valued inner product $\langle\cdot,\cdot\rangle_N$:

$$\langle X, Y\rangle_N = N\,\mathrm{Re}(\mathrm{trace}(X^*Y)). \tag{20}$$

The distribution of C_t^N is the Gaussian measure of variance $t/2$ with respect to this inner product

$$d\gamma_t(C) = d_t e^{-\langle C,C\rangle/t}\, dC,$$

where d_t is a normalization constant and dC is the Lebesgue measure on $M_N(\mathbb{C})$. This measure is a **heat kernel measure**. If we let $\mathbb{E}_t$ denote the expectation value with respect to γ_t, then we have, for any "nice" function,

$$\frac{d}{dt}\mathbb{E}_t\{f\} = \frac{1}{4}\mathbb{E}_t\{\Delta f\}, \tag{21}$$

where Δ is the Laplacian on $M_N(\mathbb{C})$ with respect to the inner product (20).

To compute more explicitly, we choose an orthonormal basis for $M_N(\mathbb{C})$ over $\mathbb{R}$ consisting of $X_1, \ldots, X_{N^2}$ and $Y_1, \ldots, Y_{N^2}$, where $X_1, \ldots, X_{N^2}$ are skew-Hermitian and where $Y_j = iX_j$. We then introduce the directional derivatives $\tilde{X}_j$ and $\tilde{Y}_j$ defined by

$$(\tilde{X}_j f)(a) = \left.\frac{d}{ds} f(a + sX_j)\right|_{s=0}; \quad (\tilde{Y}_j f)(Z) = \left.\frac{d}{ds} f(a + sY_j)\right|_{s=0}.$$

Then the Laplacian Δ is given by

$$\Delta = \sum_{j=1}^{N^2} \left((\tilde{X}_j)^2 + (\tilde{Y}_j)^2 \right).$$

We also introduce the corresponding complex derivatives, Z_j and $\bar{Z}_j$ given by

$$Z_j = \frac{1}{2}(\tilde{X}_j - i\tilde{Y}_j);$$
$$\bar{Z}_j = \frac{1}{2}(\tilde{X}_j + i\tilde{Y}_j),$$

which give

$$\frac{1}{4}\Delta = \sum_{j=1}^{N^2} \bar{Z}_j Z_j.$$

We now let C denote a matrix-valued variable ranging over $M_N(\mathbb{C})$. We may easily compute the following basic identities:

$$Z_j(C) = X_j; \quad Z_j(C^*) = 0;$$
$$\bar{Z}_j(C) = 0; \quad \bar{Z}_j(C^*) = -X_j. \tag{22}$$

(Keep in mind that X_j is skew-Hermitian.) We will also need the following elementary but crucial identity

$$\sum_{j=1}^{N^2} X_j A X_j = -\mathrm{tr}(A), \tag{23}$$

where $\mathrm{tr}(\cdot)$ is the normalized trace, given by

$$\mathrm{tr}(A) = \frac{1}{N}\mathrm{trace}(A).$$

See, for example, Proposition 3.1 in [9]. When applied to function involving a normalized trace, this will produce second trace.

Finally, we need the following formulas for differentiating matrix-valued functions of a real variable:

$$\frac{d}{ds}A(s)^{-1} = -A(s)^{-1}\frac{dA}{ds}A(s)^{-1} \tag{24}$$

$$\frac{d}{ds}\mathrm{tr}[\log A(s)] = \mathrm{tr}\left[A(s)^{-1}\frac{dA}{ds}\right]. \tag{25}$$

The first of these is standard and can be proved by differentiating the identity $A(s)A(s)^{-1} = I$. The second identity is Lemma 1.1 in [7]; it is important to emphasize that this second identity does not hold as written without the trace. One may derive (25) by using an integral formula for the derivative of the logarithm *without* the trace (see, e.g., Equation (11.10) in [27]) and then using the cyclic invariance of the trace, at which point the integral can be computed explicitly.

4.1.3 Proof of Proposition 10

We continue to let $\mathbb{E}_t$ denote the expectation value with respect to the measure γ_t, which is the distribution at time t of the Ginibre Brownian motion C_t^N, so that

$$S^{\lambda,N}(t,\varepsilon) = \mathbb{E}_t\{\mathrm{tr}[\log((C-\lambda)^*(C-\lambda)+\varepsilon)]\},$$

where the variable C ranges over $M_N(\mathbb{C})$. We apply the derivative Z_j using (25) and (22), giving

$$Z_j S^{\lambda,N}(t,\varepsilon) = \mathbb{E}_t\{\mathrm{tr}[((C-\lambda)^*(C-\lambda)+\varepsilon)^{-1}(C-\lambda)^* X_j]\}.$$

We then apply the derivative $\bar{Z}_j$ using (24) and (22), giving

$$\bar{Z}_j Z_j S^{\lambda,N}(t,\varepsilon) = -\mathbb{E}_t\{\mathrm{tr}[((C-\lambda)^*(C-\lambda)+\varepsilon)^{-1} X_j^2]\}$$
$$+\mathbb{E}_t\{\mathrm{tr}[((C-\lambda)^*(C-\lambda)+\varepsilon)^{-1} X_j (C-\lambda)((C-\lambda)^*(C-\lambda)+\varepsilon)^{-1}(C-\lambda)^* X_j]\}.$$

We now sum on j and apply the identity (23). After applying the heat equation (21) with $\Delta = \sum_j \bar{Z}_j Z_j$, we obtain

$$\begin{aligned}
&\frac{d}{dt} S^{\lambda,N}(t,\varepsilon) \\
&= \sum_j \bar{Z}_j Z_j S^{\lambda,N}(t,\varepsilon) \\
&= \mathbb{E}_t\{\mathrm{tr}[((C-\lambda)^*(C-\lambda)+\varepsilon)^{-1}]\} - \mathbb{E}_t\{\mathrm{tr}[((C-\lambda)^*(C-\lambda)+\varepsilon)^{-1}]\times \\
&\mathrm{tr}[(C-\lambda)^*(C-\lambda)((C-\lambda)^*(C-\lambda)+\varepsilon)^{-1}]\}. \qquad (26)
\end{aligned}$$

But then

$$\begin{aligned}
&(C-\lambda)^*(C-\lambda)((C-\lambda)^*(C-\lambda)+\varepsilon)^{-1} \\
&= ((C-\lambda)^*(C-\lambda)+\varepsilon-\varepsilon)((C-\lambda)^*(C-\lambda)+\varepsilon)^{-1} \\
&= 1-\varepsilon((C-\lambda)^*(C-\lambda)+\varepsilon)^{-1}.
\end{aligned}$$

Thus, there is a cancellation between the two terms on the right-hand side of (26), giving

$$\frac{\partial S^{\lambda,N}}{\partial t} = \varepsilon \mathbb{E}_t\{(\mathrm{tr}[((C-\lambda)^*(C-\lambda)+\varepsilon)^{-1}])^2\},$$

as claimed in Point 1 of the proposition.

Meanwhile, we may use again the identity (25) to compute

$$\frac{\partial}{\partial \varepsilon}\mathrm{tr}[\log((C_t^N-\lambda)^*(C_t-\lambda)+\varepsilon)]$$

to verify Point 2 3 then follows by simple algebra.

4.2 A Derivation Using Free Stochastic Calculus

4.2.1 Ordinary Stochastic Calculus

In this section, I will describe briefly how the PDE in Theorem 9 can be derived rigorously, using the tools of free stochastic calculus. We begin by recalling a little bit of ordinary stochastic calculus, for the ordinary, real-valued Brownian motion. To avoid notational conflicts, we will let ε_t denote Brownian motion in the real line. This is a random continuous path satisfying the properties proposed by Einstein in 1905, namely, that for any $0 = t_0 < t_1 < \cdots < t_k$, the increments

$$x_{t_1} - x_{t_0},\ x_{t_2} - x_{t_1}, \ldots,\ x_{t_k} - x_{t_{k-1}}$$

should be independent normal random variables with mean zero and variance $t_j - t_{j-1}$. At a rigorous level, Brownian motion is described by the Wiener measure on the space of continuous paths.

It is a famous result that, with probability one, the path x_t is nowhere differentiable. This property has not, however, deterred people from developing a theory of "stochastic calculus" in which one can take the "differential" of x_t, denoted dx_t. (Since x_t is not differentiable, we should not attempt to rewrite this differential as $\frac{dx_t}{dt}dt$.) There is then a theory of "stochastic integrals," in which one can compute, for example, integrals of the form

$$\int_a^b f(x_t)\, dx_t,$$

where f is some smooth function.

A key difference between ordinary and stochastic integration is that $(dx_t)^2$ is not negligible compared to dt. To understand this assertion, recall that the increments of Brownian motion have variance $t_j - t_{j-1}$—and therefore standard deviation $\sqrt{t_j - t_{j-1}}$. This means that in a short time interval Δt, the Brownian motion travels distance roughly Δt. Thus, if $\Delta x_t = x_{t+\Delta t} - x_t$, we may say that $(\Delta x_t)^2 \approx \Delta t$. Thus, if f is a smooth function, we may use a Taylor expansion to claim that

$$\begin{aligned} f(x_{t+\Delta t}) &\approx f(x_t) + f'(x_t)\Delta x_t + \frac{1}{2}f''(x_t)(\Delta x_t)^2 \\ &\approx f(x_t) + f'(x_t)\Delta x_t + \frac{1}{2}f''(x_t)\Delta t. \end{aligned}$$

We may express the preceding discussion in the heuristically by saying

$$(dx_t)^2 = dt.$$

Rigorously, this line of reasoning lies behind the famous Itô formula, which says that

$$df(x_t) = f'(x_t)\,dx_t + \frac{1}{2}f''(x_t)\,dt.$$

The formula means, more precisely, that (after integration)

$$f(x_b) - f(x_a) = \int_a^b f'(x_t)\,dx_t + \frac{1}{2}\int_a^b f''(x_t)\,dt,$$

where the first integral on the right-hand side is a stochastic integral and the second is an ordinary Riemann integral.

If we take, for example, $f(x) = x^2/2$, then we find that

$$\frac{1}{2}(x_b^2 - x_a^2) = \int_a^b x_t\,dx_t + \frac{1}{2}(b-a)$$

so that

$$\int_a^b x_t\,dx_t = \frac{1}{2}(x_b^2 - x_a^2) - \frac{1}{2}(b-a).$$

This formula differs from what we would get if x_t were smooth by the $b - a$ term on the right-hand side.

4.2.2 Free Stochastic Calculus

We now turn to the case of the circular Brownian motion c_t. Since c_t is a limit of ordinary Brownian motion in the space of $N \times N$ matrices, we expect that $(dc_t)^2$

will be non-negligible compared to dt. The rules are as follows; see [31, Lemma 2.5, Lemma 4.3]. Suppose g_t and h_t are processes "adapted to c_t," meaning that g_t and h_t belong to the von Neumann algebra generated by the operators c_s with $0 < s < t$. Then we have

$$dc_t\, g_t\, dc_t^* = dc_t^*\, g_t\, dc_t = \tau(g_t)\, dt \tag{27}$$

$$dc_t\, g_t\, dc_t = dc_t^*\, g_t\, dc_t^* = 0 \tag{28}$$

$$\tau(g_t\, dc_t\, h_t) = \tau(g_t\, dc_t^*\, h_t) = 0. \tag{29}$$

In addition, we have the following Itô product rule: if $a_t^1, \ldots, a_t^n$ are processes adapted to c_t, then

$$d(a_t^1 \cdots a_t^n) = \sum_{j=1}^n (a_t^1 \cdots a_t^{j-1})\, da_t^j\, (a_t^{j+1} \cdots a_t^n) \tag{30}$$

$$+ \sum_{1\leq j<k\leq n} (a_t^1 \cdots a_t^{j-1})\, da_t^j\, (a_t^{j+1} \cdots a_t^{k-1})\, da_t^k\, (a_t^{k+1} \cdots a_t^n). \tag{31}$$

Finally, the differential "d" can be moved inside the trace τ.

Suppose, for example, we wish to compute $d\tau[c_t^* c_t]$. We start by applying the product rule in (30) and (31). But by (29), there will be no contribution from the first line (30) in the product rule. We then use the second line (31) of the product rule together with (27) to obtain

$$d\tau[c_t^* c_t] = \tau[dc_t^* dc_t] = \tau(1)\, dt = dt.$$

Thus,

$$\frac{d}{dt}\tau[c_t^* c_t] = 1.$$

Since, also, $c_0 = 0$, we find that $\tau[c_t^* c_t] = t$.

4.2.3 The Proof

In the proof that follows, the Itô formula (27) plays the same role as the identity (23) plays in the heuristic argument in Section 4.1. We begin with a lemma whose proof is an exercise in using the rules of free stochastic calculus.

Lemma 11 *For each $\lambda \in \mathbb{C}$, let us use the notation*

$$c_{t,\lambda} := c_t - \lambda.$$

Then for each positive integer n, we have

$$\frac{d}{dt}\tau[(c_{t,\lambda}^* c_{t,\lambda})^n] = n\sum_{l=0}^{n-1}\tau[(c_{t,\lambda}^* c_{t,\lambda})^j]\tau[(c_{t,\lambda}c_{t,\lambda}^*)^{n-j-1}]$$

Proof We first note that $dc_{t,\lambda} = dc_t$ and $dc_{t,\lambda}^* = dc_t^*$, since λ is a constant. We then compute $d\tau[(c_{t,\lambda}^* c_{t,\lambda})^n]$ by moving the d inside the trace and then applying the product rule in (30) and (31). By (29), the terms arising from (30) will not contribute. Furthermore, by (28), the only terms from (31) that contribute are those where one d goes on a factor of $c_{t,\lambda}$ and one goes on a factor of $c_{t,\lambda}^*$.

By choosing all possible factors of $c_{t,\lambda}$ and all possible factors of $c_{t,\lambda}^*$, we get n^2 terms. In each term, after putting the d inside the trace, we can cyclically permute the factors until, say, the $dc_{t,\lambda}$ factor is at the end. There are then only n *distinct* terms that occur, each of which occurs n times. By (27), each distinct term is computed as

$$\begin{aligned}
&\tau[(c_{t,\lambda}^* c_{t,\lambda})^j\, dc_t^* c_{t,\lambda}(c_{t,\lambda}^* c_{t,\lambda})^{n-j-2}c_{t,\lambda}^*\, dc_t] \\
&= \tau[c_{t,\lambda}(c_{t,\lambda}^* c_{t,\lambda})^{n-j-2}c_{t,\lambda}^*]\tau[(c_{t,\lambda}^* c_{t,\lambda})^j]\, dt \\
&= \tau[(c_{t,\lambda}^* c_{t,\lambda})^j]\tau[c_t c_t^*(c_{t,\lambda}c_{t,\lambda}^*)^{n-j-1}]\, dt.
\end{aligned}$$

Since each distinct term occurs n times, we obtain

$$d\tau[(c_{t,\lambda}^* c_{t,\lambda})^n] = n\sum_{j=0}^{n-1}\tau[(c_{t,\lambda}^* c_{t,\lambda})^j]\tau[(c_{t,\lambda}c_{t,\lambda}^*)^{n-j-1}]\, dt,$$

which is equivalent to the claimed formula. □

We are now ready to give a rigorous argument for the PDE.

***Proof of Theorem* 9** We continue to use the notation $c_{t,\lambda} := c_t - \lambda$. We first compute, using the operator version of (25), that

$$\begin{aligned}
\frac{\partial S}{\partial \varepsilon} &= \frac{\partial}{\partial \varepsilon}\tau[\log(c_{t,\lambda}^* c_{t,\lambda} + \varepsilon)] \\
&= \tau[(c_{t,\lambda}^* c_{t,\lambda} + \varepsilon)^{-1}]. \qquad (32)
\end{aligned}$$

We note that the definition of S in (15) actually makes sense for all $\varepsilon \in \mathbb{C}$ with $\mathrm{Re}(\varepsilon) > 0$, using the standard branch of the logarithm function. We note that for $|\varepsilon| > |z|$, we have

$$\frac{1}{z+\varepsilon} = \frac{1}{\varepsilon\left(1-\left(-\frac{z}{\varepsilon}\right)\right)}$$

$$= \frac{1}{\varepsilon}\left[1 - \frac{z}{\varepsilon} + \frac{z^2}{\varepsilon^2} - \frac{z^3}{\varepsilon^3} + \cdots\right]. \tag{33}$$

Integrating with respect to z gives

$$\log(z+\varepsilon) = \log\varepsilon + \sum_{n=1}^{\infty} \frac{(-1)^{n-1}}{n}\left(\frac{z}{\varepsilon}\right)^n.$$

Thus, for $|\varepsilon| > \left\|c_t^* c_t\right\|$, we have

$$\tau[\log(c_{t,\lambda}^* c_{t,\lambda} + \varepsilon)] = \log\varepsilon + \sum_{n=1}^{\infty} \frac{(-1)^{n-1}}{n\varepsilon^n}\tau[(c_{t,\lambda}^* c_{t,\lambda})^n]. \tag{34}$$

Assume for the moment that it is permissible to differentiate (34) term by term with respect to t. Then by Lemma 11, we have

$$\frac{\partial S}{\partial t} = \sum_{n=1}^{\infty} \frac{(-1)^{n-1}}{\varepsilon^n} \sum_{j=0}^{n-1} \tau[(c_{t,\lambda}^* c_{t,\lambda})^j]\tau[(c_{t,\lambda} c_{t,\lambda}^*)^{n-j-1}]. \tag{35}$$

Now, by [5, Proposition 3.2.3], the map $t \mapsto c_t$ is continuous in the operator norm topology; in particular, $\|c_t\|$ is a locally bounded function of t. From this observation, it is easy to see that the right-hand side of (35) converges locally uniformly in t. Thus, a standard result about interchange of limit and derivative (e.g., Theorem 7.17 in [37]) shows that the term-by-term differentiation is valid.

Now, in (35), we let $k = j$ and $l = n - j - 1$, so that $n = k + l + 1$. Then k and l go from 0 to ∞, and we get

$$\frac{\partial S}{\partial t} = \varepsilon\left(\frac{1}{\varepsilon}\sum_{k=0}^{\infty} \frac{(-1)^k}{\varepsilon^k}\tau[(c_{t,\lambda}^* c_{t,\lambda})^k]\right)\left(\frac{1}{\varepsilon}\sum_{l=0}^{\infty} \frac{(-1)^l}{\varepsilon^l}\tau[(c_{t,\lambda} c_{t,\lambda}^*)^l]\right).$$

(We may check that the power of ε in the denominator is $k + l + 1 = n$ and that the power of -1 is $k + l = n - 1$.) Thus, moving the sums inside the traces and using (33), we obtain that

$$\frac{\partial S}{\partial t} = \varepsilon(\tau[(c_{t,\lambda}^* c_{t,\lambda} + \varepsilon)^{-1}])^2, \tag{36}$$

which reduces to the claimed PDE for S, by (32).

We have now established the claimed formula for $\partial S/\partial t$ for ε in the right half-plane, provided $|\varepsilon|$ is sufficiently large, depending on t and λ. Since, also, $S(0, \lambda, \varepsilon) = \log(|\lambda - 1|^2 + \varepsilon)$, we have, for sufficiently large $|\varepsilon|$,

$$S(t,\lambda,\varepsilon) = \log(|\lambda-1|^2+\varepsilon) + \int_0^t \varepsilon\tau[(c_{s,\lambda}^* c_{s,\lambda}+\varepsilon)^{-1}]\tau[(c_{s,\lambda}c_{s,\lambda}^*+\varepsilon)^{-1}]\,ds. \tag{37}$$

We now claim that both sides of (37) are well-defined, holomorphic functions of ε, for ε in the right half-plane. This claim is easily established from the standard power-series representation of the inverse:

$$\begin{aligned}(A+\varepsilon+h)^{-1} &= (A+\varepsilon)^{-1}(1+h(A+\varepsilon)^{-1})^{-1} \\ &= (A+\varepsilon)^{-1}\sum_{n=0}^{\infty}(-1)^n h^n (A+\varepsilon)^{-n},\end{aligned}$$

and a similar power-series representation of the logarithm. Thus, (37) actually holds for all ε in the right half-plane. Differentiating with respect to t then establishes the desired formula (36) for dS/dt for all ε in the right half-plane. □

5 Solving the Equation

5.1 *The Hamilton–Jacobi Method*

The PDE (16) in Theorem 9 is a first-order, nonlinear equation of Hamilton–Jacobi type. "Hamilton–Jacobi type" means that the right-hand side of the equation involves only ε and $\partial S/\partial\varepsilon$, and not S itself. The reader may consult Section 3.3 of the book [11] of Evans for general information about equations of this type. In this subsection, we describe the general version of this method. In the remainder of this section, we will then apply the general method to the PDE (16).

The Hamilton–Jacobi method for analyzing solutions to equations of this type is a generalization of the method of characteristics. In the method of characteristics, one finds certain special curves along which the solution is constant. For a general equation of Hamilton–Jacobi type, the method of characteristics is not applicable. Nevertheless, we may hope to find certain special curves along which the solution varies in a simple way, allowing us to compute the solution along these curves in a more-or-less explicit way.

We now explain the representation formula for solutions of equations of Hamilton–Jacobi type. A self-contained proof of the following result is given as the proof of Proposition 6.3 in [10].

Proposition 12 *Fix a function $H(\mathbf{x},\mathbf{p})$ defined for* $\mathbf{x}$ *in an open set $U\subset\mathbb{R}^n$ and* $\mathbf{p}$ *in $\mathbb{R}^n$. Consider a smooth function $S(t,\mathbf{x})$ on $[0,\infty)\times U$ satisfying*

$$\frac{\partial S}{\partial t} = -H(\mathbf{x},\nabla_{\mathbf{x}}S) \tag{38}$$

for $\mathbf{x} \in U$ *and* $t > 0$. *Now suppose* $(\mathbf{x}(t), \mathbf{p}(t))$ *is curve in* $U \times \mathbb{R}^n$ *satisfying Hamilton's equations:*

$$\frac{dx_j}{dt} = \frac{\partial H}{\partial p_j}(\mathbf{x}(t), \mathbf{p}(t)); \quad \frac{dp_j}{dt} = -\frac{\partial H}{\partial x_j}(\mathbf{x}(t), \mathbf{p}(t))$$

with initial conditions

$$\mathbf{x}(0) = \mathbf{x}_0; \quad \mathbf{p}(0) = \mathbf{p}_0 := (\nabla_{\mathbf{x}} S)(0, \mathbf{x}_0). \tag{39}$$

Then we have

$$S(t, \mathbf{x}(t)) = S(0, \mathbf{x}_0) - H(\mathbf{x}_0, \mathbf{p}_0)\, t + \int_0^t \mathbf{p}(s) \cdot \frac{d\mathbf{x}}{ds}\, ds \tag{40}$$

and

$$(\nabla_{\mathbf{x}} S)(t, \mathbf{x}(t)) = \mathbf{p}(t). \tag{41}$$

We emphasize that we are not using the Hamilton–Jacobi formula to *construct* a solution to the equation (38); rather, we are using the method to *analyze* a solution that is assumed ahead of time to exist. Suppose we want to use the method to compute (as explicitly as possible), the value of $S(t, \mathbf{x})$ for some fixed $\mathbf{x}$. We then need to try to choose the initial position $\mathbf{x}_0$ in (39)—which determines the initial momentum $\mathbf{p}_0 = (\nabla_{\mathbf{x}} S)(0, \mathbf{x}_0)$—so that $\mathbf{x}(t) = \mathbf{x}$. We then use (40) to get an in-principle formula for $S(t, \mathbf{x}(t)) = S(t, \mathbf{x})$.

5.2 Solving the Equations

The equation for S^λ in Theorem 9 is of Hamilton–Jacobi form with $n = 1$, with Hamiltonian given by

$$H(\varepsilon, p) = -\varepsilon p^2. \tag{42}$$

Since $S^\lambda(t, \varepsilon)$ is only defined for $\varepsilon > 0$, we take open set U in Proposition 12 to be $(0, \infty)$. That is to say, the Hamilton–Jacobi formula (40) is only valid if the curve $\varepsilon(s)$ remains positive for $0 \leq s \leq t$.

Hamilton's equations for this Hamiltonian then take the explicit form

$$\frac{d\varepsilon}{dt} = \frac{\partial H}{\partial p} = -2\varepsilon p \tag{43}$$

$$\frac{dp}{dt} = -\frac{\partial H}{\partial \varepsilon} = p^2. \tag{44}$$

Following the general method, we take an arbitrary initial position ε_0, with the initial momentum p_0 given by

$$p_0 = \frac{\partial}{\partial \varepsilon} \log(|\lambda|^2 + \varepsilon)\bigg|_{\varepsilon=\varepsilon_0} = \frac{1}{|\lambda|^2 + \varepsilon_0}. \tag{45}$$

Theorem 13 *For any $\varepsilon_0 > 0$, the solution $(\varepsilon(t), p(t))$ to* (43) *and* (44) *with initial momentum $p_0 = 1/(|\lambda|^2 + \varepsilon_0)$ exists for $0 \le t < |\lambda|^2 + \varepsilon_0$. On this time interval, we have*

$$\varepsilon(t) = \varepsilon_0 \left(1 - \frac{t}{|\lambda|^2 + \varepsilon_0}\right)^2. \tag{46}$$

The general Hamilton–Jacobi formula (40) *then takes the form*

$$S^\lambda\left(t, \varepsilon_0\left(1 - \frac{t}{|\lambda|^2 + \varepsilon_0}\right)^2\right) = \log(|\lambda|^2 + \varepsilon_0) - \frac{\varepsilon_0 t}{(|\lambda|^2 + \varepsilon_0)^2}, \quad 0 \le t < |\lambda|^2 + \varepsilon_0. \tag{47}$$

Proof Since the equation (44) for dp/dt does not involve $\varepsilon(t)$, we may easily solve it for $p(t)$ as

$$p(t) = \frac{p_0}{1 - p_0 t}.$$

We may then plug the formula for $p(t)$ into the equation (43) for $d\varepsilon/dt$, giving

$$\frac{d\varepsilon}{dt} = -2\varepsilon \frac{p_0}{1 - p_0 t}$$

so that

$$\frac{1}{\varepsilon} d\varepsilon = -2 \frac{p_0}{1 - p_0 t}\, dt.$$

Thus,

$$\log \varepsilon = 2 \log(p_0 t - 1) + c_1$$

so that

$$\varepsilon = c_2(1 - p_0 t)^2.$$

Plugging in $t = 0$ gives $c_2 = \varepsilon_0$. Recalling the expression (45) for p_0 gives the claimed formula for $\varepsilon(t)$.

Assuming $\varepsilon_0 > 0$, the solution to the system (43)–(44) continues to exist with $\varepsilon(t) > 0$ until $p(t)$ blows up, which occurs at time $t = 1/p_0 = |\lambda|^2 + \varepsilon_0$.

Finally, we work out the general Hamilton–Jacobi formula (40) in the case at hand. We note from (42) and (43) that $p(s)\frac{d\varepsilon}{ds} = -2\varepsilon(s)p(s)^2 = 2H(s)$. Since the Hamiltonian is always a conserved quantity in Hamilton's equations, we find that

$$p(s)\frac{d\varepsilon}{ds} = 2H(0) = -2\varepsilon_0 p_0^2.$$

Thus, (40) reduces to

$$\begin{aligned} S^\lambda(t, \varepsilon(t)) &= S(0, \varepsilon_0) + H(0)t \\ &= \log(|\lambda|^2 + \varepsilon_0) - \varepsilon_0 p_0^2 t. \end{aligned}$$

Using the formula (45) for p_0 gives the claimed formula (47). □

6 Letting ε Tend to Zero

Recall that the Brown measure is obtained by first evaluating

$$s_t(\lambda) := \lim_{\varepsilon \to 0^+} S^\lambda(t, 0)$$

and then taking $1/(4\pi)$ times the Laplacian (in the distribution sense) of $s_t(\lambda)$. We record the result here and will derive it in the remainder of this section.

Theorem 14 *We have*

$$s_t(\lambda) = \begin{cases} \log(|\lambda|^2) & |\lambda| \geq \sqrt{t} \\ \log t - 1 + \frac{|\lambda|^2}{t} & |\lambda| < \sqrt{t} \end{cases}. \tag{48}$$

The Brown measure is then absolutely continuous with respect to the Lebesgue measure, with density $W_t(\lambda)$ given by

$$W_t(\lambda) = \begin{cases} 0 & |\lambda| \geq \sqrt{t} \\ \frac{1}{\pi t} & |\lambda| < \sqrt{t} \end{cases}. \tag{49}$$

That is to say, the Brown measure is the uniform probability measure on the disk of radius $\sqrt{t}$ centered at the origin. The functions $s_t(\lambda)$ and $W_t(\lambda)$ are plotted for

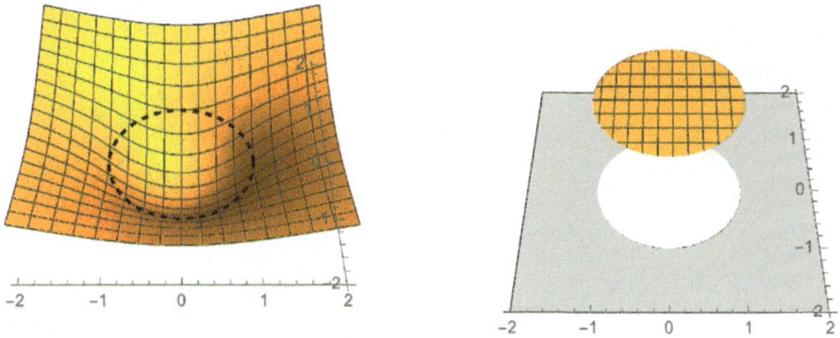

Fig. 6 Plot of $s_t(\lambda) := S^\lambda(t, 0^+)$ (left) and $\frac{1}{4\pi}\Delta s_t(\lambda)$ (right) for $t = 1$

$t = 1$ in Figure 6. On the left-hand side of the figure, the dashed line indicates the boundary of the unit disk.

6.1 *Letting ε Tend to Zero: Outside the Disk*

Our goal is to compute $s_t(\lambda) := \lim_{\varepsilon\to 0^+} S^\lambda(t, \varepsilon)$. Thus, in the Hamilton–Jacobi formalism, we want to try to choose ε_0 so that the quantity

$$\varepsilon(t) = \varepsilon_0 \left(1 - \frac{t}{|\lambda|^2 + \varepsilon_0}\right)^2 \tag{50}$$

will be very close to zero. Since there is a factor of ε_0 on the right-hand side of the above formula, an obvious strategy is to take ε_0 itself very close to zero. There is, however, a potential difficulty with this strategy: If ε_0 is small, the lifetime of the solution may be smaller than the time t we are interested in. To see when the strategy works, we take the formula for the lifetime of the solution—namely, $|\lambda|^2 + \varepsilon_0$—and take the limit as ε_0 tends to zero.

Definition 15 For each $\lambda \in \mathbb{C}$, we define $T(\lambda)$ to be the lifetime of solutions to the system (43)–(44), in the limit as ε_0 approaches zero. Thus, explicitly,

$$\begin{aligned} T(\lambda) &= \lim_{\varepsilon_0\to 0^+} (|\lambda|^2 + \varepsilon_0) \\ &= |\lambda|^2 . \end{aligned}$$

Thus, if the time t we are interested in is larger than $T(\lambda) = |\lambda|^2$, our simple strategy of taking $\varepsilon_0 \approx 0$ will not work. After all, if $t > T(\lambda)$ and $\varepsilon_0 \approx 0$, then the lifetime of the path is less than t and the Hamilton–Jacobi formula (47) is not applicable. On the other hand, if the time t we are interested in is at most $T(\lambda) = |\lambda|^2$, the simple strategy does work. Figure 7 illustrates the situation.

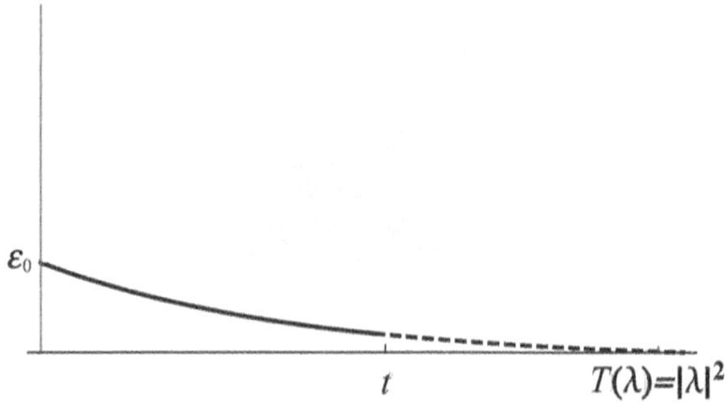

Fig. 7 If ε_0 is small and positive, $\varepsilon(s)$ will remain small and positive up to time t, *provided* that $t \leq T(\lambda) = |\lambda|^2$

Conclusion 16 *The simple strategy of letting ε_0 approach zero works precisely when $t \leq T(\lambda) = |\lambda|^2$. Equivalently, the simple strategy works when $|\lambda| \geq \sqrt{t}$, that is, when λ is outside the open disk of radius $\sqrt{t}$ centered at the origin.*

In the case that λ is outside the disk, we may then simply let ε_0 approach zero in the Hamilton–Jacobi formula, giving the following result.

Proposition 17 *Suppose $|\lambda| \geq \sqrt{t}$, that is, λ is outside the open disk of radius $\sqrt{t}$ centered at 0. Then we may let ε_0 tend to zero in the Hamilton–Jacobi formula* (47) *to obtain*

$$\begin{aligned}\lim_{\varepsilon \to 0^+} S^\lambda(t, \varepsilon) &= \lim_{\varepsilon_0 \to 0} \left(\log(|\lambda|^2 + \varepsilon_0) - \frac{\varepsilon_0 t}{(|\lambda|^2 + \varepsilon_0)^2} \right) \\ &= \log(|\lambda|^2). \end{aligned} \tag{51}$$

Since the right-hand side of (51) *is harmonic, we conclude that*

$$\Delta s_t(\lambda) = 0, \quad |\lambda| > \sqrt{t}.$$

That is to say, the Brown measure of c_t is zero outside the disk of radius $\sqrt{t}$ centered at 0.

6.2 Letting ε Tend to Zero: Inside the Disk

We now turn to the case in which the time t we are interested in is greater than the small-ε_0 lifetime $T(\lambda)$ of the solutions to (43)–(44). This case corresponds to $t > T(\lambda)^2 = |\lambda|^2$, that is, $|\lambda| < \sqrt{t}$. We still want to choose ε_0 so that $\varepsilon(t)$ will approach zero, but we cannot let ε_0 tend to zero, or else the lifetime of the solution will be less than t. Instead, we allow the *second* factor in the formula (46) for $\varepsilon(t)$ to approach zero. To make this factor approach zero, we make $|\lambda|^2 + \varepsilon_0$ approach t, that is, ε_0 should approach $t - |\lambda|^2$. Note that since we are now assuming that

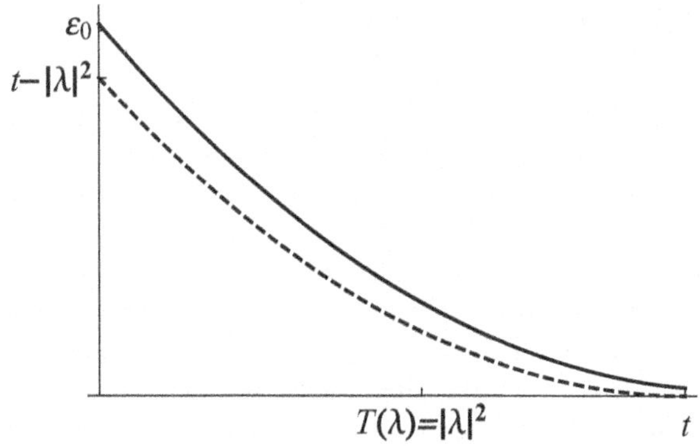

Fig. 8 If $|\lambda| < \sqrt{t}$ and we let ε_0 approach $t - |\lambda|^2$ from above, $\varepsilon(s)$ will remain positive until time t, and $\varepsilon(t)$ will approach zero

$|\lambda| < \sqrt{t}$, the quantity $t - |\lambda|^2$ is positive. This strategy is illustrated in Figure 8: When $\varepsilon_0 = t - |\lambda|^2$, we obtain $\varepsilon(t) = 0$, and if ε_0 approaches $t - |\lambda|^2$ from above, the value of $\varepsilon(t)$ approaches 0 from above.

Proposition 18 *Suppose $|\lambda| \leq \sqrt{t}$, that is, λ is inside the closed disk of radius $\sqrt{t}$ centered at 0. Then in the Hamilton–Jacobi formula* (47)*, we may let ε_0 approach $t - |\lambda|^2$ from above, and we get*

$$\lim_{\varepsilon \to 0^+} S^\lambda(t, \varepsilon) = \log t - 1 + \frac{|\lambda|^2}{t}, \quad |\lambda| \leq \sqrt{t}.$$

For $|\lambda| < \sqrt{t}$, we may then compute

$$\frac{1}{4\pi} \Delta s_t(\lambda) = \frac{1}{\pi t}.$$

Thus, inside the disk of radius $\sqrt{t}$, the Brown measure has a constant density of $1/(\pi t)$.

Proof We use the Hamilton–Jacobi formula (47). Since the lifetime of our solution is $|\lambda|^2 + \varepsilon_0$, if we let ε_0 approach $t - |\lambda|^2$ from above, the lifetime will always be at least t. In this limit, the formula (46) for $\varepsilon(t)$ approaches zero from above. Thus, we may take the limit $\varepsilon_0 \to (t - |\lambda|^2)^+$ in (47) to obtain

$$\begin{aligned}\lim_{\varepsilon \to 0^+} S^\lambda(t, \varepsilon) &= \lim_{\varepsilon_0 \to (t-|\lambda|^2)^+} \left[\log(|\lambda|^2 + \varepsilon_0) - \frac{\varepsilon_0 t}{(|\lambda|^2 + \varepsilon_0)^2} \right] \\ &= \log t - \frac{(t - |\lambda|^2)t}{t^2},\end{aligned}$$

which simplifies to the claimed formula. □

6.3 On the Boundary

Note that if $|\lambda|^2 = t$, both approaches are valid—and the two values of $s_t(\lambda) := \lim_{\varepsilon\to 0^+} S^\lambda(t,\varepsilon)$ agree, with a common value of $\log t = \log|\lambda|^2$. Furthermore, the radial derivatives of $s_t(\lambda)$ agree on the boundary: $2/r$ on the outside and $2r/t$ on the inside, which have a common value of $2/\sqrt{t}$ at $r = \sqrt{t}$. Of course, the angular derivatives of $s_t(\lambda)$ are identically zero, inside, outside, and on the boundary.

Since the first derivatives of s_t are continuous up to the boundary, we may take the distributional Laplacian by taking the ordinary Laplacian inside the disk and outside the disk and ignoring the boundary. (See the proof of Proposition 7.13 in [10].) Thus, we may compute the Laplacian of the two formulas in (48) to obtain the formula (49) for the Brown measure of c_t.

7 The Case of the Free Multiplicative Brownian Motion

7.1 Additive and Multiplicative Models

The standard GUE and Ginibre ensembles are given by Gaussian measures on the relevant space of matrices (Hermitian matrices for GUE and all matrices for the Ginibre ensemble). In light of the central limit theorem, these ensembles can be approximated by adding together large numbers of small, independent random matrices. We may therefore refer to these Gaussian ensembles as "additive" models.

It is natural to consider also "multiplicative" random matrix models, which can be approximated by *multiplying* together large numbers of independent matrices that are "small" in the multiplicative sense, that is, close to the identity. Specifically, if Z^{add} is a random matrix with a Gaussian distribution, we will consider a multiplicative version Z_t^{mult}, where the distribution of Z_t^{mult} may be approximated as

$$Z_t^{\mathrm{mult}} \sim \prod_{j=1}^{k}\left(I + i\sqrt{\frac{t}{k}}Z_j^{\mathrm{add}} - \frac{t}{k}\text{Itô}\right), \quad k \text{ large}. \tag{52}$$

Here t is a positive parameter, the Z_j^{add}s are independent copies of Z^{add}, and "Itô" is an Itô correction term. This correction term is a fixed multiple of the identity, independent of t and k. (In the next paragraph, we will identify the Itô term in the main cases of interest.) Since the factors in (52) are independent and identically distributed, the order of the factors does not affect the distribution of the product.

The two main cases we will consider are those in which Z is distributed according to the Gaussian unitary ensemble or the Ginibre ensemble. In the case that Z is distributed according to the Gaussian unitary ensemble, the Itô term is $\text{Itô} = \frac{1}{2}I$. In this case, the resulting multiplicative model may be described as *Brownian motion in the unitary group* $\mathsf{U}(N)$, which we write as U_t^N. The Itô correction is essential in this case to ensure that Z_t^{mult} actually lives in the unitary group. In the case that Z is distributed according to the Ginibre ensemble, the Itô term is zero. In this case, the resulting multiplicative model may be described as *Brownian motion in the general linear group* $\mathsf{GL}(N;\mathbb{C})$, which we write as B_t^N.

7.2 The Free Unitary and Free Multiplicative Brownian Motions

The large-N limits of the Brownian motions U_t^N and B_t^N were constructed by Biane [3]. The limits are the **free unitary Brownian motion** and the **free multiplicative Brownian motion**, respectively, which we write as u_t and b_t. The qualifier "free" indicates that the increments of these Brownian motions—computed in the multiplicative sense as $u_s^{-1}u_t$ or $b_s^{-1}b_t$—are freely independent in the sense of Section 2.3. In the case of b_t, the convergence of B_t^N to b_t was conjectured by Biane [3] and proved by Kemp [31]. In both cases, we take the limiting object to be an element of a tracial von Neumann algebra $(\mathcal{A},\tau)$.

Since u_t is unitary, we do not need to use the machinery of Brown measure, but can rather use the spectral theorem as in (7) to compute the **distribution** of u_t, denoted ν_t. We emphasize that ν_t is, in fact, the Brown measure of u_t, but it easier to describe ν_t using the spectral theorem than to use the general Brown measure construction. The measure ν_t is a probability measure on the unit circle describing the large-N limit of Brownian motion in the unitary group $\mathsf{U}(N)$. Biane computed the measure ν_t in [3] and established the following support result.

Theorem 19 *For $t<4$, the measure ν_t is supported on a proper subset of the unit circle:*

$$\operatorname{supp}(\nu_t)=\left\{e^{i\theta}\,\middle|\,|\theta|\le\frac{1}{2}\sqrt{t(4-t)}+\cos^{-1}\left(1-\frac{t}{2}\right)\right\},\qquad t<4.$$

By contrast, for all $t\ge 4$, the closed support of ν_t is the whole unit circle.

In the physics literature, the change in behavior of the support of ν_t at $t=4$ is called a *topological phase transition*, indicating that the topology of $\operatorname{supp}(\nu_t)$ changes from a closed interval to a circle.

The remainder of this article is devoted to recent results of the author with Driver and Kemp regarding the Brown measure of the free multiplicative Brownian motion b_t. We expect that the Brown measure of b_t will be the limiting empirical eigenvalue distribution of the Brownian motion B_t^N in the general linear group $\mathsf{GL}(N;\mathbb{C})$. Now, when t is small, we may take $k=1$ in (52), so that (since the Itô correction is zero in this case)

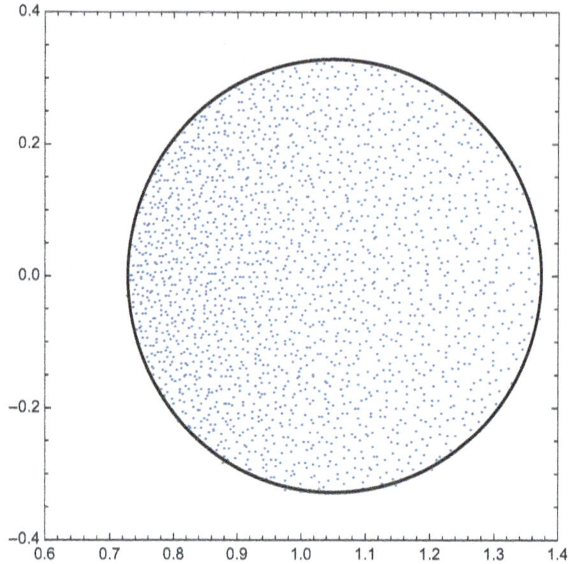

Fig. 9 The eigenvalues of B_t^N with $t = 0.1$ and $N = 2.000$

$$B_t^N \sim I + i\sqrt{\frac{t}{k}}Z, \quad t \text{ small.}$$

Thus, when t is small and N is large, the eigenvalues of B_t^N resemble a scaled and shifted version of the circular law. Specifically, the eigenvalue distribution should resemble a uniform distribution on the disk of radius $\sqrt{t}$ centered at 1.

Figure 9 shows the eigenvalues of B_t^N with $t = 0.1$ and $N = 2{,}000$. The eigenvalue distribution bears a clear resemblance to the just-described picture, with $\sqrt{t} = \sqrt{0.1} \approx 0.316$. Nevertheless, we can already see some deviation from the small-t picture: The region into which the eigenvalues are clustering looks like a disk, but not quite centered at 1, while the distribution within the region is slightly higher at the left-hand side of the region than the right. Figures 10 and 11, meanwhile, show the eigenvalue distribution of B_t^N for several larger values of t. The region into which the eigenvalues cluster becomes more complicated as t increases, and the distribution of eigenvalues in the region becomes less and less uniform. We expect that the Brown measure of the limiting object b_t will be supported on the domain into which the eigenvalues are clustering.

7.3 The Domains Σ_t

We now describe certain domains Σ_t in the plane, as introduced by Biane in [4, pp. 273–274]. It will turn out that the Brown measure of b_t is supported on Σ_t. We use here a new the description of Σ_t, as given in Section 4 of [10]. For all nonzero $\lambda \in \mathbb{C}$, we define

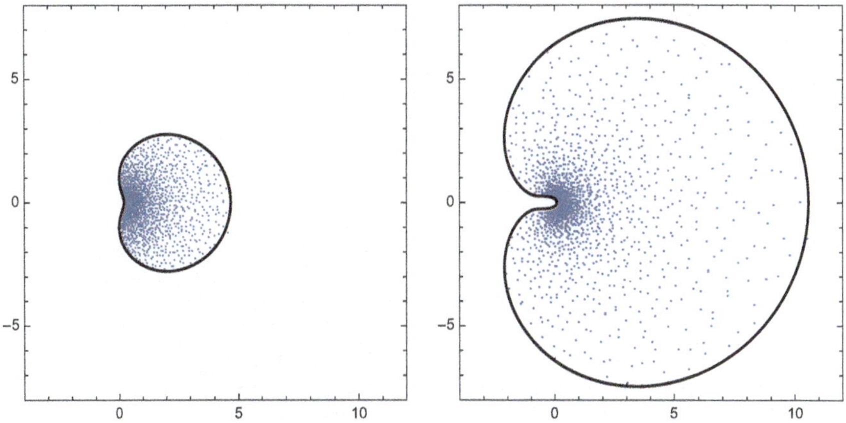

Fig. 10 Eigenvalues of B_t^N for $t = 2$ (left) and $t = 3.9$ (right), with $N = 2{,}000$

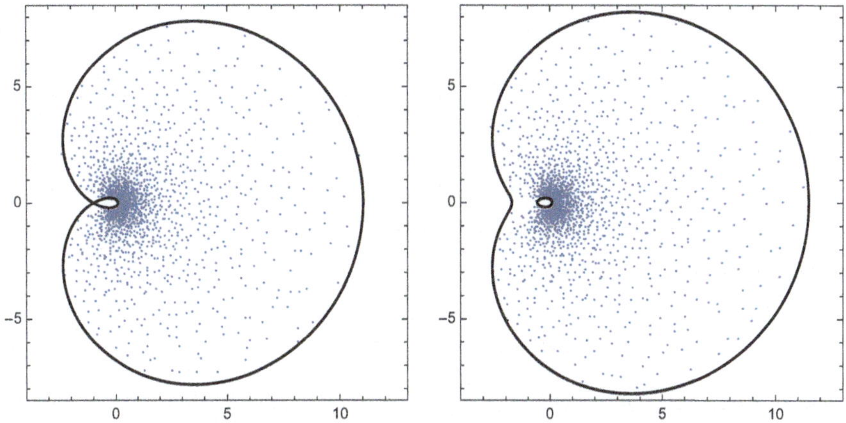

Fig. 11 Eigenvalues of B_t^N for $t = 4$ (left) and $t = 4.1$ (right), with $N = 2{,}000$

$$T(\lambda) = |\lambda - 1|^2 \frac{\log(|\lambda|^2)}{|\lambda|^2 - 1}. \tag{53}$$

If $|\lambda|^2 = 1$, we interpret $\log(|\lambda|^2)/(|\lambda|^2 - 1)$ as having the value 1 when $|\lambda|^2 = 1$, in accordance with the limit

$$\lim_{r \to 1} \frac{\log r}{r - 1} = 1.$$

See Figure 12 for a plot of this function.

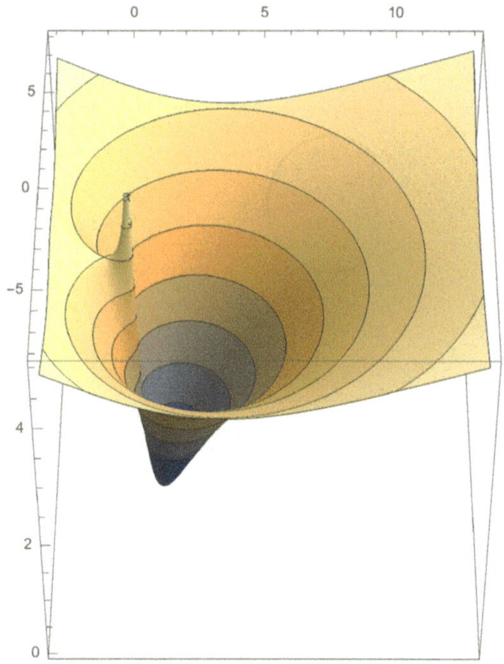

Fig. 12 A plot of the function $T(\lambda)$. The function has a minimum at $\lambda = 1$, a saddle point at $\lambda = -1$, and a singularity at $\lambda = 0$

We then define the domains Σ_t as follows.

Definition 20 For each $t > 0$, we define

$$\Sigma_t = \{\lambda \in \mathbb{C} | \, T(\lambda) < t\}.$$

Several examples of these domains were plotted already in Figures 9, 10, and 11. The domain Σ_t is simply connected for $t \leq 4$ and doubly connected for $t > 4$. The change in behavior at $t = 4$ occurs because T has a saddle point at $\lambda = -1$ and because $T(-1) = 4$. We note that a change in the topology of the region occurs at $t = 4$, which is the same value of t at which the topology of the support of Biane's measure changes (Theorem 19).

7.4 The Support of the Brown Measure of b_t

As we have noted, the domains Σ_t were introduced by Biane in [4]. Two subsequent works in the physics literature, the article [18] by Gudowska-Nowak, Janik, Jurkiewicz, and Nowak and the article [32] by Lohmayer, Neuberger, and Wettig then argued, using nonrigorous methods, that the eigenvalues of B_t^N should concentrate into Σ_t for large N. The first rigorous result in this direction was obtained by the author with Kemp [26]; we prove that the Brown measure of b_t is supported on the closure of Σ_t.

Now, we have already noted that Σ_t is simply connected for $t \leq 4$ but doubly connected for $t > 4$. Thus, the support of the Brown measure of the free *multiplicative* Brownian motion undergoes a “topological phase transition” at precisely the same value of the time parameter as the distribution of the free *unitary* Brownian motion (Theorem 19).

The methods of [26] explain this apparent coincidence, using the “free Hall transform” $\mathcal{G}_t$ of Biane [4]. Biane constructed this transform using methods of free probability as an infinite-dimensional analog of the Segal–Bargmann transform for $\mathsf{U}(N)$, which was developed by the author in [21]. More specifically, Biane's definition $\mathcal{G}_t$ draws on the stochastic interpretation of the transform in [21] given by Gross and Malliavin [17]. Biane conjectured (with an outline of a proof) that $\mathcal{G}_t$ is actually the large-N limit of the transform in [21]. This conjecture was then verified by in independent works of Cébron [8] and the author with Driver and Kemp [9]. (See also the expository article [25].)

Recall from Section 7.2 that the distribution of the free unitary Brownian motion is Biane's measure ν_t on the unit circle, the support of which is described in Theorem 19. A key ingredient in [26] is the function f_t given by

$$f_t(\lambda) = \lambda e^{\frac{t}{2}\frac{1+\lambda}{1-\lambda}}. \tag{54}$$

This function maps the complement of the closure of Σ_t conformally to the complement of the support of Biane's measure:

$$f_t : \mathbb{C} \setminus \overline{\Sigma}_t \to \mathbb{C} \setminus \mathrm{supp}(\nu_t). \tag{55}$$

(This map f_t will also play a role in the results of Section 7.5; see Theorem 23.)

The key computation in [26] is that for λ outside $\overline{\Sigma}_t$, we have

$$\mathcal{G}_t^{-1}\left(\frac{1}{z-\lambda}\right) = \frac{f_t(\lambda)}{\lambda}\frac{1}{u - f_t(\lambda)}, \quad \lambda \notin \overline{\Sigma}_t. \tag{56}$$

See Theorem 6.8 in [26]. Properties of the free Hall transform then imply that for λ outside $\overline{\Sigma}_t$, the operator $b_t - \lambda$ has an inverse. Indeed, the noncommutative L^2 norm of $(b_t - \lambda)^{-1}$ equals to the norm in $L^2(S^1, \nu_t)$ of the function on the right-hand side of (56). This norm, in turn, is finite because $f_t(\lambda)$ is outside the support of ν_t whenever λ is outside $\overline{\Sigma}_t$. The existence of an inverse to $b_t - \lambda$ then shows that λ must be outside the support of μ_{b_t}.

An interesting aspect of the paper [26] is that we not only *compute* the support of μ_{b_t} but also that we *connect* it to the support of Biane's measure ν_t, using the transform $\mathcal{G}_t$ and the conformal map f_t.

We note, however, that none of the papers [18, 32], or [26] says anything about the distribution of μ_{b_t} within Σ_t; they are only concerned with identifying the region Σ_t. The actual computation of μ_{b_t} (not just its support) was done in [10].

7.5 The Brown Measure of b_t

We now describe the main results of [10]. Many of these results have been extended by Ho and Zhong [29] to the case of the free multiplicative Brownian motion with an arbitrary unitary initial distribution.

The first key result in [10] is the following formula for the Brown measure of b_t (Theorem 2.2 of [10]).

Theorem 21 *For each $t > 0$, the Brown measure μ_{b_t} is zero outside the closure of the region Σ_t. In the region Σ_t, the Brown measure has a density W_t with respect to Lebesgue measure. This density has the following special form in polar coordinates:*

$$W_t(r, \theta) = \frac{1}{r^2} w_t(\theta), \quad re^{i\theta} \in \Sigma_t,$$

for some positive continuous function w_t. The function w_t is determined entirely by the geometry of the domain and is given as

$$w_t(\theta) = \frac{1}{4\pi}\left(\frac{2}{t} + \frac{\partial}{\partial\theta}\frac{2r_t(\theta)\sin\theta}{r_t(\theta)^2 + 1 - 2r_t(\theta)\cos\theta}\right),$$

where $r_t(\theta)$ is the "outer radius" of the region Σ_t at angle θ.

See Figure 13 for the definition of $r_t(\theta)$, Figure 14 for plots of the function $w_t(\theta)$, and Figure 15 for a plot of W_t. The simple explicit dependence of W_t on r is a major surprise of our analysis. See Corollary 22 for a notable consequence of the form of W_t.

Using implicit differentiation, it is possible to compute $dr_t(\theta)/d\theta$ explicitly as a function of $r_t(\theta)$. This computation yields the following formula for w_t, which does not involve differentiation:

$$w_t(\theta) = \frac{1}{2\pi t}\omega(r_t(\theta), \theta),$$

where

$$\omega(r, \theta) = 1 + h(r)\frac{\alpha(r)\cos\theta + \beta(r)}{\beta(r)\cos\theta + \alpha(r)}, \tag{57}$$

and

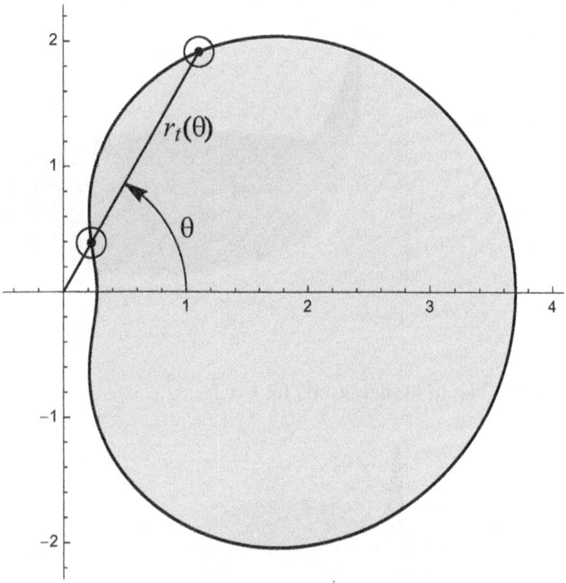

Fig. 13 The quantity $r_t(\theta)$ is the larger of the two radii at which the ray of angle θ intersects the boundary of Σ_t

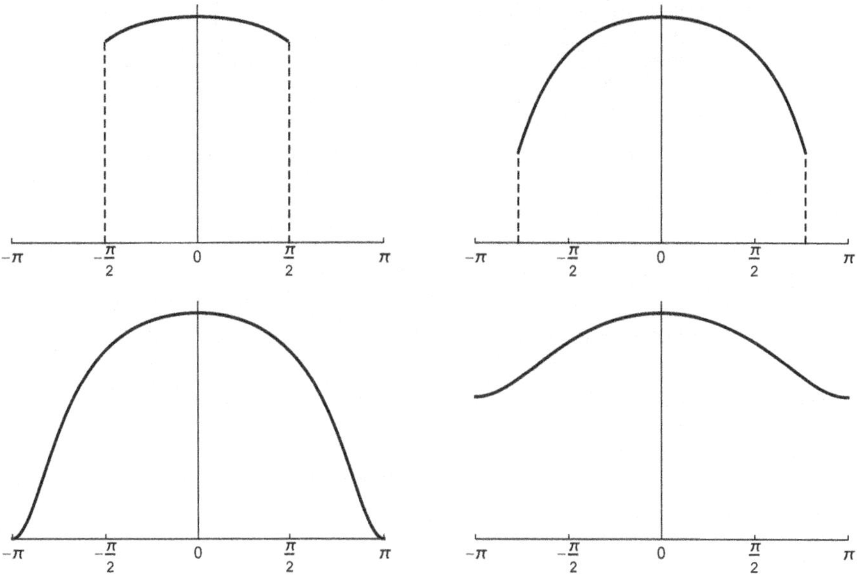

Fig. 14 Plots of $w_t(\theta)$ for $t = 2$, 3.5, 4, and 7

$$h(r) = r\frac{\log(r^2)}{r^2 - 1}; \quad \alpha(r) = r^2 + 1 - 2rh(r); \quad \beta(r) = (r^2 + 1)h(r) - 2r.$$

See Proposition 2.3 in [10].

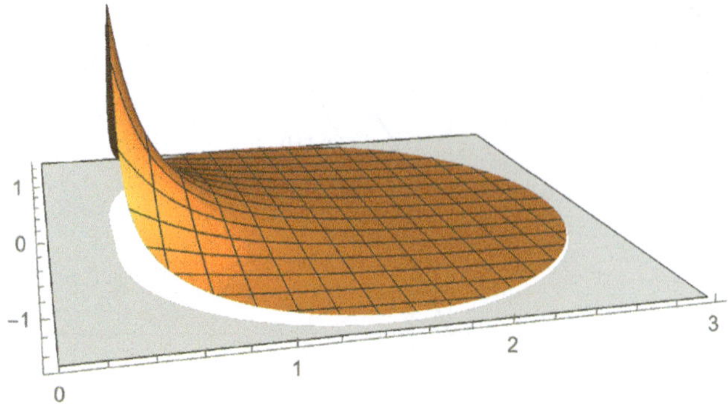

Fig. 15 Plot of the density W_t for $t = 1$

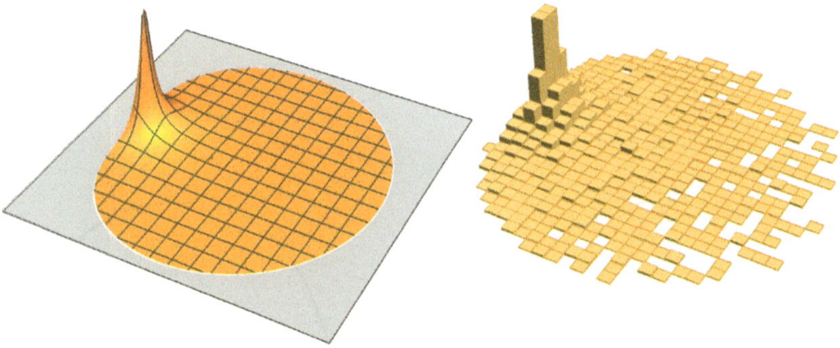

Fig. 16 The density W_t (left) and a histogram of the eigenvalues of B_t^N (right), for $t = 1$ and $N = 2{,}000$

We expect that the Brown measure of b_t will coincide with the limiting empirical eigenvalue distribution of the Brownian motion B_t^N in $\mathsf{GL}(N;\mathbb{C})$. This expectation is supported by simulations; see Figure 16.

We note that the Brown measure (inside Σ_t) can also be written as

$$\begin{aligned} d\mu_{b_t} &= \frac{1}{r^2} w_t(\theta)\, r\, dr\, d\theta \\ &= w_t(\theta)\, \frac{1}{r}\, dr\, d\theta \\ &= w_t(\theta)\, d\log r\, d\theta. \end{aligned}$$

Since the complex logarithm is given by $\log(re^{i\theta}) = \log r + i\theta$, we obtain the following consequence of Theorem 21.

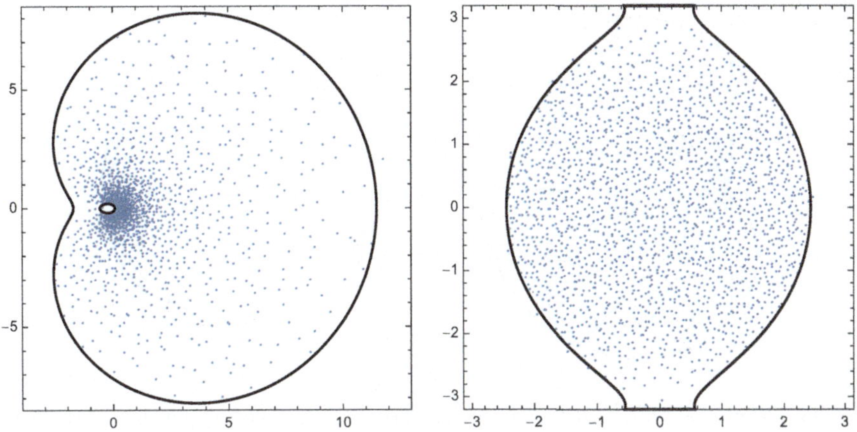

Fig. 17 The eigenvalues of B_t^N for $t = 4.1$ and $N = 2{,}000$ (left) and the logarithms thereof (right). The density of points on the right-hand side of the figure is approximately constant in the horizontal direction

Corollary 22 *The push-forward of the Brown measure μ_{b_t} under the complex logarithm has density that is constant in the horizontal direction and given by w_t in the vertical direction.*

In light of this corollary, we expect that for large N, the logarithms of the eigenvalues of B_t^N should be approximately uniformly distributed in the horizontal direction. This expectation is confirmed by simulations, as in Figure 17.

We conclude this section by describing a remarkable connection between the Brown measure μ_{b_t} and the distribution ν_t of the free unitary Brownian motion. Recall the holomorphic function f_t in (54) and (55). This map takes the boundary of Σ_t to the unit circle. We may then define a map

$$\Phi_t : \overline{\Sigma}_t \to S^1$$

by requiring (a) that Φ_t should agree with f_t on the boundary of Σ_t and (b) that Φ_t should be constant along each radial segment inside $\overline{\Sigma}_t$, as in Figure 18. (This specification makes sense because f_t has the same value at the two boundary points on each radial segment.) We then have the following result, which may be summarized by saying that *the distribution ν_t of free unitary Brownian motion is a "shadow" of the Brown measure of b_t.*

Theorem 23 *The push-forward of the Brown measure of b_t under the map Φ_t is Biane's measure ν_t on S^1. Indeed, the Brown measure of b_t is the* unique *measure μ on $\overline{\Sigma}_t$ with the following two properties:* (1) *the push-forward of μ by Φ_t is ν_t,* and (2) *μ is absolutely continuous with respect to Lebesgue measure with a density W having the form*

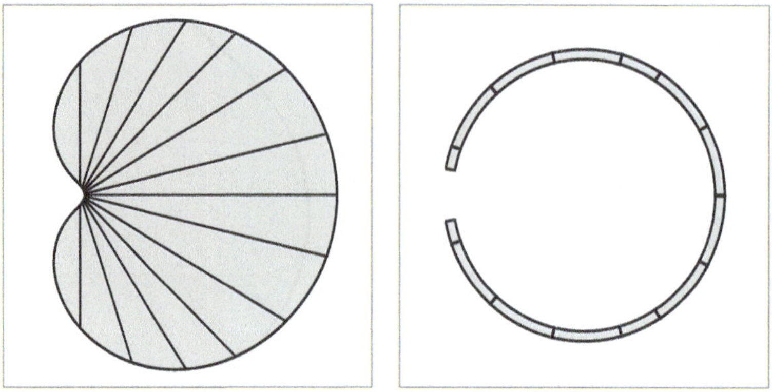

Fig. 18 The map Φ_t maps $\overline{\Sigma}_t$ to the unit circle by mapping each radial segment in $\overline{\Sigma}_t$ to a single point in S^1

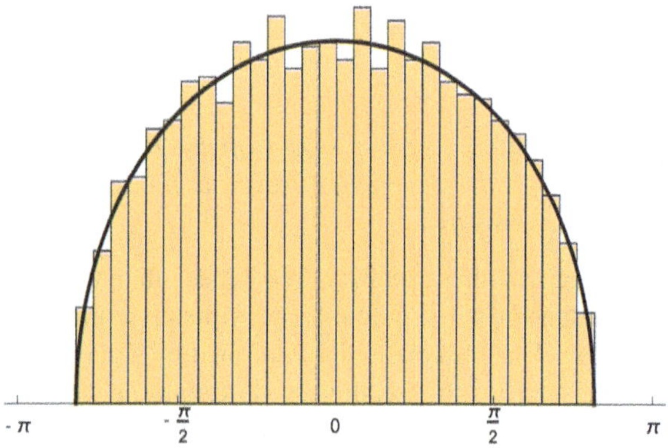

Fig. 19 The eigenvalues of B_t^N, mapped to the unit circle by Φ_t, plotted against the density of Biane's measure ν_t. Shown for $t = 2$ and $N = 2{,}000$

$$W(r, \theta) = \frac{1}{r^2} g(\theta)$$

in polar coordinates, for some continuous function g.

This result is Proposition 2.6 in [10]. Figure 19 shows the eigenvalues for B_t^N after applying the map Φ_t, plotted against the density of Biane's measure ν_t. We emphasize that we have computed the eigenvalues of the Brownian B_t^N motion in $\mathsf{GL}(N; \mathbb{C})$ (in the two-dimensional region Σ_t) and then mapped these points to the unit circle. The resulting histogram, however, looks precisely like a histogram of the eigenvalues of the Brownian motion in $\mathsf{U}(N)$.

7.6 *The PDE and Its Solution*

We conclude this article by briefly outlining the methods used to obtain the results in the previous subsection.

7.6.1 The PDE

Following the definition of the Brown measure in Theorem 7, we consider the function

$$S(t, \lambda, \varepsilon) := \tau[\log((b_t - \lambda)^*(b_t - \lambda) + \varepsilon)]. \quad (58)$$

We then record the following result [10, Theorem 2.8].

Theorem 24 *The function S in* (58) *satisfies the following PDE:*

$$\frac{\partial S}{\partial t} = \varepsilon \frac{\partial S}{\partial \varepsilon}\left(1 + (|\lambda|^2 - \varepsilon)\frac{\partial S}{\partial \varepsilon} - a\frac{\partial S}{\partial a} - b\frac{\partial S}{\partial b}\right), \quad \lambda = a + ib, \quad (59)$$

with the initial condition

$$S(0, \lambda, \varepsilon) = \log(|\lambda - 1|^2 + \varepsilon). \quad (60)$$

Recall that in the case of the circular Brownian motion (the PDE in Theorem 9), the complex number λ enters only into the initial condition and not into the PDE itself. By contrast, the right-hand side of the PDE (59) involves differentiation with respect to the real and imaginary parts of λ.

On the other hand, the PDE (59) is again of Hamilton–Jacobi type. Thus, following the general Hamilton–Jacobi method in Section 5.1, we define a Hamiltonian function H from (the negative of) the right-hand side of (59), replacing each derivative of S by a corresponding momentum variable:

$$H(a, b, \varepsilon, p_a, p_b, p_\varepsilon) = -\varepsilon p_\varepsilon(1 + (a^2 + b^2)p_\varepsilon - \varepsilon p_\varepsilon - ap_a - bp_b). \quad (61)$$

We then consider Hamilton's equations for this Hamiltonian:

$$\frac{da}{dt} = \frac{\partial H}{\partial p_a}; \quad \frac{db}{dt} = \frac{\partial H}{\partial p_b}; \quad \frac{d\varepsilon}{dt} = \frac{\partial H}{\partial p_\varepsilon};$$
$$\frac{dp_a}{dt} = -\frac{\partial H}{\partial a}; \quad \frac{dp_b}{dt} = -\frac{\partial H}{\partial b}; \quad \frac{dp_\varepsilon}{dt} = -\frac{\partial H}{\partial \varepsilon}. \quad (62)$$

Then, after a bit of simplification, the general Hamilton–Jacobi formula in (40) then takes the form

$$S(t,\lambda(t),\varepsilon(t)) = \log(|\lambda_0 - 1|^2 + \varepsilon_0) - \frac{\varepsilon_0 t}{(|\lambda_0 - 1|^2 + \varepsilon_0)^2} + \log|\lambda(t)| - \log|\lambda_0|. \tag{63}$$

(See Theorem 6.2 in [10].)

The analysis in [10] then proceeds along broadly similar lines to those in Sections 5 and 6. The main structural difference is that because λ is now a variable in the PDE, the ODE's in (62) now involve both x and λ and the associated momenta. (That is to say, the vector $\mathbf{x}$ in Proposition 12 is equal to $(\lambda, \varepsilon) \in \mathbb{C} \times \mathbb{R} \cong \mathbb{R}^3$.) The first key result is that the system of ODE's associated to (59) can be solved explicitly; see Section 6.3 of [10]. Solving the ODE's gives an implicit formula for the solution to (59) with the initial conditions (60).

We then evaluate the solution in the limit as ε tends to zero. We follow the strategy in Section 6. Given a time t and a complex number λ, we attempt to choose initial conditions ε_0 and λ_0 so that $\varepsilon(t)$ will be very close to zero and $\lambda(t)$ will equal λ. (Recall that the initial momenta in the system of ODE's are determined by the positions by (39).)

7.6.2 Outside the Domain

As in the case of the circular Brownian motion, we use different approaches for λ outside Σ_t and for λ in Σ_t. For λ outside Σ_t, we allow the initial condition ε_0 in the ODE's to approach zero. As it turns out, when ε_0 is small and positive, $\varepsilon(t)$ remains small and positive for as long as the solution to the system exists. Furthermore, when ε_0 is small and positive, $\lambda(t)$ is approximately constant. Thus, our strategy will be to take $\varepsilon_0 \approx 0$ and $\lambda_0 \approx \lambda$.

A key result is the following.

Proposition 25 *In the limit as ε_0 tends to zero, the lifetime of the solution to* (62) *with initial conditions λ_0 and ε_0—and initial moment determined by* (39)*—approaches $T(\lambda_0)$, where T is the same function* (53) *that enters into the definition of the domain Σ_t.*

This result is Proposition 6.13 in [10]. Thus, the strategy in the previous paragraph will work—meaning that the solution continues to exist up to time t—provided that $T(\lambda_0) \approx T(\lambda)$ is greater than t. The condition for success of the strategy is, therefore, $T(\lambda) > t$. In light of the characterization of Σ_t in Definition 20, we make have the following conclusion.

Conclusion 26 *The simple strategy of taking $\varepsilon_0 \approx 0$ and $\lambda_0 \approx \lambda$ is successful precisely if $T(\lambda) > t$ or, equivalently, if λ is outside $\overline{\Sigma}_t$.*

When this strategy works, we obtain a simple expression for $\lim_{\varepsilon \to 0^+} S(t, \lambda, \varepsilon)$, by letting ε_0 approach zero and λ_0 approach λ in (63). Since $\lambda(t)$ approaches λ in this limit [10, Proposition 6.11], we find that

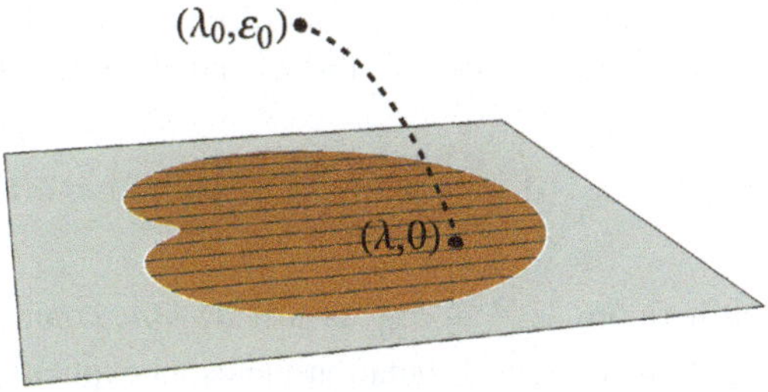

Fig. 20 For each λ in Σ_t, there exists $\varepsilon_0 > 0$ and $\lambda_0 \in \Sigma_t$ such that with these initial conditions, we have $\varepsilon(t) = 0$ and $\lambda(t) = \lambda$

$$\lim_{\varepsilon \to 0^+} S(t, \lambda, \varepsilon) = \log(|\lambda - 1|^2), \quad \lambda \notin \overline{\Sigma}_t. \tag{64}$$

This function is harmonic (except at $\lambda = 1$, which is always in the domain Σ_t), so we conclude that *the Brown measure of* b_t *is zero outside* $\overline{\Sigma}_t$. See Section 7.2 in [10] for more details.

7.6.3 Inside the Domain

For λ inside Σ_t, the simple approach in the previous subsection does not work, because when λ is outside Σ_t and ε_0 is small, the solutions to the ODE's (62) will cease to exist prior to time t (Proposition 25). Instead, we must prove a "surjectivity" result: For each $t > 0$ and $\lambda \in \Sigma_t$, there exist—in principle—$\lambda_0 \in \mathbb{C}$ and $\varepsilon_0 > 0$ giving $\lambda(t) = \lambda$ and $\varepsilon(t) = 0$. See Figure 20. Actually the proof shows that λ_0 again belongs to the domain Σ_t; see Section 6.5 in [10].

We then make use of the second Hamilton–Jacobi formula (41), which allows us to compute the derivatives of S directly, without having to attempt to differentiate the formula (63) for S. Working in logarithmic polar coordinates, $\rho = \log|\lambda|$ and $\theta = \arg\lambda$, we find an amazingly simple expression for the quantity

$$\frac{\partial s_t}{\partial \rho} = \lim_{\varepsilon \to 0^+} \frac{\partial S}{\partial \rho}(t, \lambda, \varepsilon),$$

inside Σ_t, namely,

$$\frac{\partial s_t}{\partial \rho} = \frac{2\rho}{t} + 1, \quad \lambda \in \Sigma_t. \tag{65}$$

(See Corollary 7.6 in [10].) This result is obtained using a certain constant of motion of the system of ODE's, namely, the quantity

$$\Psi = \varepsilon p_\varepsilon + \frac{1}{2}(a p_a + b p_b)$$

in [10, Proposition 6.5].

If we evaluate this constant of motion at a time t when $\varepsilon(t) = 0$, the $\varepsilon p_\varepsilon$ term vanishes. But if $\varepsilon(t) = 0$, the second Hamilton–Jacobi formula (41) tells us that

$$\left(a\frac{\partial S}{\partial a} + b\frac{\partial S}{\partial b}\right)(t, \lambda(t), 0) = a(t)p_a(t) + b(t)p_b(t).$$

Furthermore, $a\frac{\partial S}{\partial a} + b\frac{\partial S}{\partial b}$ is just $\partial S/\partial\rho$, computed in rectangular coordinates. A bit of algebraic manipulation yields an explicit formula for $a\frac{\partial S}{\partial a} + b\frac{\partial S}{\partial b}$, as in [10, Theorem 6.7], explaining the formula (65). To complete the proof (65), it still remains to address certain regularity issues of $S(t, \lambda, \varepsilon)$ near $\varepsilon > 0$, as in Section 7.3 of [10].

Once (65) is established, we note that the formula for $\partial s_t/\partial\rho$ in (65) is independent of θ. It follows that

$$\frac{\partial}{\partial\rho}\frac{\partial s_t}{\partial\theta} = \frac{\partial}{\partial\theta}\frac{\partial s_t}{\partial\rho} = 0,$$

that is, that $\partial s_t/\partial\theta$ is independent of ρ inside Σ_t. Writing the Laplacian in logarithmic polar coordinates, we then find that

$$\begin{aligned}\Delta s_t(\lambda) &= \frac{1}{r^2}\left(\frac{\partial^2 s_t}{\partial\rho^2} + \frac{\partial^2 s_t}{\partial\theta^2}\right)\\ &= \frac{1}{r^2}\left(\frac{2}{t} + \frac{\partial}{\partial\theta}\left(\frac{\partial s_t}{\partial\theta}\right)\right), \quad \lambda \in \Sigma_t,\end{aligned} \tag{66}$$

where $2/t$ term in the expression comes from differentiating (65) with respect to ρ. Since $\partial s_t/\partial\theta$ is independent of ρ, we can understand the structure of the formula in Theorem 21.

The last step in the proof of Theorem 21 is to compute $\partial s_t/\partial\theta$. Since $\partial s_t/\partial\theta$ is independent of ρ—or, equivalently, independent of $r = |\lambda|$—inside Σ_t, the value of $\partial s_t/\partial\theta$ at a point λ in Σ_t is the same as its value as we approach the boundary of Σ_t along the radial segment through λ. We show that $\partial s_t/\partial\theta$ is continuous over the whole complex plane, even at the boundary of Σ_t. (See Section 7.4 of [10].) Thus, on the boundary of Σ_t, the function $\partial s_t/\partial\theta$ will agree with the angular derivative of $\log(|\lambda - 1|^2)$, namely

$$\begin{aligned}\frac{\partial}{\partial\theta}\log(|\lambda - 1|^2) &= \frac{2\,\mathrm{Im}\,\lambda}{|\lambda - 1|^2}\\ &= \frac{2r\sin\theta}{r^2 + 1 - 2r\cos\theta}.\end{aligned} \tag{67}$$

Thus, to compute $\partial s_t/\partial\theta$ at a point λ in Σ_t, we simply evaluate (67) at either of the two points where the radial segment through λ intersects $\partial\Sigma_t$. (We get the same value at either point.)

One such boundary point is the point with argument $\theta = \arg\lambda$ and radius $r_t(\theta)$, as in Figure 13. Thus, inside Σ_t, we have

$$\frac{\partial s_t}{\partial\theta} = \frac{2r_t(\theta)\sin\theta}{r_t(\theta)^2 + 1 - 2r_t(\theta)\cos\theta}.$$

Plugging this expression into (66) gives the claimed formula in Theorem 21.

References

1. Z.D. Bai, Circular law. Ann. Probab. **25**, 494–529 (1997)
2. P. Biane, On the free convolution with a semi-circular distribution. Indiana Univ. Math. J. **46**, 705–718 (1997)
3. P. Biane, Free Brownian motion, free stochastic calculus and random matrices, in *Free Probability Theory*, Waterloo, 1995. Fields Institute Communications, vol. 12 (American Mathematical Society, Providence, 1997), pp. 1–19
4. P. Biane, Segal–Bargmann transform, functional calculus on matrix spaces and the theory of semi-circular and circular systems. J. Funct. Anal. **144**, 232–286 (1997)
5. P. Biane, R. Speicher, Stochastic calculus with respect to free Brownian motion and analysis on Wigner space. Probab. Theory Related Fields **112**, 373–409 (1998)
6. P. Bourgade, J.P. Keating, Quantum chaos, random matrix theory, and the Riemann ζ-function, in *Chaos*. Progress in Mathematical Physics, vol. 66 (Birkhäuser/Springer, Basel, 2013), pp. 125–168
7. L.G. Brown, Lidskiĭ's theorem in the type II case, in *Geometric Methods in Operator Algebras*, Kyoto, 1983. Pitman Research Notes in Mathematics Series, vol. 123 (Longman Scientific & Technical, Harlow, 1986), pp. 1–35
8. G. Cébron, Free convolution operators and free Hall transform. J. Funct. Anal. **265**, 2645–2708 (2013)
9. B.K. Driver, B.C. Hall, T. Kemp, The large-N limit of the Segal–Bargmann transform on $\mathbb{U}_N$. J. Funct. Anal. **265**, 2585–2644 (2013)
10. B.K. Driver, B.C. Hall, T. Kemp, The Brown measure of the free multiplicative Brownian motion, preprint arXiv:1903.11015 [math.PR] (2019)
11. L.C. Evans, *Partial Differential Equations*. Graduate Studies in Mathematics, vol. 19, 2nd edn. (American Mathematical Society, Providence, 2010), xxii+749pp
12. O. Feldheim, E. Paquette, O. Zeitouni, Regularization of non-normal matrices by Gaussian noise. Int. Math. Res. Not. IMRN **18**, 8724–8751 (2015)
13. B. Fuglede, R.V. Kadison, On determinants and a property of the trace in finite factors. Proc. Nat. Acad. Sci. U. S. A. **37**, 425–431 (1951)
14. B. Fuglede, R.V. Kadison, Determinant theory in finite factors. Ann. Math. (2) **55**, 520–530 (1952)
15. J. Ginibre, Statistical ensembles of complex, quaternion, and real matrices. J. Math. Phys. **6**, 440–449 (1965)
16. V.L. Girko, The circular law. (Russian) Teor. Veroyatnost. i Primenen. **29**, 669–679 (1984)
17. L. Gross, P. Malliavin, Hall's transform and the Segal–Bargmann map, in *Itô's Stochastic Calculus and Probability Theory*, ed. by N. Ikeda, S. Watanabe, M. Fukushima, H. Kunita (Springer, Tokyo, 1996), pp. 73–116
18. E. Gudowska-Nowak, R.A. Janik, J. Jurkiewicz, M.A. Nowak, Infinite products of large random matrices and matrix-valued diffusion. Nuclear Phys. B **670**, 479–507 (2003)
19. A. Guionnet, P.M. Wood, O. Zeitouni, Convergence of the spectral measure of non-normal matrices. Proc. Am. Math. Soc. **142**, 667–679 (2014)

20. M.C. Gutzwiller, *Chaos in Classical and Quantum Mechanics*. Interdisciplinary Applied Mathematics, vol. 1 (Springer, New York, 1990)
21. B.C. Hall, The Segal–Bargmann "coherent state" transform for compact Lie groups. J. Funct. Anal. **122**, 103–151 (1994)
22. B.C. Hall, Harmonic analysis with respect to heat kernel measure. Bull. Am. Math. Soc. (N.S.) **38**, 43–78 (2001)
23. B.C. Hall, *Quantum Theory for Mathematicians*. Graduate Texts in Mathematics, vol. 267 (Springer, New York, 2013)
24. B.C. Hall, *Lie Groups, Lie Algebras, and Representations. An Elementary Introduction*. Graduate Texts in Mathematics, vol. 222, 2nd edn. (Springer, Cham, 2015)
25. B.C. Hall, The Segal–Bargmann transform for unitary groups in the large-N limit, preprint arXiv:1308.0615 [math.RT] (2013)
26. B.C. Hall, T. Kemp, Brown measure support and the free multiplicative Brownian motion. Adv. Math. **355**, article 106771, 1–36 (2019)
27. N.J. Higham, *Functions of Matrices. Theory and Computation*. (Society for Industrial and Applied Mathematics (SIAM), Philadelphia, 2008)
28. C.-W. Ho, The two-parameter free unitary Segal-Bargmann transform and its Biane-Gross-Malliavin identification. J. Funct. Anal. **271**, 3765–3817 (2016)
29. C.-W. Ho, P. Zhong, Brown Measures of free circular and multiplicative Brownian motions with probabilistic initial point, preprint arXiv:1908.08150 [math.OA] (2019)
30. N.M. Katz, P. Sarnak, Zeroes of zeta functions and symmetry, Bull. Am. Math. Soc. (N.S.) **36**, 1–26 (1999)
31. T. Kemp, The large-N limits of Brownian motions on $\mathbb{GL}_N$. Int. Math. Res. Not. **2016**, 4012–4057 (2016)
32. R. Lohmayer, H. Neuberger, T. Wettig, Possible large-N transitions for complex Wilson loop matrices. J. High Energy Phys. **2008**(11), 053, 44pp (2008)
33. M.L. Mehta, *Random Matrices*. Pure and Applied Mathematics (Amsterdam), vol. 142, 3rd edn. (Elsevier/Academic Press, Amsterdam, 2004)
34. J.A. Mingo, R. Speicher, *Free Probability and Random Matrices*. Fields Institute Monographs, vol. 35 (Springer/Fields Institute for Research in Mathematical Sciences, New York/Toronto, 2017)
35. H.L. Montgomery, The pair correlation of zeros of the zeta function. Analytic number theory. (Proceedings of Symposia in Pure Mathematics, vol. XXIV, St. Louis University, St. Louis, 1972) (American Mathematical Society, Providence, 1973), pp. 181–193
36. A. Nica, R. Speicher, *Lectures on the Combinatorics of Free Probability*. London Mathematical Society Lecture Note Series, vol. 335 (Cambridge University Press, Cambridge, 2006)
37. W. Rudin, *Principles of Mathematical Analysis*. International Series in Pure and Applied Mathematics, 3rd edn. (McGraw-Hill Book Co., New York/Auckland/Düsseldorf, 1976)
38. P. Śniady, Random regularization of Brown spectral measure. J. Funct. Anal. **193**, 291–313 (2002)
39. H.-J. Stöckmann, *Quantum Chaos. An Introduction* (Cambridge University Press, Cambridge, 1999)
40. T. Tao, *Topics in Random Matrix Theory*. Graduate Studies in Mathematics, vol. 132 (American Mathematical Society, Providence, 2012)
41. D. Voiculescu, Symmetries of some reduced free product C^*-algebras, in *Operator Algebras and Their Connections with Topology and Ergodic Theory*, Buşteni, 1983. Lecture Notes in Mathematics, vol. 1132 (Springer, Berlin, 1985), pp. 556–588
42. D. Voiculescu, Limit laws for random matrices and free products. Invent. Math. **104**, 201–220 (1991)
43. E. Wigner, Characteristic vectors of bordered matrices with infinite dimensions. Ann. Math. (2) **62**, 548–564 (1955)

Structure and Optimisation in Computational Harmonic Analysis: On Key Aspects in Sparse Regularisation

Anders C. Hansen and Bogdan Roman

Abstract Computational harmonic analysis has a rich history spanning more than half a century, where the last decade has been strongly influenced by sparse regularisation and compressed sensing. The theory has matured over the last years, and it has become apparent that the success of compressed sensing in fields like magnetic resonance imaging (MRI), and imaging in general, is due to specific structures beyond just sparsity. Indeed, structured sampling and the structure of images represented in X-lets, for example, sparsity in levels, are key ingredients. The field relies on the crucial assumption that one can easily compute minimisers of convex optimisation problem. This assumption is false in general. One can typically easily compute the objective function of convex optimisation problems, but not minimisers. However, due to the specific features in compressed sensing, one can actually compute the desired minimisers fast and reliably to sufficient precision. In short, as we demonstrate here: the success of sparse regularisation and compressed sensing is due to specific key structures that allow for a beneficial interaction between harmonic analysis and optimisation.

1 Introduction

Compressed sensing (CS) and sparse regularisation [1–4] concern the recovery of an object (e.g. a signal or image) from an incomplete set of linear measurements. In a discrete setting, this can be formulated as the linear system

$$y = Ax,$$

where $y \in \mathbb{C}^m$ is the vector of measurements, $x \in \mathbb{C}^N$ is the object to recover and $A \in \mathbb{C}^{m \times N}$ is the so-called measurement matrix. In practice, the number

A. C. Hansen (✉) · B. Roman
DAMTP, Centre for Mathematical Sciences, University of Cambridge, Cambridge, UK
e-mail: a.hansen@damtp.cam.ac.uk; abr28@cam.ac.uk

M. Th. Rassias (ed.), *Harmonic Analysis and Applications*, Springer Optimization and Its Applications 168, https://doi.org/10.1007/978-3-030-61887-2_6

of measurements m is often substantially smaller than the dimension N, making the recovery of x from y generally impossible. To overcome this, compressed sensing leverages two key properties: first, *sparsity* of the vector x, and second, the *incoherence* of the measurement vectors (rows of the matrix A). The first property asserts that x should have at most $s \leq m$ significant components, with the remainder being small. The second property asserts that the rows of A should be (in a sense that can be rigorously defined) spread out, rather than concentrated around a small number of entries [5]. Taking noise into account, we can say that $y = Ax + e$. Recovering x from y can be achieved by a number of recovery approaches, including ℓ^1 minimisation of basis pursuit, where one seeks

$$x \in \underset{z}{\operatorname{argmin}} \|z\|_1 \text{ subject to } \|Az - y\| \leq \eta, \qquad \eta \geq 0, \tag{1}$$

where $\|e\|_2 \leq \eta$. The key to this is the assumed structure of x. We say that x is s-sparse if it has at most s non-zero entries, regardless of their locations. The CS literature now contains a wealth of results giving sufficient and/or necessary conditions for recovering an s-sparse vector from y by solving either (1) or other appropriate algorithms.

The key observation, in the context of this paper, is that there is a mismatch between the sparsity model and the class of recovered signals for many applications such as magnetic resonance imaging (MRI), computerised tomography (CT), electron microscopy (EM), fluorescence microscopy (FM), radio interferometry (RI) and others. Specifically, one does not actually recover all sparse vectors, but only a small subset of sparse vectors which in fact possess far more structure than sparsity alone. It is possible to observe this phenomenon through the so-called flip test [6, 7], which we shall detail Sect. 2.2.

This observation gives rise to the aforementioned question about what kind of structured signals does CS actually recover, which in turn can be approached from two linked perspectives:

(a) *Given a sampling mechanism, does one recover an arbitrary sparse signal?*
(b) *If not, what kind of sparsity structure does one recover?*

Question (b) is highly non-trivial. The typical approach is to conjecture a structured sparsity model and then ask:

(c) *Given a sampling mechanism and a structured sparsity model, does one recover an arbitrary signal in this class?*

We designed a numerical test, a generalised flip test, that allows one to investigate questions (a) and (c) above.

In addition, in many applications of CS such as those listed above, the sampling mechanism is itself not just random, but also highly structured [6–13], and so it is perhaps not too surprising that one recovers only signals with a particular structure. This raises a second fundamental question: if there was complete freedom to design the sampling operator (this is impossible or restricted in many applications, e.g.

MRI, but we ask this as a basic question), should one use incoherent sampling, e.g. sub-Gaussian random matrices, which can recover any s-sparse signal, or should one choose or design a sampling operator and/or strategy that recovers only structured sparse signals, such as natural images? In other words,

Does structured sampling outperform incoherent sampling?

Section 3 endeavours to answer this question both numerically and mathematically, and the answer is: yes, provided the signal is structured.

Given that compressed sensing is based on the assumption that (1) can be solved accurately and efficiently, it is legitimate to ask whether this is actually possible. Indeed, convex optimisation is a very well-established field; however, it is mostly concerned about computing the objective function in such optimisation problems. Compressed sensing and sparse recovery are based on computing the actual minimisers. Thus, we ask the following:

Do there exist algorithms that can compute minimisers to general problems of the form (1)*? If not, how come compressed sensing works well in practical imaging problem?*

These questions will be discussed in Sect. 4 and onwards.

2 What Is the Correct Model?

In the classic CS theory, the restricted isometry property (RIP) [3, 4] has been widely used and allows to obtain various recovery guarantees. The RIP states that, given any s-sparse vector $x \in \mathbb{C}^N$ and a matrix $A \in \mathbb{C}^{m\times N}$, then A has the RIP with constant δ_s if for every submatrix $A_s \in \mathbb{C}^{m\times s}$ of A we have

$$(1-\delta_s)\,\|x\|_2^2 \le \|A_s x\|_2^2 \le (1+\delta_s)\,\|x\|_2^2,$$

which can be understood as saying that any submatrix of A acting on s-sparse vectors will behave close to an isometry, preserving vector norms (energy), up to a given constant. The smaller the constant δ_s, the more isometric the submatrix. This powerful condition, while allowing to obtain theoretical recovery guarantees for a number of matrix classes (e.g. sub-Gaussian), makes no further assumptions regarding the structure of x besides being s-sparse. For example, the location (indices) of the s non-zero vector entries is not important.

A slightly weaker property is the *robust nullspace property (rNSP) of order s* [4, 14], which is implied by the RIP. We say that a matrix $A \in \mathbb{C}^{m\times N}$ satisfies the ℓ^2 rNSP if there is a $\rho \in (0, 1)$ and a $\tau > 0$ such that

$$\|v_S\|_2 \le \frac{\rho}{\sqrt{s}}\|v_{S^c}\|_1 + \tau\|Av\|_2,$$

for all sets $S \subset \{1, \dots, N\}$ with $|S| \leq s$ (where $|\cdot|$ denotes set cardinality) and vectors $v \in \mathbb{C}^N$. The notation v_S means that the coefficients of v on S^c are set to zero and the coefficients on S are kept as is. A key CS result is the following recovery guarantee:

Theorem 1 (rNSP implies stable and robust recovery) *Suppose that $A \in \mathbb{C}^{m\times N}$ has the rNSP of order s with constants $0 < \rho < 1$ and $\tau > 0$. Let $x \in \mathbb{C}^N$ and $y = Ax + e \in \mathbb{C}^m$, where $\|e\|_{\ell^2} \leq \eta$ for some $\eta \geq 0$. Then any minimiser $\hat{x} \in \mathbb{C}^N$ of (1) satisfies*

$$\|\hat{x} - x\|_{\ell^1} \leq C_1\sigma_s(x)_{\ell^1} + C_2\sqrt{s}\eta,$$

and

$$\|\hat{x} - x\|_{\ell^2} \leq C_3\frac{\sigma_s(x)_{\ell^1}}{\sqrt{s}} + C_4\eta,$$

where the constants C_1, C_2, C_3, C_4 depend on ρ and τ only, and $\sigma_s(y)_1 = \min_{z\in\Sigma_s}\|y - z\|_1$,, where Σ_s denotes the set of all s-sparse vectors.

The above theorem is a mainstay in CS theory. Thus, in view of question (1.a) above, a reasonable question that we will address below is whether the rNSP, the RIP and the standard sparsity model are satisfied in the many areas where CS can be used.

In many practical applications, CS is performed using various types of sensing matrices. A popular choice is to couple the matrix A with a sparsifying matrix W, such as a wavelet transform. This is commonly performed because most natural signals, such as a brain image in MRI, are not sparse themselves, but are sparse in some appropriate basis like wavelets. In practice, A is often a random matrix, with entries drawn independently from the same distribution (e.g. a random Gaussian matrix), or obtained by randomly selecting (undersampling) m rows from a known $N \times N$ matrix, indexed by a (random) subset $\Omega \subseteq \{1, \dots, N\}$ called the sampling pattern or sampling map. In this case, we have $A = P_\Omega\Psi$ where P_Ω is the is the diagonal projection matrix with $|P_\Omega| = m$ and jth entry 1 if $j \in \Omega$ and 0 otherwise. Here Ψ is called the raw *sampling operator*, as it models the actual sampling device which acquires samples (A would be the undersampling operator taking the measurements $y = Ax = P_\Omega\Psi x$), and we call W the *sparsifying operator*, as it renders the signal x sparse, i.e. Wx is s-sparse, and not x. In this context, (1) becomes

$$x \in \underset{z}{\operatorname{argmin}}\, \|z\|_1 \text{ subject to } \|y - P_\Omega U z\| \leq \eta, \qquad \eta \geq 0, \tag{2}$$

The resulting operator $U = \Psi W^*$ will have the RIP if $A = P_\Omega\Psi$ is universal – we say that a random matrix $A \in \mathbb{C}^{m\times N}$ is universal if for any isometry $W \in \mathbb{C}^{N\times N}$, the matrix AW has the RIP with high probability. However, in practice, Ψ may be imposed, like the Fourier operator is imposed in MRI, and so A may have a very weak RIP (i.e. very large δ_s) or may not possess the RIP. We call such operators *structured operators* because they tend to reveal specific features of s-sparse vectors, as opposed to matrices that possess the RIP. One particular such feature that is worth investigating and which helps in answering questions (1.a) and (1.c) is whether such matrices are sensitive to further structure in the input vector x. A first question would be: is the location of the s non-zero entries important?

We designed a simple test for this paradigm, which we termed flip test [6]. In brief, we *flip* (reverse the order of) the wavelet coefficients of x, hence preserving its sparsity but changing its structure. Any permutation sufficiently far from the identity could in fact be used. We then perform the same CS experiment as for the original x, and finally reverse the permutation of the wavelet coefficients. If the original reconstruction and the flipped reconstruction have (visibly) different reconstruction qualities, then it means that the position of the coefficients is important, and that the operator $U = AW^*$ does not possess the RIP, or that the RIP and sparsity alone do not explain the recovery for this class of operators and signals.

The flip test. Let $x \in \mathbb{C}^N$ be a vector, and $U \in \mathbb{C}^{N \times N}$ a sensing matrix. We sample according to some pattern $\Omega \subseteq \{1, \dots, N\}$ with $|\Omega| = m$ and solve (1) for x, i.e. $\min \|z\|_1$ s.t $P_\Omega U z = P_\Omega U x$ to obtain a reconstruction $z = \alpha$. Now we *flip* x to obtain a vector x' with reverse entries, $x'_i = x_{N-i}, i = 1, \dots, N$ and solve (2) for x' using the same U and Ω, i.e. $\min \|z\|_1$ s.t. $P_\Omega U z = P_\Omega U x'$. Assuming z to be a solution, then by flipping z, we obtain a second reconstruction α' of the original vector x, where $\alpha'_i = z_{N-i}$. Assume Ω is a sampling pattern for recovering x using α. If sparsity alone dictates the reconstruction quality, then α' must yield thef:flip-test-univ same reconstruction quality (since x' has the same sparsity as x, being merely a permutation of x).

Let us first perform the flip test for a universal operator, such as a Bernoulli matrix B. A Bernoulli matrix has random Bernoulli numbers as its entries, and is a sub-Gaussian matrix, hence it has the RIP and is also universal, i.e. BW also has the RIP for any isometry W. As described earlier, let us choose W to be the wavelet basis which renders most natural signals sparse. Given $U = BW^*$ has the RIP, we expect the two reconstructions to have the same quality, and indeed this is the case, as illustrated in Fig. 1. This is yet another confirmation that a matrix possessing the RIP is not sensitive to changes in the sparsity structure of x. We note that it is possible to obtain universal-like operators deterministically, for example, by randomly permuting columns of known matrices such as Fourier or Hadamard [15, 16], Kronecker products of random matrix stencils [17], or even fully orthogonal matrices such as the Sum-To-One (STOne) matrix [18]; these yield an operator that behaves similar to a random matrix in the CS context but allows for fast transforms.

Let us now perform the flip test for structured operators that are commonly used in practice, all of which are highly nonuniversal. These include the Fourier operator, the Hadamard operator, the Radon operator and others. Here we use $U = P_\Omega \Psi W^*$. The results are shown in Fig. 2. It is clear that the reconstruction from original wavelet coefficients and the one from flipped wavelet coefficients are strikingly different for the same operator U. This shows that the RIP does not explain the reconstructions seen in practice (left column of Fig. 2). Furthermore, it suggests that

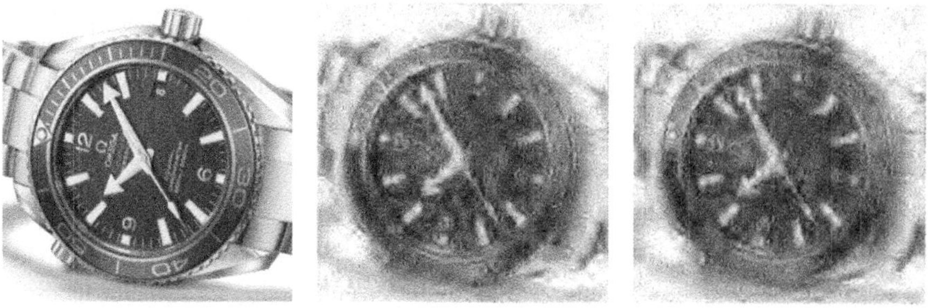

Fig. 1 *The flip test for a universal operator.* Recovery from original versus flipped wavelet coefficients for a 256×256 image. The sampling matrix is a Bernoulli matrix $B \in \{0, 1\}^{9831 \times 256^2}$ (15% subsampling). The sparsity matrix $W \in \mathbb{R}^{256^2 \times 256^2}$ is Daubechies-4. *Left:* Original image. *Middle:* Recovery from original wavelet coefficients. *Right:* Recovery from flipped wavelet coefficients

the class of signals that these operators recover must in fact be much reduced, as posed in questions (1.a) and (1.c).

One remark which we expand later on is that using universal operators in practice, even when presented with the option, is in fact undesirable in most cases. The reason is that they offer inferior reconstruction performance compared to structured operators. However, we note that this doesn't hold when the signal has unknown properties, specifically when we don't know of a representation that can render it sparse (luckily, most signals encountered in practice do offer such prior knowledge).

2.1 Generalised Flip Test

We now generalise the flip test to allow for any sparsity model, any signal class, any sampling operator and any recovery algorithm. The original flip test presented earlier is then just a special case. The purpose behind designing such a test is to allow probing some of the properties chosen for (or imposed by) the problem, e.g. whether the sparsity model chosen is too crude. To exemplify its applicability, we apply it to three existing sparsity models when considering Fourier and Hadamard measurements and X-lets as the sparsity representation: the *classic sparsity* model [1, 2], the *weighted sparsity* model [19] and the *sparsity in levels* model [6].

The generalised flip test is as follows:

(i) Signal model. Decide on the type of signals of interest in, e.g. 1D piecewise smooth functions, 2D images, smooth functions, etc. Create a discrete vector x coming from the discretisation of this desired signal model.

(ii) Sparse transform. Choose a sparsifying transform W such that Wx is sufficiently sparse. See more details in step (vi).

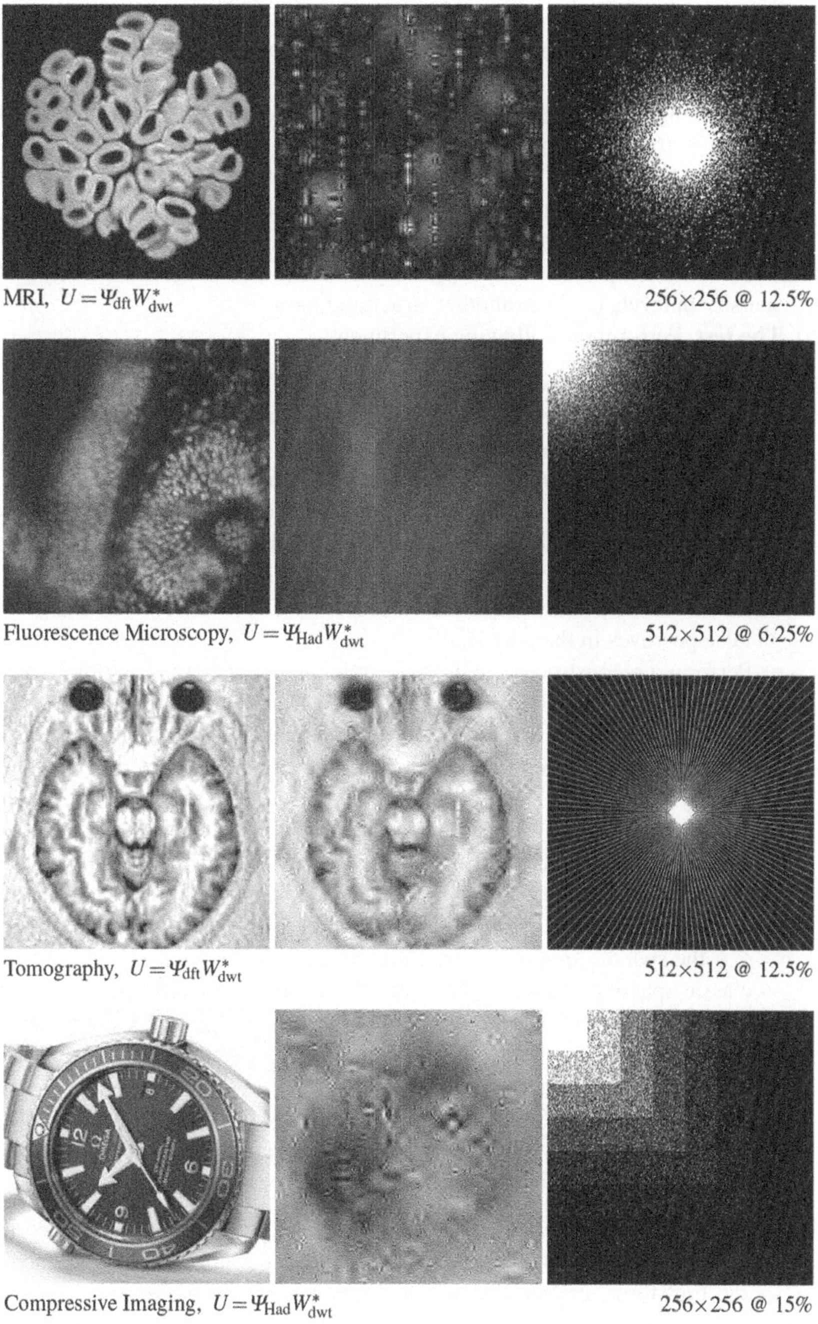

Fig. 2 *Flip tests for structured operators.* Recovery from original versus flipped wavelet coefficients for various images of sizes 256×256 and 512×512. The percentage shown is the subsampled fraction of coefficients, i.e. $|P_\Omega|/N$. The sparsity matrix W is Daubechies-4. *Left:* Recovery from original wavelet coefficients. *Middle:* Recovery from flipped wavelet coefficients. *Right:* The subsampling map P_Ω

(iii) **Measurements and sampling strategy.** Choose a measurement operator $A \in \mathbb{C}^{m \times N}$ e.g. Fourier, Hadamard, Gaussian, etc. For orthogonal operators, such as Fourier, Hadamard, DCT and others, also specify the subsampling strategy for selecting the m rows, e.g. uniform random, power law [9], half-half [2, 20], multilevel [6, 7], etc.

(iv) **Recovery algorithm.** Choose a recovery algorithm $\Delta : \mathbb{C}^m \to \mathbb{C}^N$. For example, ℓ^1 minimisation, i.e. solving $\min_{z \in \mathbb{C}^N} ||z||_1$ s.t. $AW^{-1}z = Ax$.

(v) **Sparsity structure model.** Choose the model of the sparsity structure to be tested, for example, the classic sparsity model [1–4], weighted sparsity [19], sparsity in levels [6, 7] or another structured model.

(vi) **The test.** Perform the following experiment:

- Generate a signal x_0 from the model in (i) and obtain Wx_0 using the sparsifying transform from (ii), and then threshold Wx_0 so that it is perfectly sparse (see next step for how to determine the threshold level). If the sparsity model from (v) does not depend on the magnitudes of the entries in Wx_0 (e.g. models such as sparsity, weighted sparsity and sparsity in levels), then set all the non-zero entries in Wx_0 to a positive constant α, i.e. obtain an x so that $(Wx)_i = \alpha$ for all $i \in \text{supp}(Wx_0)$. This is to make all non-zeros equally important in order to avoid small coefficients giving false positives in the next step.
- Perform a reconstruction with measurements Ax using the sampling operator from (iii) and the recovery algorithm from (iv). If x is not recovered exactly (within a low tolerance), then decrease the thresholding level for Wx_0 in the above step to obtain a sparser x and repeat until x is recovered exactly.
- Create several new signals $x_1, x_2, \dots$ such that Wx_j are in the same structured sparsity class from (iv) as Wx is. Specifically, ensure that all vectors Wx and Wx_j give the same value under the sparsity measure defined in the model from (iv), with non-zero entries set to some positive α if the sparsity model does not involve magnitudes. For example, for the classic sparsity model, all Wx_j must have the same number of non-zero entries as Wx and magnitudes equal to some positive α; for the weighted sparsity model, all Wx_j must have the same weighted ℓ^0 norm and non-zero entries set to some positive α; etc.
- Obtain a recovery for each x_j using the same operator A and algorithm Δ. Ensure the recovery is consistent by averaging over several trials if A entails any kind of randomisation.

(vii) **Interpreting test results.** First, if any of the recovery tests failed, then the structured sparsity model chosen in (v) is not appropriate for the signal model, sparse transform, measurements and recovery algorithm chosen in (i)–(iv), respectively, and the conjectured model can be ruled out. Second, if the recovery of sufficiently many and different x_j is successful, this suggests that the structured sparsity model could be correct – though this is never a complete validation of the model.

We shall now test two sparsity models for structured 1D and 2D signals using structured operators. Figures 3 and 4 show the signals x_0, Wx_0, Wx (thresholded) and their reconstructions via ℓ^1 minimisation of (1), which is the recovery approach we shall use. Throughout all examples, $\eta = 0$ in (1).

2.2 *The Classic Sparsity Model and X-Lets*

The classic sparsity model is the first CS model [1–4] and states that the location of the non-zero entries of Wx is not important, only the sparsity measure $s = |\operatorname{supp}(Wx)|$ is important. In this model, operators that satisfy the RIP with appropriate constant can recover all s-sparse vectors when using ℓ^1 minimisation. The flip test for this model is the original flip test introduced in [6]. Specifically, we take A to be the Fourier operator, W to be any discrete wavelet transform and x to be a natural image, and the new signal x_j is generated so that Wx_j is the flipped version of Wx, i.e. $(Wx_j)_k = (Wx)_{N-k+1}$, thus being in the same sparsity model as the sparsity measure is preserved since $|\operatorname{supp}(Wx)| = |\operatorname{supp}(Wx_j)| = s$.

Figures 5 and 6 show a generalised flip test for this model. The marked differences between the two recoveries demonstrate that the classic sparsity model and RIP are not appropriate for this class of operators and signals. Simply put, one does not recover all s-sparse vectors, and the location of the non-zero entries in Wx is actually important when dealing with structured operators and structured signals. The same conclusion was reached when we repeated this experiment for a large number of various natural images and combinations of Fourier, Hadamard and DCT measurements (see [6, 7]).

2.3 *The Sparsity in Levels Model and X-Lets*

Another structured sparsity model is sparsity in levels [6, 7]. It is motivated by the fact that any X-lets have a particular level structure according to their scales. As seen below, this model defines a vector $\{s_k\}$ of local sparsities and level boundaries $\{M_k\}$ in order to capture sparsity in a more refined manner. As such, it is expected to contain a smaller class of signals. This model also has its own RIP and robust nullspace variant.

Definition 1 (Sparsity in levels) Let $y \in \mathbb{C}^N$. For $r \in \mathbb{N}$ let $\mathbf{M} = (M_1, \dots, M_r) \in \mathbb{N}^r$ and $\mathbf{s} = (s_1, \dots, s_r) \in \mathbb{N}^r$, with $s_k \leq M_k - M_{k-1}$, $k = 1, \dots, r$, where $M_0 = 0$. We say that y is $(\mathbf{s}, \mathbf{M})$-sparse if, for each $k = 1, \dots, r$, we have $|\Delta_k| \leq s_k$, where

$$\Delta_k := \operatorname{supp}(y) \cap \{M_{k-1} + 1, \dots, M_k\}.$$

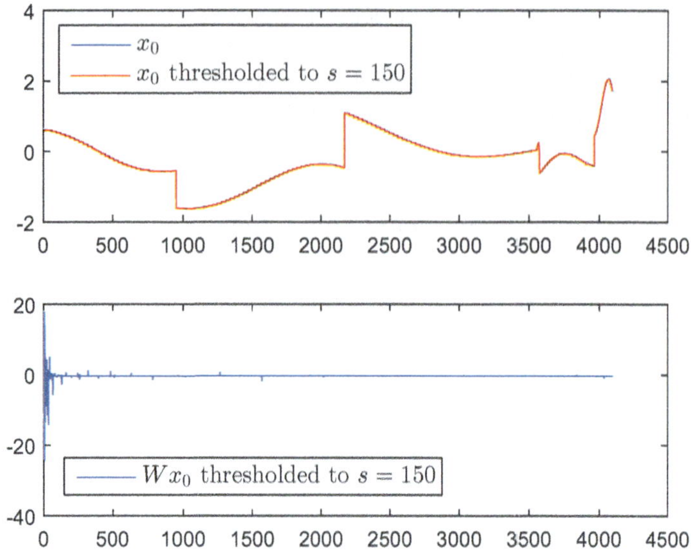

x_0 **(top) and thresholded** Wx_0 **(bottom)**

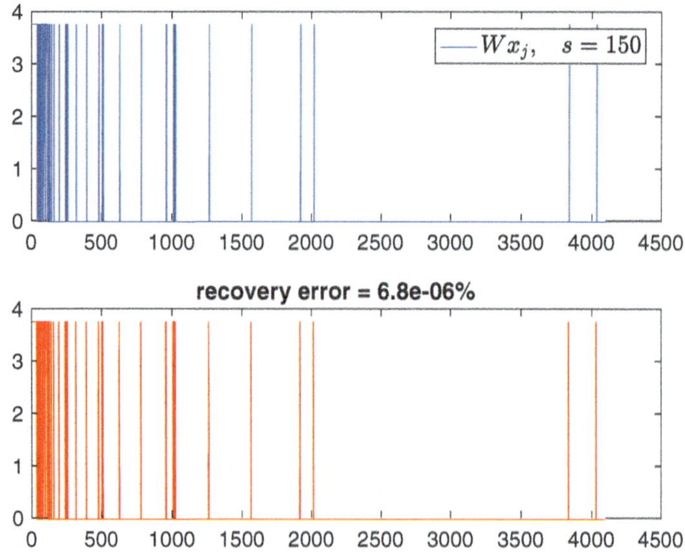

Wx **(top) and its** ℓ^1 **recovery (bottom)**

Fig. 3 Piecewise smooth 1D signal x_0 and recovery into Haar wavelets from $m = 1000$ Hadamard samples taken using a half-half scheme (first $m/2$ samples taken fully from the lower ordered rows and the other $m/2$ uniformly at random from the remaining rows), which is known to be a good all-round strategy [2, 20]. Here Wx was thresholded to $s = 150$ Haar coefficients

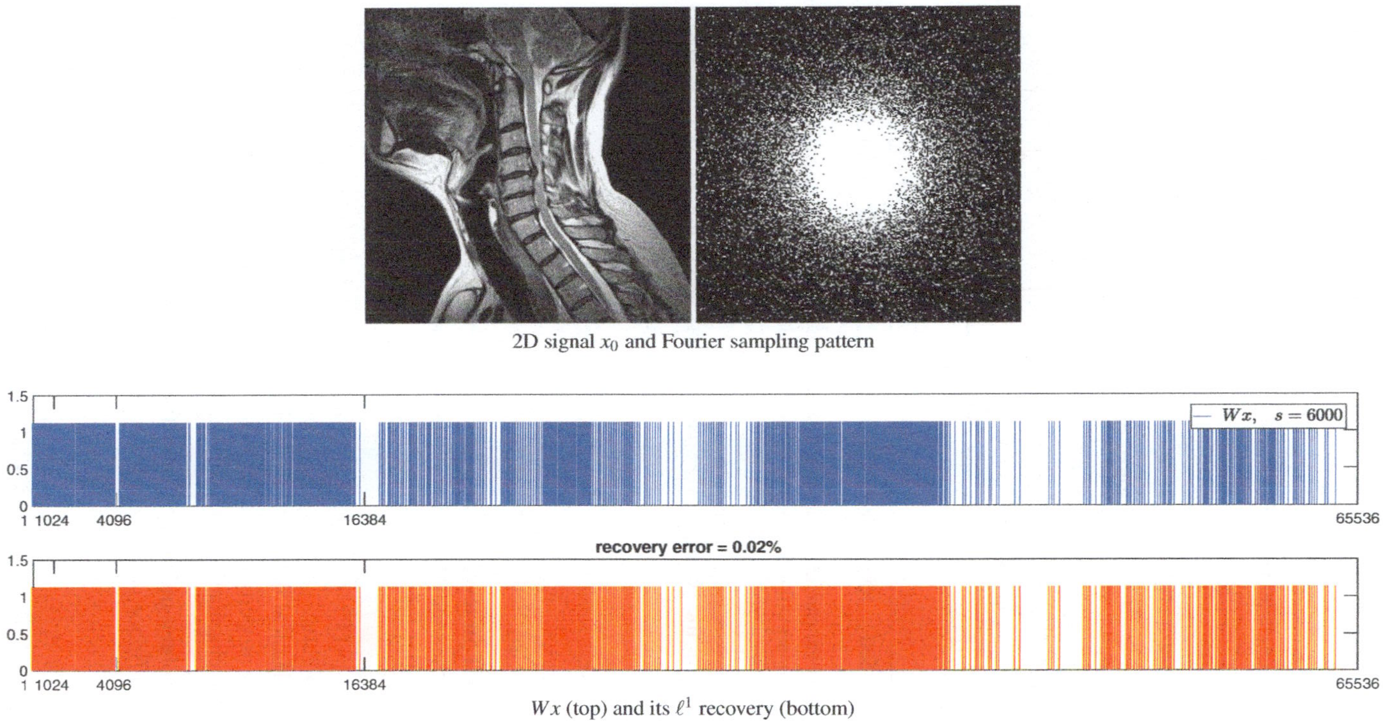

Fig. 4 Natural 2D signal at 256×256 resolution and recovery into Haar wavelets from $m = 18{,}000$ Fourier samples taken using a power law $\sim (k_1 + k_2)^{-3/2}$ as in [9]. Here Wx was thresholded to $s = 6000$ Haar coefficients

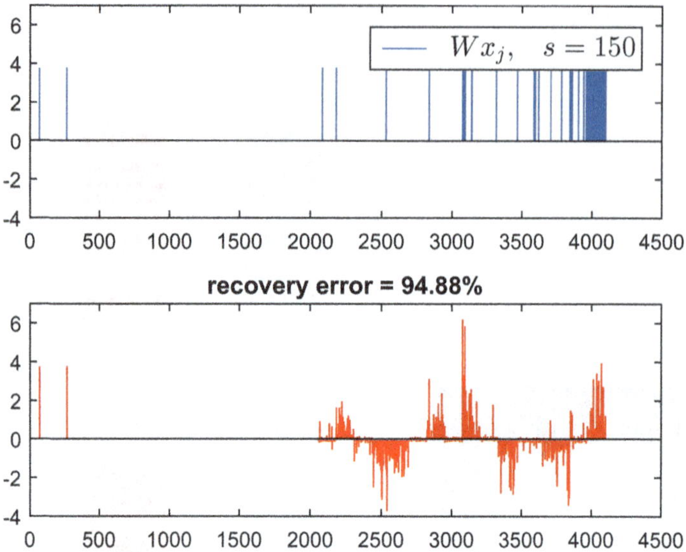

Fig. 5 Flip test for the classic sparsity model with piecewise smooth 1D signals. Wx_j (top) and its ℓ^1 recovery (bottom). All test elements, except x_j, are identical to those used in Fig. 3

We write $\Sigma_{\mathbf{s},\mathbf{M}}$ for the set of $(\mathbf{s}, \mathbf{M})$-sparse vectors and define the best $(\mathbf{s}, \mathbf{M})$-term approximation as

$$\sigma_{\mathbf{s},\mathbf{M}}(y)_1 = \min_{z \in \Sigma_{\mathbf{s},\mathbf{M}}} \|y - z\|_1.$$

Moreover, we say that $\cup_{k=1}^r \Delta_k$ is an $(\mathbf{s}, \mathbf{M})$-sparse set of integers.

Definition 2 (RIP in levels) Given an r-level sparsity pattern $(\mathbf{s}, \mathbf{M})$, where $M_r = N$, we say that the matrix $U \in \mathbb{C}^{m \times N}$ satisfies the *RIP in levels* (RIP_L) with RIP_L constant $\delta_{\mathbf{s},\mathbf{M}} \geq 0$ if for all $y \in \Sigma_{\mathbf{s},\mathbf{M}}$, we have

$$(1 - \delta_{\mathbf{s},\mathbf{M}})\|y\|_2^2 \leq \|Uy\|_2^2 \leq (1 + \delta_{\mathbf{s},\mathbf{M}})\|y\|_2^2.$$

Just as the classical definition of the RIP can be extended to RIP in levels, the robust nullspace property has a an extension called the robust nullspace property in levels. As in the classical case, the RIP in levels is a stronger assumption.

Definition 3 (Robust nullspace property in levels) A matrix $U \in \mathbb{C}^{m \times n}$ satisfies the ℓ^2 *robust nullspace property in levels of order* $(\mathbf{s}, \mathbf{M})$ if there is a $\rho \in (0, 1)$ and a $\tau > 0$ such that

$$\|v_S\|_2 \leq \frac{\rho}{\sqrt{\tilde{s}}}\|v_{S^c}\|_1 + \tau\|Uv\|_2 \tag{3}$$

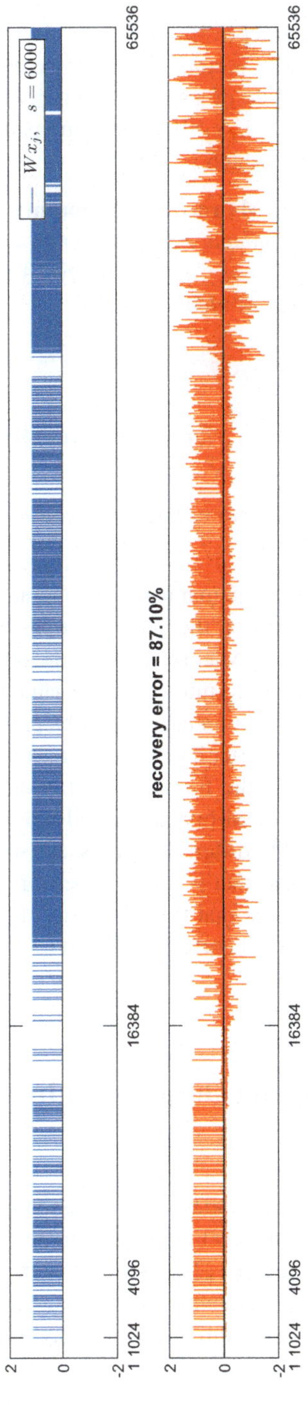

Fig. 6 Flip test for the classic sparsity model with natural 2D signals. Wx_j (top) and its ℓ^1 recovery (bottom). All test elements, except x_j, are identical to those used in Fig. 4

for all $(\mathbf{s}, \mathbf{M})$ -sparse sets S and vectors $v \in \mathbb{C}^n$.

Similar to the classical RIP concepts, the RIP in levels and robust NSP in levels imply recovery of all $(\mathbf{s}, \mathbf{M})$ sparse vectors [21, 22] (see also [23]). However, we need the concept of the *ratio constant* of a sparsity pattern $(\mathbf{s}, \mathbf{M})$, which we denote by $\eta_{\mathbf{s},\mathbf{M}}$, is given by $\eta_{\mathbf{s},\mathbf{M}} := \max_{i,j} s_i/s_j$. If the sparsity pattern $(\mathbf{s}, \mathbf{M})$ has r levels and there is a $j \in \{1, 2, \ldots, r\}$ for which $s_j = 0$, then we write $\eta_{\mathbf{s},\mathbf{M}} = \infty$.

Theorem 2 (ℓ^2 rNSP of order $(\mathbf{s}, \mathbf{M})$ recovery theorem) *Suppose that a matrix $A \in \mathbb{C}^{m\times n}$ satisfies the ℓ^2 robust nullspace property of order $(\mathbf{s}, \mathbf{M})$ with constants $\rho \in (0, 1)$ and $\tau > 0$. Let $x \in \mathbb{C}^n$ and $y \in \mathbb{C}^m$ satisfy $\|Ax - y\|_2 < \eta$. Then any minimiser $\tilde{x}$ of the ℓ^1 minimisation problem (1) satisfy*

$$\|\tilde{x} - x\|_1 \leq C_1\sigma_{\mathbf{s},\mathbf{M}}(x)_1 + C_2\eta\sqrt{\tilde{s}} \tag{4}$$

$$\|\tilde{x} - x\|_2 \leq \frac{\sigma_{\mathbf{s},\mathbf{M}}(x)_1}{\sqrt{\tilde{s}}}\left(C_3 + C_4\sqrt[4]{r\eta_{\mathbf{s},\mathbf{M}}}\right) + 2\eta\left(C_5 + C_6\sqrt[4]{r\eta_{\mathbf{s},\mathbf{M}}}\right) \tag{5}$$

where $\tilde{s} = s_1 + \ldots + s_r$, r is the number of levels and the constants C_j only depend on ρ and τ.

We shall test this model in the same manner, with an interest towards structured operators and structured signals. Let the sparsity transform W be a wavelet transform and let the level boundaries $\mathbf{M}$ correspond to the wavelet scale boundaries. For this model, the *flipped* signals Wx_j must have the same $(\mathbf{s}, \mathbf{M})$ sparsity as Wx, i.e. the local sparsities and the level boundaries must be preserved. In other words, we can move coefficients within wavelet levels, but not across levels.

Figures 7 and 8 show results of the flip test. We ran the same test for various other structured signals and structured operators, and the results were consistent, e.g. Fig. 9 shows the result of 1000 different signals. As suggested by the results, sparsity in levels seems to be a class that is actually recovered. As previously stated though, the flip test cannot guarantee that this is true for the entire class, since the flip test cannot entirely prove a model correct (unless it tests all signals in the class, which is infeasible).

3 Does Structured Sampling Outperform Incoherent Sampling?

Being interested in structured signals, and having discussed sparsity models, we turn our attention to problems where one has the freedom to design the sampling mechanism, such as the single-pixel camera [24], lensless camera [25] or fluorescence microscopy [7, 20], which can implement either incoherent matrices (e.g. random sub-Gaussian, expanders, etc.) or structured matrices (e.g. Hadamard or DCT).

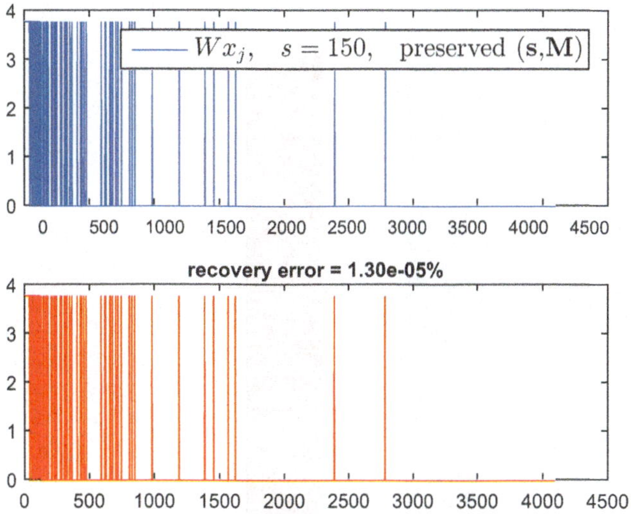

Fig. 7 Flip test for the sparsity in levels model with piecewise smooth 1D signals. Wx_j (top) and its ℓ^1 recovery (bottom). All test components, except x_j, are identical to those used in Fig. 3

Can we outperform incoherent sampling? At first that appears to be difficult as one typically needs $m \gtrsim s\log(N)$ samples to recover all s-sparse vectors from incoherent measurements. This bound is optimal for recovering sparse vectors, so it seems hard to believe that one can do better. However, the context changes when we restrict the class of s-sparse signals to signals that have substantially more structure, such as natural images. As shown empirically and discussed in [7], when dealing with such signals, using a variable density sampling procedure and either Fourier or Hadamard measurements, one can substantially outperform sampling with incoherent matrices whenever the sparsifying transform consists of wavelets, X-lets or total variation. In fact, [7] showed that using standard ℓ^1 recovery and structured sampling, one can substantially outperform incoherent sampling even when structured recovery algorithms are used, i.e. algorithms that exploit signal structure during the recovery phase, such as model-based CS [26], TurboAMP [27], Bayesian CS [28], etc.

Here we provide a theoretical justification for this phenomenon, followed by numerical results. We commence by defining the sampling scheme.

Definition 4 (Multilevel random sampling) Let $r \in \mathbb{N}$, $\mathbf{N} = (N_1, \ldots, N_r) \in \mathbb{N}^r$ with $1 \leq N_1 < \ldots < N_r$, $\mathbf{m} = (m_1, \ldots, m_r) \in \mathbb{N}^r$, with $m_k \leq N_k - N_{k-1}$, $k = 1, \ldots, r$, and suppose that $\Omega_k \subseteq \{N_{k-1}+1, \ldots, N_k\}$, $|\Omega_k| = m_k$, $k = 1, \ldots, r$, are chosen uniformly at random, where $N_0 = 0$. We refer to the set $\Omega = \Omega_{\mathbf{N},\mathbf{m}} = \Omega_1 \cup \ldots \cup \Omega_r$ as an $(\mathbf{N}, \mathbf{m})$-multilevel sampling scheme.

First, we define the discrete Fourier transform U_{dft}. Let $x = \{x(t)\}_{t=0}^{N-1} \in \mathbb{C}^N$ be a signal and the Fourier transform of x be $\mathscr{F}x(\omega) = N^{-1/2}\sum_{t=1}^{N} x(t)e^{2\pi i\omega t/N}$,

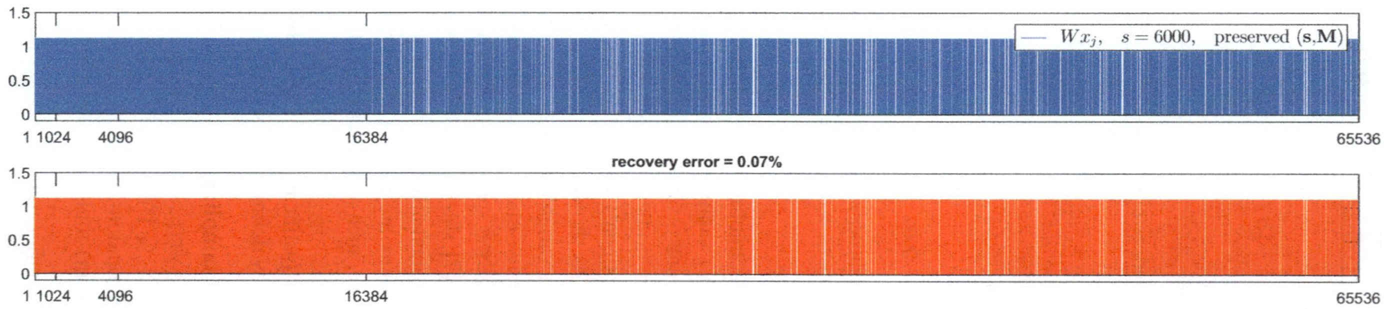

Fig. 8 Flip test for the sparsity in levels model with natural 2D signals. Wx_j (top) and its ℓ^1 recovery (bottom). All test elements, except x_j, are identical to those used in Fig. 4

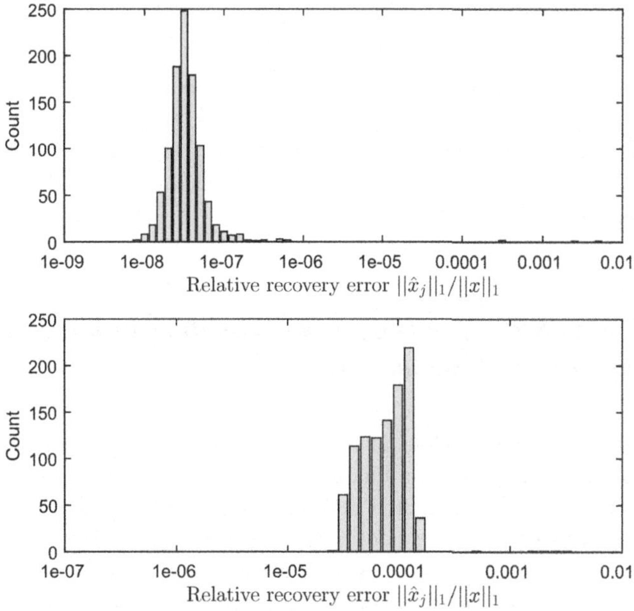

Fig. 9 Histogram showing recovery of 1000 different x_j randomised signals that have the same $(\mathbf{s}, \mathbf{M})$ sparsity as the original signal x. *Top:* 1D signal x from Fig. 3. *Bottom:* 2D signal x from Fig. 4. All test elements, except x_j, are identical to those used in Figs. 3 and 4, respectively

with $\omega \in \mathbb{R}$, and then write $F \in \mathbb{C}^{N\times N}$ for the corresponding matrix, so that $Fx = \{\mathscr{F}x(\omega)\}_{\omega=-N/2+1}^{N/2}$. We then let U_{dft} be the row permuted version of F where frequencies are reordered according to the bijection $\theta : \mathbb{Z} \to \mathbb{N}$ defined by $\theta(0) = 1, \theta(1) = 2, \theta(-1) = 3$, etc.

Theorem 3 (Fourier to Haar) *Let* $\epsilon \in (0, \mathrm{e}^{-1}]$ *and* $U = U_{\text{dft}} V_{\text{dwt}}^{-1} \in \mathbb{C}^{N\times N}$, *where* V_{dwt} *denotes the discrete Haar transform. Let* $x \in \mathbb{C}^N$. *Suppose that* $\Omega = \Omega_{\mathbf{N},\mathbf{m}}$ *is a multilevel sampling scheme and* $(\mathbf{s}, \mathbf{M})$ *is a multilevel sparsity structure as described above where* $\mathbf{M} = \mathbf{N}$ *correspond to levels defined by the wavelet scales (where potentially several scales could be combined into one level). Moreover, suppose that* $s_1 = M_1$ *and* $s_1 \leq s_2$. *If*

$$m_1 = M_1, \qquad m_j \gtrsim \left(s_j + \sum_{l=2, l\neq j}^{r} 2^{-\frac{|j-l|}{2}} s_l \right) \log(\epsilon^{-1}) \log(N), \qquad j = 2, \dots, r, \tag{6}$$

then any minimiser z *of* (1) *with* $A = P_\Omega U$ *satisfies*

$$\|z - x\| \leq C \left(\delta \sqrt{D}(1 + E\sqrt{s}) + \sigma_{\mathbf{s},\mathbf{M}}(x) \right), \tag{7}$$

with probability exceeding $1-s\epsilon$*, where* $s = s_1+\ldots+s_r$*, C is a universal constant,* $D = 1 + \frac{\sqrt{\log_2(6\epsilon^{-1})}}{\log_2(4EN\sqrt{s})}$ *and* $E = \max_{j=1,\ldots,r}\{(N_j - N_{j-1})/m_j\}$*. In particular, the total number of measurements* $m = M_1 + m_2 + \ldots + m_r$ *satisfies*

$$m \geq M_1 + G\,(s_2 + \ldots + s_r)\log(\epsilon^{-1})\log(N), \tag{8}$$

where G is a universal constant.

3.1 Exploiting Structure During the Sampling Procedure: Discussion and Comparison

The universality of many random matrices is a reason of their popularity in CS literature. As mentioned earlier, a random matrix $A \in \mathbb{C}^{m\times N}$ is universal if for any isometry $W \in \mathbb{C}^{N\times N}$, the matrix AW has the RIP with high probability. This allows the usage of sparsifying transforms such as wavelets to render signals sparse, coupled with a random (fat) matrix to sample the signal, which would then allow the usage of the classic CS theory to obtain performance guarantees.

These operators are agnostic to the signal structure and tend to *spread out* the signal information in each sample, i.e. tend to make every sample carry the same amount of information about the signal. This is good when there is no prior information about the signal, as one does not need to care *which* samples to take, but *only how many* samples to take, but what if the signal does possess further structure?

Typical signals in practice exhibit far more structure than sparsity alone. For starters, natural signals tend to possess asymptotic structures in their wavelet representation. Using a universal operator to sample these signals cannot exploit this structure during the sampling procedure. However, it is possible to leverage this structure during the reconstruction procedure. This is what we call *structured recovery* algorithms, which include model-based CS [26], Bayesian CS [28], TurboAMP [27] and others. These aim to exploit an assumed structure of the signal during the reconstruction procedure, after the signal has been sampled. These algorithms typically exploit the connected structure of the wavelet coefficients, known as the connected tree, a dependency model between the wavelet coefficients of most real-world signals, which is a stronger assumption than asymptotic sparsity (see also *Note 2*). A typical drawback of such algorithms is that they generally also assume a prior probability distribution of the information within the samples (performance may suffer if the signal or the sampling operator gives a different distribution). Another drawback is that they may assume a specific class of sparsity bases, e.g. wavelets (replacing the sparsifying operator can yield poor results).

A sampling strategy that leverages the asymptotic sparsity behaviour is variable density sampling. This is applied to sample rows from structured sampling operators in a more systematic fashion, in an attempt to extract more relevant information from the signal during the sampling stage. In what follows, we use a variant of

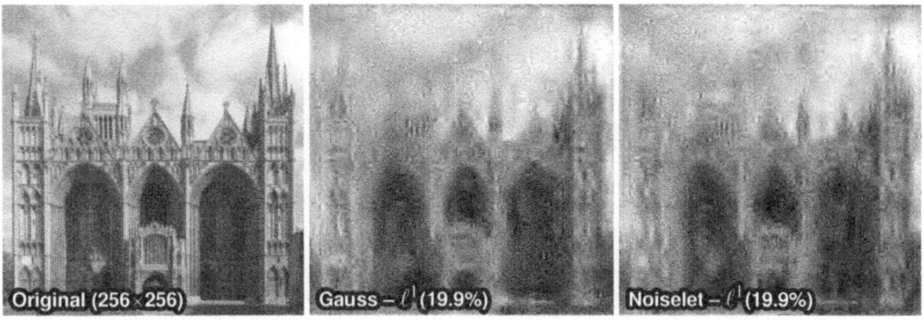

Fig. 10 12.5% subsampling at 256 × 256 (8192 samples). Unstructured (incoherent) sampling using random Gaussian measurements, as well as noiselets, with unstructured recovery (standard ℓ^1). The sparsifying operator was Daubechies-4 wavelets. The percentages shown are the relative ℓ_2 errors with respect to the original image

variable density sampling which we called *multilevel sampling* in Definition 4 [6, 7]. In short, the sampling space is split into r adjacent levels, with boundaries not necessarily matching the incoherence levels, where each level samples a fixed but decreasing fraction of samples p_k [7]. This ensures that there is non-zero sampling in all sampling levels, which is important. In essence, we want to make it *easier* for the reconstruction algorithm to approach the original signal, by sampling more relevant information about the signal to begin with (see also *Note 1*). This sampling structure allowed to obtain the theoretical recovery guarantees in [6] and to explain the success of variable density seen previously in practice.

In Figs. 10, 11, 12, 13 and 14, we perform a series of reconstructions with both unstructured (universal) and structured sampling, as well as unstructured and structured recovery. A key takeaway is that exploiting structure during the sampling procedure allows more headroom during reconstruction, and thus better performance is expected whenever the signal has an asymptotic sparsity structure in wavelets (luckily, the majority of real-world images possess this property). Essentially, it allows the reconstruction to better highlight the important wavelet coefficients.

Figure 10 shows the typical performance when sampling with a universal random matrix and reconstructed with classic ℓ^1 minimisation. For natural images, this is bound to be of very poor performance. Speed or storage is not really an issue, as one can obtain statistics similar to sub-Gaussian random matrices in the limit by using deterministic matrices with specific properties (e.g. noiselets, permuted Fourier, etc.). Figure 11 exemplifies performance of some structured algorithms that exploit the connected wavelet tree when the signal is sampled using a random matrix. It is interesting to observe that not all such algorithms achieve a better reconstruction than standard ℓ^1 reconstruction. There are several reasons for this, e.g. the samples distribution happens to be less well aligned with the algorithm's prior, the sampling amount is too low for the algorithm to perform, etc.

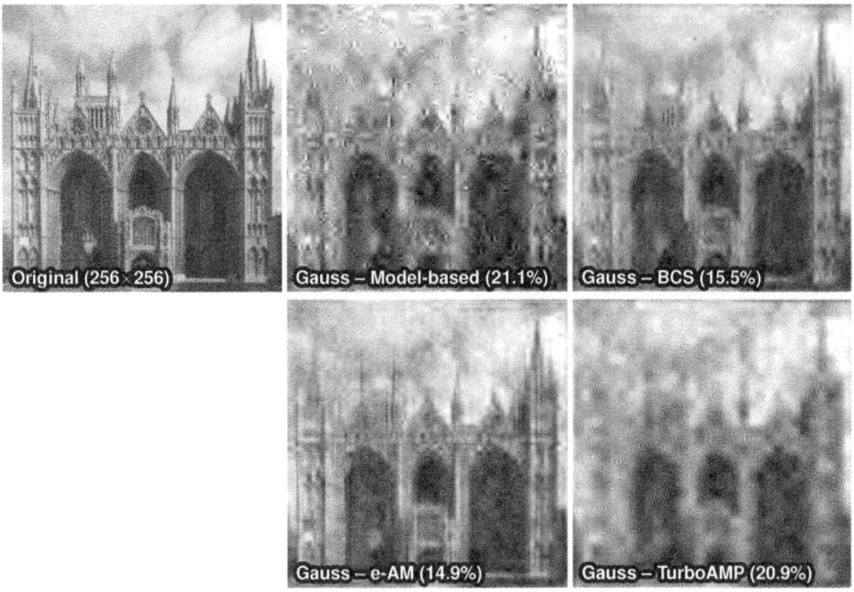

Fig. 11 12.5% subsampling at 256 × 256 (8192 samples). Unstructured (incoherent) sampling with random Gaussian matrices and structured recovery (algorithms that exploit the wavelet tree structure during reconstruction). The sparsifying operator in all cases was Daubechies-4 wavelets. The percentages shown are the relative ℓ_2 errors with respect to the original image

Figure 12 shows that even a partly structured sampling procedure can offer visible gains over unstructured sampling. Here we undersample rows from the discrete cosine transform (DCT), making sure that all the low ordered (low frequency) rows are included – this is based on the sensible assumption that for natural images, much of the signal's energy is concentrated in those rows.

In Fig. 13, we use fully structured sampling, where the rows of the DCT matrix are sampled using variable density sampling (multilevel sampling). The results are visibly superior, for reasons described above. It also shows that using a simple and general reconstruction algorithm (standard ℓ^1 minimisation) allows us to use any different types of sparsifying operators: we show results using wavelets, dual-tree wavelets [29], curvelets [30] and shearlets [31]. These all share the property that the coefficients of the signal representation have an asymptotic sparsity structure; hence, using variable density sampling works efficiently for all of them. In this context, the holy grail question is: *what is the optimum sampling map?* In other words, which m rows of the DCT matrix should one sample to maximise reconstruction quality. Sadly, although the matrices are fully known in advance, the signal is not. The optimum sampling map depends on the sparsity structure of Wx, i.e. of the signal representation within the sparsity basis – see also Note 1.

Figure 14 illustrates that simply pairing structured recovery algorithms that exploit the wavelet connected tree with a structured sampling operator can fail.

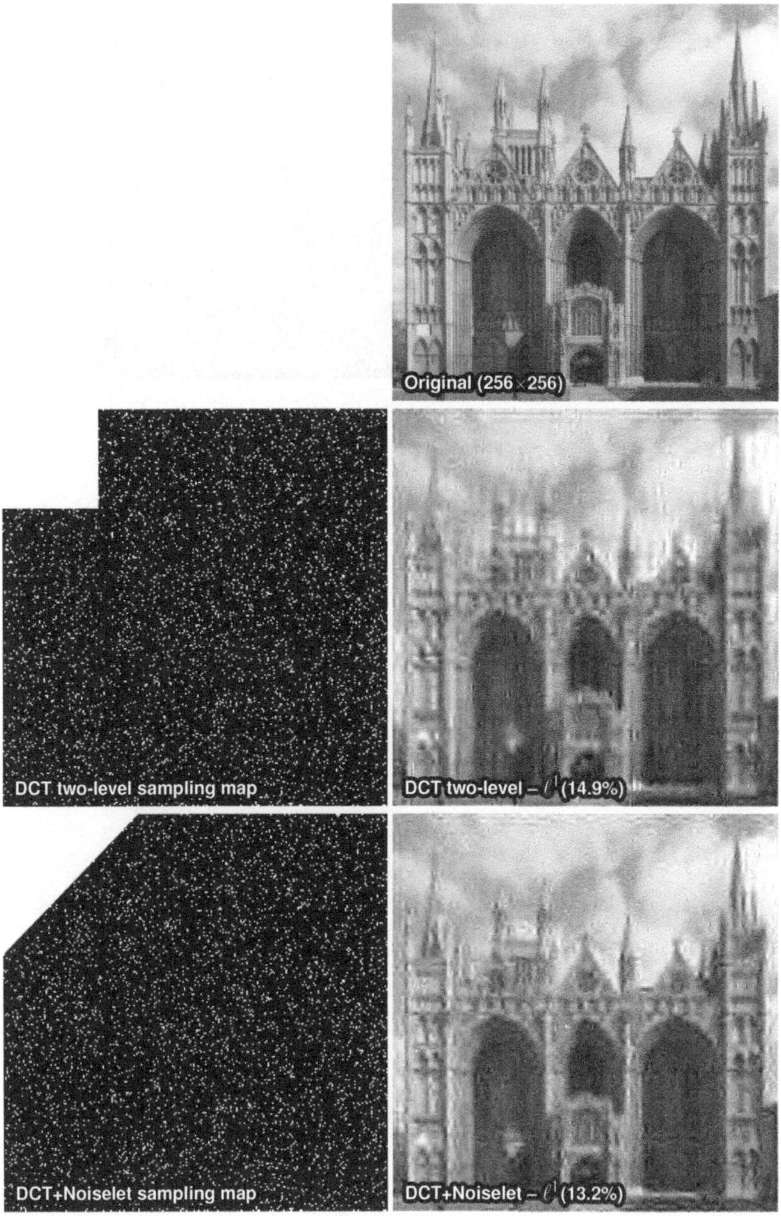

Fig. 12 12.5% subsampling at 256 × 256 (8192 samples). Partly structured sampling using DCT with unstructured recovery. The DCT + noiselet sampling takes part of the samples using the DCT matrix, while the rest are taken using noiselet (this is close to sub-Gaussian in the limit). The sparsifying operator in all cases was Daubechies-4 wavelets. The percentages shown are the relative ℓ_2 errors with respect to the original image

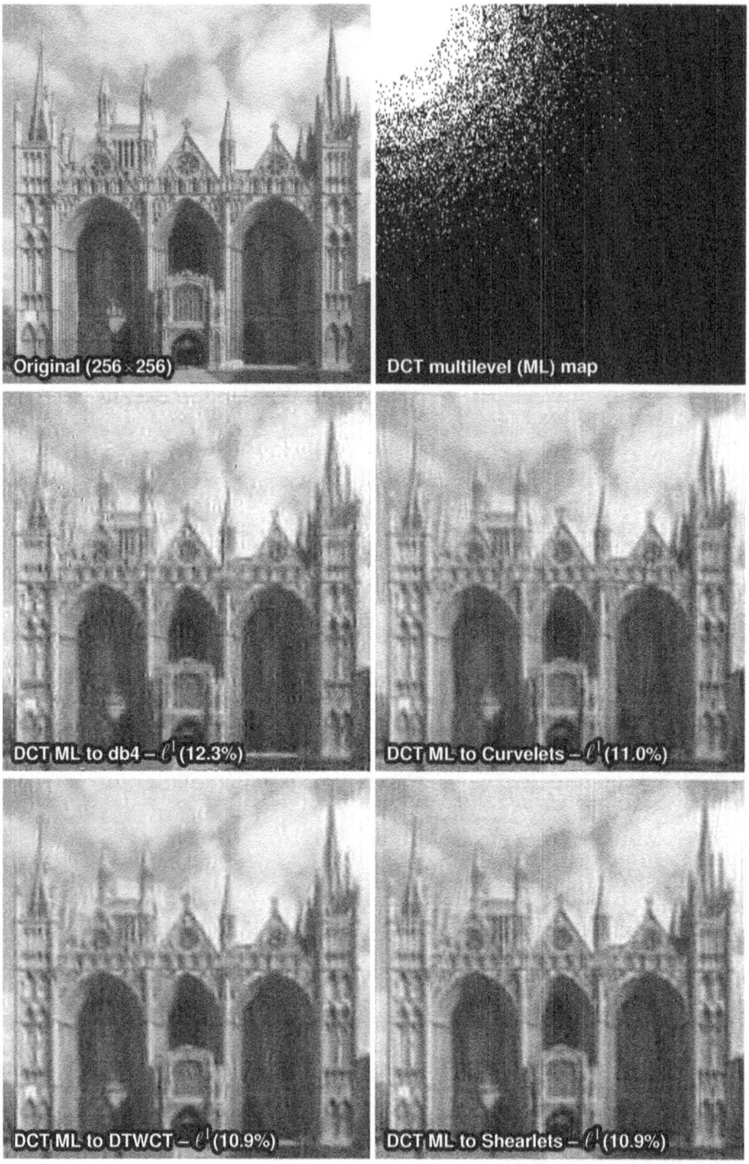

Fig. 13 12.5% subsampling at 256 × 256 (8192 samples). Fully structured sampling (DCT multilevel) and unstructured recovery (standard ℓ^1) with different sparsifying operators (Daubechies-4 wavelets, curvelets, dual-tree complex wavelets, shearlets, TV). These show that properly exploiting sparsity structure during the sampling procedure via a fully structured sampling operator is key, and using standard ℓ^1 recovery is sufficient for a visibly superior reconstruction quality compared to unstructured (incoherent) sampling or partly structured sampling seen in Figs. 10, 11 and 12, respectively. The percentages shown are the relative ℓ_2 errors with respect to the original image

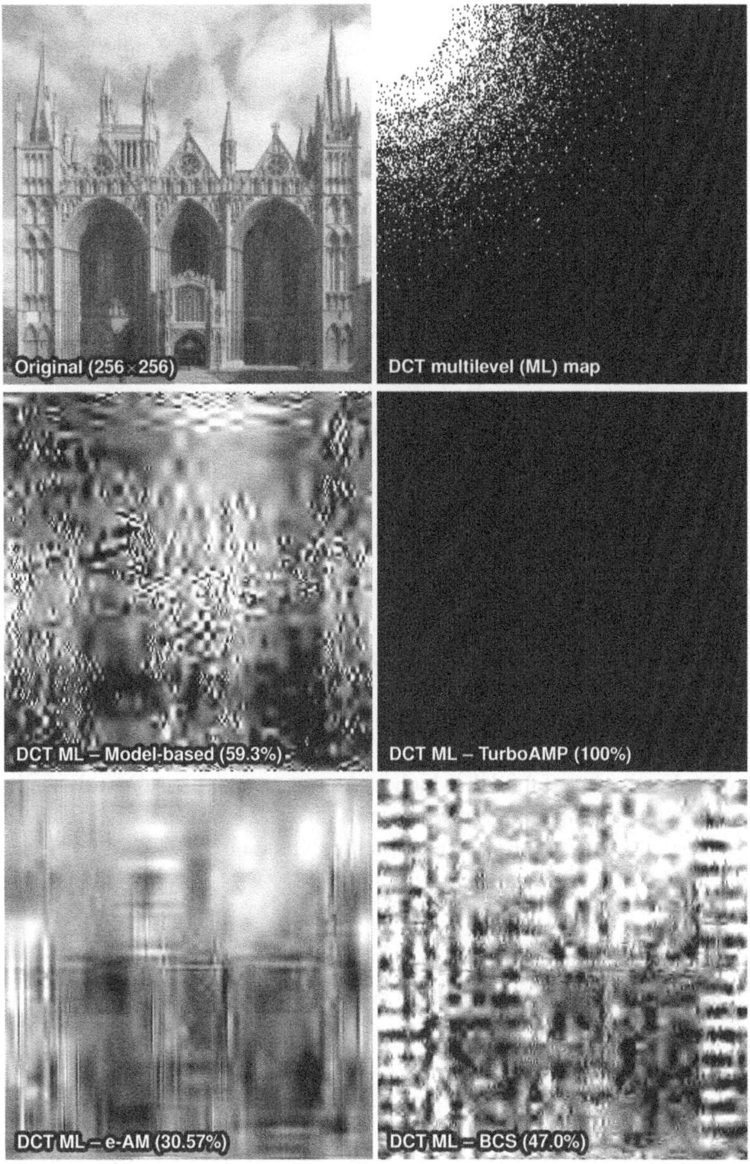

Fig. 14 12.5% subsampling at 256 × 256 (8192 samples). Fully structured sampling (DCT multilevel) and structured recovery algorithms. The failure of these algorithms to recover the image is due to the fact that they assume the sampling operator to be incoherent, such as sub-Gaussian, which in this case it is not (it is a structured DCT operator). This example shows that blindly combining structured sampling operators with structured recovery algorithms can in fact break down rather than improve the result. The sparsifying operator in all cases was Daubechies-4 wavelets. The percentages shown are the relative ℓ_2 errors with respect to the original image

In all the cases shown, these algorithms expect measurements to be taken using an incoherent operator, such as a random matrix, whereas the undersampled DCT gives a very different measurements distribution, and thus the algorithms fail to converge to the most important wavelet coefficients. It is theoretically possible to successfully couple such a structured sampling operator with a structured recovery algorithm, which is the subject of further research.

***Note 1:** Optimal sampling.* Sampling using a structured operator like Fourier, DCT or Hadamard does not mean that one should seek to maximise the *sampled* signal energy. That would be the classic compression case and is a common misconception in practice when it comes to CS. In CS, this is in fact not optimal. For example, if we sample with a Fourier matrix F and we are restricted to maximum m samples, then choosing P_Ω such that $||P_\Omega F x||$ is maximised (or some other vector norm) does not in fact give the best CS reconstruction when we recover in, say, a wavelet basis. This is because the wavelet (sparsity) structure is different than that of Fourier, and the CS reconstruction algorithm leverages the sparsity of the wavelet representation, not that of the Fourier representation. The takeaway message is that, even if we could sample the 'best' Fourier samples (which, in practice, we cannot), we should not aim to do that.

***Note 2:** Asymptotic sparsity vs. wavelet tree.* We note that there are fundamental differences between the asymptotic sparsity model we presented and the connected tree of wavelet coefficients. The latter assumes that wavelet coefficients live on a connected tree, a known phenomenon of wavelet signal representation stemming from the 'persistence across scales' phenomenon [32], where large wavelet coefficients have large child coefficients. The asymptotic sparsity model is a much more general and relaxed model, which makes no assumption about dependencies between wavelet coefficients. It only assumes different local sparsities s_k in a level-based structure, though the levels need not be dyadic or correspond to the wavelet levels.

4 From Linear to Non-linear: New Computational Challenges

In the previous sections, we have discussed how sparse regularisation and compressed sensing have become mainstays in modern computational harmonic analysis and how structure is the key to make this approach work. However, sparse regularisation has also, through the optimisation problems that have to be solved, incorporated non-linear techniques as new standard tools in signal and image processing. This means that there are computational challenges that are very different from more traditional problems using linear techniques. In particular, classical computational hurdles in harmonic analysis have been related to creating fast transforms such as fast Fourier transforms, wavelet transforms or in general X-lets transforms. However, non-linear techniques mean different numerical problems, and so far we have only treated these non-linear problems as computational tasks that can be computed by a 'black box' providing a minimiser to a convex optimisation problem. In the following sections, we will discuss the following:

(i) One cannot treat the optimisation problems needed in sparse regularisation as "black box" problems that can be solved by standard software packages. Indeed, as the goal is to compute minimisers and not the objective function, it is easy to make standard packages in, for example, MATLAB, fail and not even produce one correct digit, even if the input is well-conditioned.

(ii) The failure of standard algorithms and software packages (documented in Sect. 4.1.4) on simple well-conditioned problems can be explained by theorems from the foundations of computational mathematics such as Theorems 4 and 6 in Sect. 6. These theorems reveal several intricate phenomena suggesting how the success of computing minimisers of problems occurring in sparse regularisation is a delicate matter.

(iii) The success of computing minimisers used in sparse regularisation cannot be described by standard optimisation theory. The success can only be guaranteed under very specific conditions found in sparse regularisation, and this is linked to Smale's ninth problem (from the list of mathematical problems for the twenty-first century) [33] and its extensions [34]. Moreover, the specific conditions are exactly those established through the structure in sparse regularisation described in the previous sections.

The results presented are based on recent developments in the theory of the solvability complexity index (SCI) hierarchy [34–38], and the proofs can be found in [34].

4.1 Key Problems in Modern Computational Harmonic Analysis

The key assumption in order to apply, for example, compressed sensing successfully in practice is that one can easily compute minimisers to the basis pursuit (BP) problem:

$$x \in \underset{z}{\operatorname{argmin}} \|z\|_1 \text{ subject to } \|Az - y\| \leq \eta, \qquad \eta \geq 0, \tag{9}$$

where $A \in \mathbb{C}^{m \times N}$ and $y \in \mathbb{C}^m$. Note that in the real case, the BP problem (9) can be recasted into a linear program when $\eta = 0$, in particular, the problem of finding

$$z \in \underset{x}{\operatorname{argmin}} \langle x, c \rangle \text{ such that } Ax = y, \quad x \geq 0, \tag{10}$$

when given the input $A \in \mathbb{R}^{m\times N}$, $c \in \mathbb{R}^N$ and $y \in \mathbb{R}^m$. Other popular and related optimisation problems are unconstrained Lasso (UL)

$$x \in \underset{z}{\operatorname{argmin}} \|Az - y\|_2^2 + \lambda\|z\|_1, \qquad \lambda > 0, \tag{11}$$

and constrained Lasso (CL)

$$x \in \underset{z}{\operatorname{argmin}} \|Az - y\|_2 \text{ subject to } \|z\|_1 \leq \tau, \qquad \tau > 0. \tag{12}$$

The three problems above are quite similar in the way that they may share minimisers for different values of δ, λ and τ. However, they are not equivalent as computational problems. Indeed, BP may not always have a solution, whereas both UL and CL always will. But although they are not equivalent as computational problems one may often, in practice, approximate BP by solving either UL or CL trying to estimate the values λ and τ to get close to solutions of BP.

Note that constrained Lasso and unconstrained Lasso are also key problems in statistical estimation and imaging and are therefore of interest beyond, for example, compressed sensing. The problems above may have multivalued solutions in certain cases. Whenever this occurs, the computational problem is to compute any of these solutions. We will throughout the paper use the notation

$$\Xi : \Omega \to \mathcal{M}, \tag{13}$$

where, for an input $\iota \in \Omega$, $\Xi(\iota)$ represents the set of all solutions of any of the problems above. Moreover, Ω is the domain of the function, and $\mathcal{M}$ denotes the metric space where the appropriate metric will be specified. Typically, this is $\mathbb{R}^N$ or $\mathbb{C}^N$ equipped with the $\|\cdot\|_2$ norm; however, any norm will suffice.

4.1.1 Computing Minimisers: Not the Objective Function

It is tempting to treat the optimisation problems above as already solved, meaning that the rich classical literature [39–45] in optimisation will cover these convex problems, and hence the task is reduced to finding ones favourite numerical solver. Here is where we have to be careful. Note that classical optimisation is traditionally concerned with computing the objective function. Indeed, in many classical problems in operational research that have provided much motivation for convex optimisation, the objective function typically has a physical meaning. For example, one may be interested in minimising time or energy consumption in order to optimise a certain procedure. Thus, most of the classical optimisation theory is devoted to finding

$$f(x^*) = \min\{f(x) : x \in \mathcal{X}\},$$

where $f : \mathbb{R}^n \to \mathbb{R}$ is some convex function, $\mathscr{X} \subset \mathbb{R}^n$ is some convex set and $x^* \subset \mathscr{X}$ is a minimiser. Numerically, one wants to compute, for any $\epsilon > 0$, a $x_\epsilon \in \mathscr{X}$ such that

$$f(x_\epsilon) - f(x^*) \leq \epsilon.$$

Note, however, that x_ϵ may have very little to do with any minimiser x^*. In particular, $f(x_\epsilon) - f(x^*) \leq \epsilon$ does not mean that $\|x_\epsilon - x^*\| \leq \epsilon$.

The objective of computing the objective function changes sharply with the introduction of new methods in the 1990s and early 2000s in mathematics of information, in particular in statistical estimation, image processing, compressed sensing and machine learning. In these areas, where convex optimisation is used, one almost exclusively focuses on the minimisers, and not the objective function. As an example, we may consider the developments in the previous sections and realise that in compressed sensing, the objective function is typically the ℓ^1 norm of the wavelet coefficients of an image. This is a quantity that is rather uninteresting; moreover, its physical meaning is hard to interpret in any meaningful way. In fact, when doing computation in compressed sensing and sparse regularisation, one rarely ever concern oneself with the objective function. It is the minimiser that is the focus, and this changes also the computational challenges.

Warning! Computing a minimiser $x \in \mathbb{C}^N$ of any of the problems above is much more difficult than computing the minimum value $f(x^*) \in \mathbb{R}_+$. The key is that one will rarely work with exact numerical representations of numbers, and this fact has consequences when it comes to computing minimisers. Moreover, as we will see, this issue is much more subtle than just stability analysis.

The key is to understand how to deal with inexact input, and to motivate this, we use two simple examples from linear programming and linear systems. Note that, as mentioned above, BP can be recast as a linear program in the real case. Thus, linear programs go hand in hand with one of the key optimisation problems of compressed sensing.

4.1.2 Question on Existence of Algorithms in Optimisation

The change of focus from the objective function to the minimisers, as discussed above, leaves the following fundamental question that does not have an obvious answer in the standard optimisation literature.

Q1 *Given any of the following problems: linear programming (LP), basis pursuit (BP), unconstrained Lasso (UL) and constrained Lasso (CL), does there exist an algorithm such that for any $K \in \mathbb{N}$ and any input with real (or complex) numbers, the algorithm can produce a minimiser that has at least K correct digits?*

Remark 1 (Real (or complex) input) The reader may ask why this is not known in the optimisation literature. In particular, how does this relate to the well-established statement: LP is in P? This statement is only true for rational inputs with L digits, and the 'in P' statement means that there is a polynomial time algorithm (polynomial in the number of variables n and L) for the problem. The key here is that we require the algorithm to work not just with rational inputs, but other (computable) real numbers such as $\sqrt{2}$, $e^{2/5\pi i}$, etc. These numbers can never be represented exactly; however, they can be approximated arbitrarily well. Thus, to produce K correct digits in the solution, the algorithm can use as many correct digits L in its input as it wants (it may even choose L adaptively). However, it must guarantee K correct digits in its output (see also Sect. 4.2.1).

4.1.3 Computing with Inexact Input

Computing with exact numbers is a luxury that is rarely enjoyed in scientific computing. There is a variety of reasons for this. The most obvious reason is that numbers like $\sqrt{2}$, $e^{2\pi i/5}$ or cos(3) can never be computed exactly, and we have to resort to an approximate decimal representation. What is important is that such cases happen in compressed sensing all the time. In particular, we use the discrete Fourier transform, the discrete cosine transform, discrete wavelet transforms and random Gaussian matrices on a daily basis. All of these contain irrational numbers. However, the numerical representation can be arbitrarily close to the number it approximates.

Another reason for inexact representation is that most of the popular programming languages used for numerical calculations such as MATLAB, C++, Python, Fortran, etc. are based on floating point arithmetic. This means that even rational numbers may be represented inexactly. For example, $1/3$ is represented as a base-2 approximation to $1/3$. The number of decimals is dependent on the machine epsilon ϵ_{mach}, which in IEEE standard double precision is $2^{-52} \approx 2.22 \times 10^{-16}$. However, many modern programming languages (MATLAB, Mathematica, etc.) allow variable precision that, depending on computer memory, can be arbitrarily small. To be able to analyse the computations, the issue of inexact input must be taken into account. Indeed, the following quote from the list of mathematical problems for the twenty-first century illustrates the problem in a very accurate way:

> But real number computations and algorithms which work only in exact arithmetic can offer only limited understanding. Models which process approximate inputs and which permit round-off computations are called for.
>
> — S. Smale (from the list of mathematical problems for the twenty-first century [33])

Smale's argument highlights the traditional dichotomy between the discrete and continuous. Problems with finite input are pervasive in practical computing. However, the world of scientific computing is not finite. Given the widespread encounter of approximated real numbers in operators and transforms used in practice, a key question we can ask is, therefore, how does this impact the computation of the key problems we want to compute?

4.1.4 Can We Compute Minimisers of Linear Programs? A Simple Test

We will consider MATLAB's standard solver for LP, namely, `linprog`. However, there are many other commercial packages for solving LPs and other convex optimisation problems. They will all suffer from the same issues as we present below.

Example 1 (MATLAB's `linprog` for linear programming) The `linprog` command offers three different solvers: 'dual-simplex' (default), interior-point-legacy' and 'interior-point'. It returns a minimiser of the linear program

$$z \in \underset{x}{\operatorname{argmin}} \langle x, c \rangle \text{ such that } Ax = y, \quad x \geq 0, \tag{14}$$

when given the input $A \in \mathbb{R}^{m \times N}$, $c \in \mathbb{R}^N$ and $y \in \mathbb{R}^m$. Note that regardless of the solver, the `linprog` command has an `EXIT FLAG` parameter that determines if MATLAB can certify the computed solution as correct. This parameter can take the following values:

```
 1  linprog converged to a solution X.
 0  Maximum number of iterations reached.
-2  No feasible point found.
-3  Problem is unbounded.
-4  NaN value encountered during execution
     of algorithm.
-5  Both primal and dual problems are infeasible.
-7  Magnitude of search direction became too
     small; no further progress can be made.
     The problem is ill-posed or badly conditioned.
```

In the following example, we provide an intriguing test of the performance of `linprog`.

Example 2 (Testing MATLAB's `linprog` (edition R2019a)) Let

$$A = [1, 1 - x],\ x > 0, \quad c = [1, 1], \quad y = 1,$$

where we observe that the exact solution is given by `exact_soln` $= [1, 0]^T$ for all $x > 0$. We test all three of the built-in solvers: 'dual-simplex', interior-point-legacy' and 'interior-point' in this order. The following snippet tests both `linprog` and the `EXIT FLAG`:

```
c = [1,1]; y = 1; exact_soln = [1;0];
for k = 1:10
```

```
            x = 10^(-k); A = [1,1-x];
            [computed_soln, FVAL, EXIT FLAG] =
            linprog(c,[],[],A,y,[0,0],[100,100], options);
            error(k) = norm(computed_soln-exact_soln);
            flag(k) = EXIT FLAG;
        end
```

The `options` parameter in `linprog` allows for choosing the solvers 'dual-simplex', 'interior-point-legacy' and 'interior-point'. The other parameters in `options` are the default settings, and the results are as follows:

```
'dual-simplex'

error =  0 0 0 0 0 0 0 1.4 1.4 1.4

flag = 1 1 1 1 1 1 1 1 1 1
---
'interior-point-legacy'

error =  8.7e-11 1.2e-12 3.0e-07 8.7e-12 6.7e-6
         7.0e-7 7.1e-7 0.2 0.6 0.7

flag = 1 1 1 1 1 1 1 1 1 1
---
'interior-point'

error =  2.0e-9 1.8e-7 3.2e-07 3.4e-07 3.5e-4
         0.7 0.7 1.4 1.4 1.4

flag = 1 1 1 1 1 1 1 1 1 1
```

The experiment reveals two important issues:

- (Failure of the algorithm) Modern commercial software fails when attempting to produce minimisers of very basic problems and struggles to produce even one correct digit despite running double precision.
- (Failure of the exit flag) The software fails to recognise an incorrect output and wrongly certifies it as correct even though the first digit is incorrect.

There are several reasons why this experiment is of interest. First, it raises basic questions on why this can happen, and we will address those later on. Second, it is clear that we have given MATLAB a 'bad' problem, but why is this a bad problem? Moreover, the fact that there might be 'bad' problems is well known in numerical analysis in general, most notably when solving linear systems. Indeed, this is the case and is typically connected to condition numbers. In view of the failure `linprog`, it is tempting to see if MATLAB can spot a 'bad' problem in the context of solving a linear system.

Example 3 (Linear systems) In this case, we will also give MATLAB a 'bad' problem, an ill-conditioned matrix that is. In particular, we let

$$A = \begin{pmatrix} 1 & 0 \\ 0 & 2 \cdot 10^{-16} \end{pmatrix}, \qquad y = (1, 1)^T.$$

The result of the MATLAB computation is as follows:

```
computed_soln = A\y
Warning: Matrix is close to singular or badly scaled.
Results may be inaccurate. RCOND =  2.0000e-16.

computed_soln =  [1; 5.0000e15]
```

Note that MATLAB does discover that this is an ill-conditioned problem and indeed warns that the result may be inaccurate. This is in stark contrast to the linear programming case where it not only produces wrong results but also certifies them as correct.

Examples 2 and 3 raise several fundamental questions about computing minimisers of basic convex optimisation problems. The most immediate questions are as follows:

(1) Is the problem of failure specific to the solvers provided by MATLAB, or is this a global problem for any algorithm?
(2) How can one trust any 'black box' algorithm providing minimisers for compressed sensing and sparse regularisation problems?
(3) Why did we get a warning about a potential inaccurate result in the case of a linear system, but not for linear programming?
(4) Why did MATLAB certify the nonsensical solution as correct in the case of linear programming?
(5) Can the issue be resolved by using standard notions of condition numbers in optimisation?

It turns out that the answers to these questions are rather deep and one has to utilise the theory from the foundations of computational mathematics to provide the analysis. The rest of this paper is devoted to answering these questions; however, we will provide some short answers and remarks here.

Remark 2 (Is the problem of failure specific to the MATLAB solvers?) The short answer is no. Failure will happen for any algorithm. In particular, we have the somewhat paradoxical result that there does not exist any algorithm that can compute minimisers of linear programs or any of the problems in (9), (10), (11) and (12) given inputs of irrational numbers. In the case of linear programming, one can compute minimisers if the input is rational; however, the fact that almost all modern programming languages (such as MATLAB) rely on floating point arithmetic means that in practice most algorithms will fail on certain examples even when given rational inputs (just as in Example 2). This will be explained in Sect. 6, where Example 4 provides the main result.

Remark 3 (Can one trust standard algorithms for compressed sensing?) Despite the rather paradoxical result mentioned in Remark 2, one can compute solutions to compressed sensing problems reliably and fast. However, there is a striking phenomenon that must be handled with care. Indeed, there are standard compressed sensing problems for which no algorithm can produce arbitrary accuracy even when allowed arbitrary precision. In fact, there are problems for which no algorithm can produce five-digit accuracy; however, there exists an algorithm that can produce four-digit accuracy fast (in $\mathcal{O}(n^{3.5})$ time, where n is the total number of variables). This is discussed in Example 8. In imaging problems, however, four-digit accuracy is sufficient as the human eye cannot distinguish differences in pixel values in the fourth digit.

Remark 4 (Why can MATLAB produce a warning for linear systems?) One can always compute the solution to a linear system to arbitrary accuracy as long as one knows that the matrix is invertible and one can utilise arbitrary high precision. An example of an algorithm that could do this would roughly be as follows. One first estimates the condition number. Then based on that, decide the precision needed for solving the linear system. This is in some sense what MATLAB is doing, except it does not try to refine the precision, but rather gives a warning when the condition number is too big.

Remark 5 (Why does MATLAB certify a wrong solution as correct?) Contrary to the linear system case, it is impossible to compute a correct exit flag for an algorithm trying to solve linear programs with irrational inputs. In fact, it is strictly harder to compute an exit flag of an algorithm than computing a minimiser of a linear program. Thus, the `EXIT FLAG` parameter in `linprog` is impossible to compute. As described in Example 6, this is universal and has nothing to do with the particular implementation in MATLAB.

Remark 6 (Can the issue be resolved by using standard notions of condition?) The short answer is no. There are four standard condition numbers in optimisation: (1) the standard condition number of a matrix, (2) the condition number of the solution map (as a non-linear mapping), (3) the feasibility primary condition numbers and (4) the condition number capturing the distance to inputs that would provide several minimisers. As we will see in Example 4, there are examples of classes of inputs where we have finite and known condition numbers, yet there does not exist any algorithm that can compute minimisers. However, as we will see in Example 8, there are cases where the condition numbers may be infinite, yet one can find algorithms that can compute minimisers to arbitrary accuracy with runtime bounded by $p(n, K)$, where p is a polynomial, n is the total number of variables and K is the number of correct digits in the computed solution.

4.2 The Reason Behind the Failure: Computing with Inexact Input

The root to all the issues about failure of algorithms, as discussed in the previous sections, is the fact that we compute on a daily basis with inexact input. A key question is, therefore, how does this impact the computation of the key problems we want to compute? To illustrate this question, it is convenient to take a look at the standard complexity theory.

4.2.1 Standard Complexity Theory: LP is in P, Right?

Standard complexity theory shows that linear programming (LP) is in P (the set of computable problems that can be computed in polynomial time given a deterministic Turing machine). This is one of the most celebrated positive results in complexity theory of optimisation, and when the result was announced, it reached the front page of *The New York Times* [46]. Moreover, the standard estimate is that there exist algorithms, for example Karmarkar's algorithm, that can compute a minimiser of a LP with maximum runtime bounded by

$$\mathcal{O}(n^{3.5}L^2 \cdot \log L \cdot \log\log L), \tag{15}$$

where n denotes the number of variables and L is the number of bits or digits required in the representation of the inputs.

However, what if the input matrix A contains rows from the discrete cosine transform? What does the estimate (15) tell us? In order to answer these questions, we need a reformulation so that we can ask a mathematically precise question. Indeed, we may ask:

(i) Suppose that the input matrix $A \in \mathbb{N}^{m \times N}$ in the LP in (10) contains the rows of the discrete cosine transform. Suppose also that the inputs $y \in \mathbb{R}^m$ and $c \in \mathbb{R}^N$ have rational coefficients with, say, five digits. Given $K \in \mathbb{N}$, can one compute a minimiser to the LP with the runtime bounded by $p(n, K)$, where p is a polynomial, n is the total number of variables and K is the number of correct digits in the computed solution?
(ii) Is it obvious that there exists an algorithm that can compute K correct digits?
(iii) If such an algorithm exists, what should L (the number of correct digits in the approximation of the irrational numbers in the matrix A) be, given that we want K correct digits in the output?

To answer these questions, we first note that we cannot use Karmarkar's algorithm (or any of the known polynomial time algorithms) and (15) directly. Indeed, $L = \infty$ because we have irrational inputs, and thus (15) does not become very helpful. However, if we want K correct digits, maybe we could set $L = K$ and then apply the above algorithm.

As we will see in Sect. 6, the situation is much more complex. One cannot simply set $L = K$ and use standard algorithms. In fact, regarding answering Question (i), it turns out that the standard complexity for finite size inputs can say very little. Moreover, not only is it impossible to find an algorithm that can compute K digits with runtime bounded by a polynomial $p(n, K)$, but also it is impossible to find an algorithm that can compute K digits even when $K = 1$. Thus, we end up with the following rather paradoxical result: computing a minimiser of LP is in P when given rational inputs with L digits; however, it is impossible to compute K correct digits if the input is irrational, such as for the discrete cosine transform.

It should be noted that asking the questions above in relation to the discrete cosine transform is deliberate from an applications point of view. Indeed, discrete cosine and Fourier transforms are used on a daily basis in a wide range of applications from signal processing, via medical imaging, to astronomy. Thus, given that sparse regularisation can be used in all these areas, the examples motivating the questions are far from contrived; they occur daily.

5 Existence of Algorithms and Condition Numbers

As Example 2 suggests and the discussion in the previous sections has alluded to, there does not exist algorithm that can compute minimisers of some of the basic key optimisation problems when given inputs such as the discrete Fourier transform. However, we must make such a statement precise. In order to ask the question does there exist an algorithm that can compute an approximate solution to the problem, one must define what an algorithm is. Moreover, when asking about the existence of an algorithm, this is typically done in connection with condition. In particular, in scientific computing, one traditionally considers well-conditioned problems and ill-conditioned problems. And, typically, ill-conditioned problems could be hard to compute. Thus, it is reasonable to ask about existence of algorithms for well-conditioned problems.

5.1 The Basic Model: What Is an Algorithm?

There are two main models of computation that formally define an algorithm. The first is the Turing model where the basic concept of an algorithm is a Turing machine [47]. A Turing machine only works with integers and hence only with rational numbers, a feature that makes it very different compared to the BSS-machine [48]. Indeed, the second approach is the Blum-Shub-Smale (BSS) model where an algorithm is defined as a BSS-machine that allows for arbitrary real numbers. Since the two models of computation are not equivalent, one must be careful when making a statement of the form: 'there does not exist an algorithm'.

In order to be able to make universal statements, we will use a theoretical framework that is based on the solvability complexity index (SCI) hierarchy [34–38] that encompasses any model of computation. However, we mention in passing that both the Turing and the BSS model do suffer from the same issue when computing minimisers of optimisation problems, where the matrix is, for example, the discrete Fourier transform. Indeed, in either model, the function

$$x \mapsto e^x \qquad x \in \mathbb{Q} \text{ (Turing model)}, \quad x \in \mathbb{R} \text{ (BSS model)}$$

can only be computed up to a finite, yet arbitrary, small precision. Thus, both models will have to deal with an approximate input.

Hence, following Smale's demand for an extended model, suppose now that the algorithm (Turing or Blum-Shub-Smale (BSS) machine) that should solve any of the problems in (9), (10), (11), and (12) is equipped with an oracle that can produce the input to any precision $\hat{\epsilon}$. Moreover, the oracle computes the input in polynomial time in $|\log(\hat{\epsilon})|$ (this is a common assumption; see, e.g. Lovász [49, p. 36]). One may think of this model in the following way. We are given a domain $\Omega \subset \mathbb{C}^n$ of inputs; however, for $\iota \in \Omega$, the algorithm cannot access ι but rather $\tilde{\iota}$ such that, for any $k \in \mathbb{N}$, $\tilde{\iota}(k) \in \mathbb{C}^n$ and $\|\tilde{\iota}(k) - \iota\|_\infty \leq 2^{-k}$. In particular, the algorithm can access $\tilde{\iota}(k)$ for any k, and the time cost of accessing $\tilde{\iota}(k)$ is polynomial in k. The key is that the algorithm must work with any such approximate representation $\tilde{\iota}$.

5.2 *Questions on Existence of Algorithms for Compressed Sensing*

Suppose now that the algorithm (Turing or Blum-Shub-Smale machine) that should solve LP is equipped with an oracle that can produce the input to any precision $\hat{\epsilon}$ and the oracle computes the input polynomially in $|\log(\hat{\epsilon})|$ as described in Sect. 5.1. The natural question would be:

Question 1 (Is the problem in P?) *Given any $\epsilon > 0$, does there exist an algorithm that has a uniform bound on the runtime $T(\epsilon)$ such that*

$$T(\epsilon) \leq P(n, K),$$

where P is a polynomial, $K = |\log(\epsilon)|$ is the number of correct digits in the computed solution and n is the number of variables in the input?

The model, where one measures the computational cost in the number of variables n and the error (the number of correct digits K), is indeed well established (see Blum, Cucker, Shub and Smale [50, p. 29], Grötschel, Lovász and Schrijver [46, p. 34] and Valiant [51, p. 131]). However, before one can address Question 1, one needs to answer the following questions:

Question 2 (Existence of algorithms: deterministic or randomised) *Consider any of the problems in* (9), (10), (11), (12), *where the input may be given with some inaccuracy controlled by* $\hat{\epsilon} > 0$.

(i) *Does there exist an algorithm that can compute an approximate solution, such that, for an arbitrary* $\epsilon > 0$, *the output will be no further than* ϵ *away from a true solution? The algorithm can choose* $\hat{\epsilon}$ *to be as small as desired (as a function of* ϵ *and the input) to produce the output.*
(ii) *Does there exist an algorithm that has a uniform bound on the runtime* $T(\epsilon)$ *(depending on an arbitrary* $\epsilon > 0$*) for all inputs in its domain when the output is at least* ϵ*-accurate?*

When considering randomised algorithms, we can ask questions (i)–(iii) as well, where the only difference is that we require that the algorithm produces an output that is at least ϵ*-accurate with probability* $p > 1/2$.

A problem, for which the answer to Question 2 (i) is negative, is referred to by Turing as non-computable. Note that, clearly, (i) implies no on (ii) and yes on (ii) implies yes on (i). The immediate reaction from an expert in computational mathematics will be that Questions 1 and 2 must be related to condition. Indeed, there is a well-established literature on condition of computational problems.

5.2.1 Condition Numbers in Optimisation

Condition numbers help in understanding the behaviour of mappings under perturbations and are crucial for the analysis of performance of algorithms and complexity theory. We want to highlight the pioneering work (mentioned in order of publication dates) by Renegar [52, 53], Blum, Cucker, Shub and Smale [50] and Burgisser and Cucker [54] where their recent book gives a thorough up-to-date account of the field. See also [55]. We recall the basic definitions here. The classical condition number of a matrix A is given by

$$\mathrm{Cond}(A) = \|A\| \|A^{-1}\|. \tag{16}$$

For different types of condition numbers related to a mapping $\Xi : \Omega \subset \mathbb{C}^n \to \mathbb{C}^m$, we need to establish what types of perturbations we are interested in. For example, if Ω denotes the set of diagonal matrices, we may not be interested in perturbations in the off-diagonal elements as they will always be zero. In particular, we may only be interested in perturbations in the coordinates that are varying in the set Ω. Thus, given $\Omega \subset \mathbb{C}^n$, we define the active coordinates of Ω to be $\mathcal{A}(\Omega) = \{j \mid \exists\, x, y \in \Omega, x_j \neq y_j\}$. Moreover, for $\nu > 0$, we define

$$\tilde{\Omega}_\nu = \{x \mid \exists\, y \in \Omega \text{ such that } \|x - y\|_\infty \leq \nu, x_{\mathcal{A}^c} = y_{\mathcal{A}^c}\}$$

In other words, $\tilde{\Omega}_\nu$ is the set of ν-perturbations along the nonconstant coordinates of elements in Ω. We can now recall some of the classical condition numbers from the literature [54].

(1) *Condition of a mapping*: Let $\Xi : \Omega \subset \mathbb{C}^n \to \mathbb{C}^m$ be a linear or non-linear mapping, and suppose that Ξ is also defined on $\tilde{\Omega}_\nu$ for some $\nu > 0$. Then,

$$\mathrm{Cond}(\Xi) = \sup_{x\in\Omega} \lim_{\epsilon\to 0^+} \sup_{\substack{x+z\in\tilde{\Omega}_\nu \\ 0<\|z\|\leq\epsilon}} \left\{ \frac{\mathrm{dist}(\Xi(x+z), \Xi(x))}{\|z\|} \right\}, \tag{17}$$

where we allow for multivalued functions by defining

$$\mathrm{dist}(\Xi(x), \Xi(z)) = \min_{\tilde{x}\in\Xi(x), \tilde{z}\in\Xi(z)} \|\tilde{x} - \tilde{z}\|.$$

(2) *Distance to infeasibility – the feasibility primal condition number*: If Ξ denotes the solution map to any of the problems in (9), (10), (11), and (12) with domain Ω, we define, for $(A, y) \in \Omega$

$$\begin{aligned} &\rho(A, y) \\ &= \sup\big\{\delta \mid \|\hat{A}\|, \|\hat{y}\| \leq \delta, (A + \hat{A}, y + \hat{y}) \in \tilde{\Omega}_1 \Rightarrow (A + \hat{A}, y + \hat{y}) \\ &\qquad \text{are feasible inputs}\big\}, \end{aligned}$$

and this yields the *feasibility primal* (FP) condition number

$$C_{\mathrm{FP}}(A, y) := \frac{\|A\| \vee \|y\|}{\rho(A, y)}, \tag{18}$$

where we use the standard $\vee$, $\wedge$ max/min notation.

(3) *Distance to solution with several minimisers – the RCC condition number*: If Ξ denotes the solution map to any of the problems in (9), (10), (11), and (12) with domain Ω then we define, for $(A, y) \in \Omega$,

$$\begin{aligned} \varrho(A, y) = \sup\big\{\delta : \|\hat{A}\|, \|\hat{y}\| \leq \delta, (A + \hat{A}, y + \hat{y}) \in \tilde{\Omega}_1 \\ \Rightarrow (A + \hat{A}, y + \hat{y}) \text{ yields at most one solution}\big\}, \end{aligned}$$

and this yields the *RCC condition number*

$$C_{\mathrm{RCC}}(A, y) := \frac{\|A\|_2 \vee \|y\|_2}{\varrho(A, y)}. \tag{19}$$

The above condition numbers are the standard ones used in optimisation. A potential surprise is that they do not give any insight in the existence of algorithms for the key computational problems such as linear programming, basis pursuit or Lasso. We will discuss the details in Sect. 6.1.1.

6 Paradoxical Results in Optimisation

The following theorems may come as a surprise and provide answers to Questions 1 and 2 and insight into the rather intricate phenomenon of existence and non-existence of algorithms. The result can be summed up as follows. There is an infinite classification theory of problems according to the accuracy for which one can compute a solution. This is a phenomenon that is not covered by standard complexity and computability theory. In particular, complexity theory strictly considers only computable problems; however, as the next theorem demonstrates, paradoxically, there is a complexity theory for non-computable problems.

6.1 Determining the Boundaries of Computation: Why Things Work and Fail

The following examples help in shedding light on why algorithms may fail completely on getting a particular accuracy (as suggested in the example in Sect. 4.1.4); however, some algorithms may be able to get some accuracy fast. For specific examples in application such as compressed sensing and sparse regularisation, see Example 8. The following example is a short summary of some of the results in [34]. The statements are made deliberately non-technical in order to be reader friendly.

Example 4 (Determining the boundaries of optimisation) Let Ξ denote the solution map (as in (13)) to any of the problems (9), (10), (11), and (12), and let $K > 2$ be an integer. There exists a class Ω of inputs for Ξ so that we have the following:

(i) No algorithm, even randomised, can produce K correct digits for all inputs in Ω (with probability greater than $p > 1/2$ in the randomised case).

(ii) There does exist an algorithm that will provide $K - 1$ correct digits for all inputs in Ω. However, any algorithm will need an arbitrarily long time to reach $K - 1$ correct digits. In particular, there is an $\Omega' \subset \Omega$, with inputs of fixed dimensions m, N, such that for any $T > 0$ and any algorithm Γ, there exists an input $\iota \in \Omega'$ such that either $\Gamma(\iota)$ does not approximate $\Xi(\iota)$ with $K - 1$ correct digits or the runtime of Γ on ι is greater than T. Moreover, for any randomised algorithm Γ^{ran} and $p = (0, 1/2)$ there exists an input $\iota \in \Omega'$ such that

$$\mathbb{P}\big(\Gamma^{\mathrm{ran}}(\iota) \text{ does not approximate } \Xi(\iota) \text{ with } K - 1 \text{ correct digits}$$
$$\text{or the runtime of } \Gamma \text{ on } \iota \text{ is } > T\big) > p.$$

(iii) The problem of producing $K - 2$ correct digits for inputs in Ω is in P (can be solved in polynomial time in n, the number of variables).

(iv) If one only considers (i)–(iii), Ω can be chosen with any fixed dimensions $m < N$ with $N \geq 4$. Moreover, if one only considers (i), then K can be chosen to be one.

The statements above are true even when we require the input to be well-conditioned and bounded from above and below, in particular, for any input $\iota = (y, A) \in \Omega$ ($\iota = (y, c, A)$ in the case of LP), we have $\mathrm{Cond}(AA^*)$, $C_{\mathrm{FP}}(\iota)$, $\mathrm{Cond}(\Xi) \leq c_1$, $c_2 \leq \|\iota\| \leq c_1$ for some constants $c_1, c_2, > 0$.

There is traditionally a sharp divide in foundations between problems that are computable, according to Turing's definition, and those that are not. Moreover, complexity theory is normally only considered for problems that are computable. Turing's definition of computability means that one can compute an approximation to any accuracy. Although this is a natural definition, it deems many key problems non-computable despite that they are computed on a daily basis. The key is that these problems are computed to a sufficient precision needed. Arbitrary precision, which as discussed in Example 4 may be impossible, is typically not needed.

Example 5 (High precision is not needed in imaging sciences) Take any black and white (the example is similar for colour images) image at any resolution where the pixel values are in $[0, 1]$. Zoom in to any pixel of the image, and perturb the value of the pixel with the number 10^{-4}. Compared visually with the original pixel, will you see a difference? The answer will simply be no.

The above example illustrates the delicate issue of Turing's definition of computability. Indeed, if the human eye cannot distinguish between images with perturbations in the fourth digit, it may seem like a worthless effort to compute images with six correct digits. One may argue that what is important is that the accuracy needed in the particular application is important and that Turing's definition should be extended. For example, one could define the concept of K-computability (in the desired metric), if one can compute K correct digits, and ∞-computability would be the same as Turing's current concept of computability. The key is that the traditional definition of computability does not capture vast areas of problems in computational science that are non-computable, yet form the basis of many everyday computations in modern data science.

Remark 7 (Linear programming vs. linear systems) Note that, in contrast, the answer to Question 2 (i) is *yes* for solving linear systems $Ax = y$ even when the only information available is that A is invertible. In particular, if we let Ξ_1 denote the solution map for linear systems, the problem may become arbitrarily unstable ($\mathrm{Cond}(\Xi_1) = \infty$), yet for linear systems, the answer to Question 2 (i) is yes. However, if we let Ξ_2 denote the solution map for linear programming, as discussed in Example 4, the answer to Question 2 (i) is *no* for linear programming, and one cannot even get one correct digit, even when $\mathrm{Cond}(\Xi_2) \leq 2$, and the norm of the input is bounded.

6.1.1 Surprises on Condition in Optimisation

Condition numbers have been crucial in numerical analysis since the dawn of the field; however, it is equally crucial in foundations of computation and complexity theory. Indeed, the standard wisdom is that well-conditioned problems should be easy to compute, whereas ill-conditioned problems may be troublesome. Nevertheless, Example 4 demonstrates a rather different scenario. Combined with Theorem 8 on the extended Smale's ninth problem and compressed sensing, we end up with the following take-home message:

- (**Bounded condition numbers, yet no algorithm exists**) All the condition numbers such as the classical condition number of a matrix (16), the condition number of the solution map (17) and the feasibility primal condition number (18) may be bounded with known bounds in addition to known bounds on the input, yet there are cases where one cannot find an algorithm that will produce even one correct digit.
- (**Infinite RCC condition number, yet there do exist algorithms**) The RCC condition number can be ∞, yet there exist algorithms that can compute the problem to any precision. Moreover, for most problems in compressed sensing, the RCC condition number is ∞. In particular, problems including Bernoulli and subsampled Hadamard matrices always have infinite RCC condition number. However, many of these problems are even in P.

Thus, the classical notions of condition are not that helpful when answering questions on the existence of algorithms. For example, requiring that the RCC condition number should be finite may guarantee the existence of algorithms in certain cases; however, such an assumption would exclude whole fields such as compressed sensing and sparse regularisation. Hence, a new concept of condition numbers is needed in optimisation in order to capture the delicate issues of existence and non-existence of accurate algorithms.

6.1.2 Why the EXIT FLAG Cannot Be Computed

Example 4 provides a justification for the first part of the experiment in Sect. 4.1.4. However, the next example explains why MATLAB could not compute a correct exit flag in its `linprog` routine. The following example is written with a deliberately reader friendly jargon; for precise statements, see [34].

Example 6 (Impossibility of computing the exit flag) Let Ξ denote the solution map (as in (13)) to any of the problems (9), (10), (11), and (12), and let $K \in \mathbb{N}$. For any fixed dimensions $m < N$ with $N \geq 4$, there exists a class of inputs Ω for Ξ such that if Γ' is an algorithm, for the computational problem of approximating Ξ with K correct digits, we have the following:

(i) No algorithm, even randomised with access to an exact solution oracle, can compute the exit flag of Γ' (with probability greater than $p > 1/2$ in the randomised case).
(ii) The problem of computing the exit flag of Γ' is strictly harder than computing a K correct digit approximation to Ξ, the original problem.
(iii) For linear programming and basis pursuit, however, there exists a class of inputs $\Omega^{\sharp} \neq \Omega$ such that no algorithm, even randomised with non-zero probability of not halting, can compute the exit flag of Γ' (with probability greater than $p > 1/2$ in the randomised case), yet one can compute the exit flag with a deterministic algorithm with access to an exact solution oracle.

Statements (i) and (ii) are true even when we require the input to be well-conditioned and bounded, in particular, for any input $\iota = (y, A) \in \Omega$ ($\iota = (y, c, A)$ in the case of LP), we have $\mathrm{Cond}(AA^*)$, $C_{\mathrm{FP}}(\iota)$, $\mathrm{Cond}(\Xi) \leq c$, $\|\iota\| \leq c$ for some constant $c > 0$.

6.2 A Practical Consequence: How Do You Set Halting Criteria?

Examples 4 and 6 conclude that we cannot design algorithms and software that can compute minimisers of arbitrary linear programs, basis pursuit problems or constrained/unconstrained Lasso when working with inexact input due to irrational numbers or floating point arithmetic. Moreover, one cannot detect if an algorithm produces the wrong answer. Hence, a pertinent question to the reader is therefore:

> How does one set the halting criterion in ones code, and how does one guarantee an accurate computation?

We have just seen in Sect. 4.1.4 how a popular tool such as MATLAB fails on basic LPs, and Examples 4 and 6 demonstrate that this is not a particular problem for MATLAB, but rather a universal phenomenon. As an overwhelming amount of problems in computational harmonic analysis, signal processing, imaging and modern data science is based around computing minimisers of the problems covered by Examples 4 and 6, how do we make sure that examples used in scientific publishing and indeed in practical issues are computed accurately? Moreover, how can we separate between methodological and algorithmic errors?

6.2.1 Methodological vs. Algorithmic Error

In order to show how Examples 4 and 6 imply a rather delicate issue regarding how to determine the difference between methodological and algorithmic errors, we will begin with an example.

Example 7 Suppose we run two synthetic experiments modelling MRI, where we use the same sampling pattern, i.e. the output of the data from the sampling is the same for both experiments. More precisely, the sampling data is given by $y = P_\Omega U_{\text{df}} x$, where $U_{\text{df}} \in \mathbb{C}^{m \times N}$ denotes the discrete Fourier transform, $x \in \mathbb{C}^N$ is a vectorised version of a medical image and P_Ω is the same in both experiments. However, we use two different reconstruction techniques. In experiment (1), we compute

$$x_1 = V_{\text{dw}}^{-1} x, \qquad x \in \underset{z \in \mathbb{C}^N}{\operatorname{argmin}} \|z\|_1 \text{ subject to } \|P_\Omega U_{\text{df}} V_{\text{dw}}^{-1} z - y\| \leq \delta, \tag{20}$$

where V_{dw}^{-1} denotes the discrete wavelet transform (DB4), whereas in experiment (2), we compute

$$x_2 \in \underset{z \in \mathbb{C}^N}{\operatorname{argmin}} \|z\|_{\text{TV}} \text{ subject to } \|P_\Omega U_{\text{df}} z - y\| \leq \delta. \tag{21}$$

Note that these two approaches represent two different methods for reconstruction. The goal is now to compare the results and deduce which method, at least on one image, will give the best image quality most suited for radiologists to provide accurate diagnostics.

Note that this is not a contrived example, but an issue that is a daily encounter when methods are compared in scientific publications. And here lies the problem. In order to test the methods, we will have to compute minimisers to the above problems (20) and (21) for which Examples 4 and 6 demonstrate the impossibility of controlling the error in general. Thus, we end up with the key question regarding methodological and algorithmic errors:

> How can one deduce that method (1) is better than method (2) (or the other way), if we cannot control the algorithmic error that is introduced when we compute an approximation to the minimisers?

In particular, without any control of the error, the exercise of plotting two images produced with the two methods in order to deduce a winner is rather pointless. Note that the human eye has problems distinguishing colour differences at pixel level given a difference in the fourth digit (base 10). Thus, for the purpose of looking at images, then four-digit accuracy in the l^∞ norm is sufficient. But how can we secure four-digit accuracy? The following halting criterion may seem familiar:

halt $= no$
while halt $= no$
Tweak with the parameters in the code until the image looks better
    if you think the image looks good enough **then**
        halt $= yes$
    endif
end

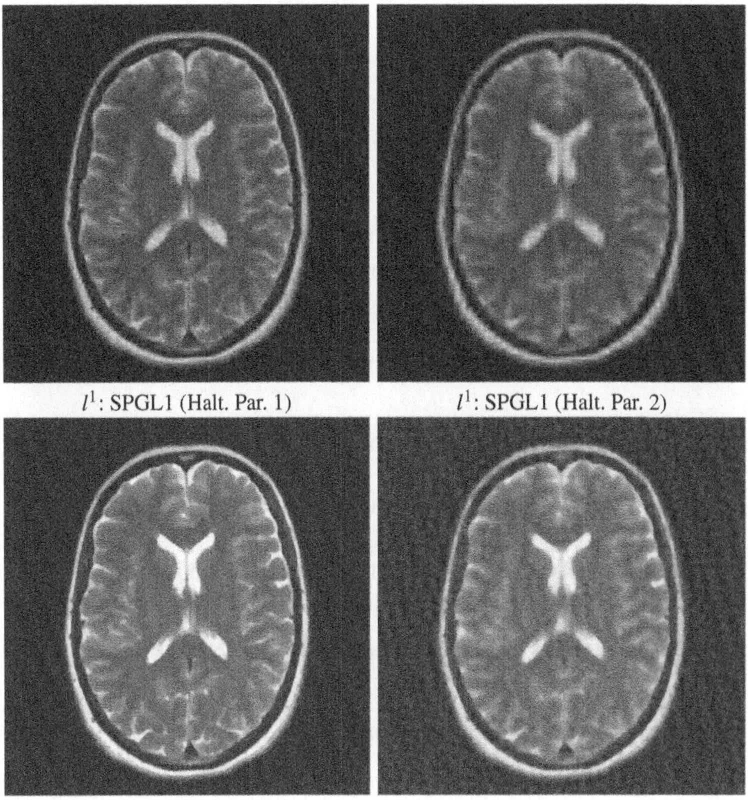

Fig. 15 Four reconstructions using standard packages with different halting criteria for solving (20) and (21)

Or in order to demonstrate that one's method is better than a competitor's method:

```
halt = no
while halt = no
Tweak with the parameters in the code until the image looks better
    if you think the image looks better than the competitor's then
        halt = yes
    endif
end
```

Indeed, the reason why such halting criterion may be the only choice is suggested in Example 4. Moreover, Example 6 demonstrates that it is impossible to compute an exit flag in order to verify if the computation is correct.

To illustrate the issue, we have displayed different approximations to minimisers of (20) and (21) using standard algorithms with different halting parameters in Fig. 15. As is evident from Fig. 15, the results vary greatly and make it very hard to deduce which of these figures actually represent a reasonable approximation to the minimiser one seeks.

6.2.2 Computations in Compressed Sensing: Why Things Often Work in Practice

In view of Example 4, it is legitimate to ask if the standard assumptions in compressed sensing will imply the existence of fast algorithms for the optimisation problems solved in practice. In particular, let $K \in \mathbb{N} \cup \{\infty\}$ and consider

$$[z]_K, \quad z \in \operatorname*{argmin}_x \|x\|_1 \text{ subject to } \|Ax - y\|_2 \leq \eta, \qquad \eta \geq 0, \tag{22}$$

where the jth coordinate of $[z]_K$ is given by

$$([z]_K)_j = \lfloor 10^K z_j \rfloor 10^{-K}.$$

In particular, for $K \in \mathbb{N}$, the problem is to compute a solution that has K correct digits in the $\|\cdot\|_\infty$ norm. The following then becomes a key question:

> Is the BP problem (22) in P when A satisfies the robust nullspace property of order s with parameters ρ and τ, and where $y = Ax$ where x is s-sparse?

As we will see below, this is a rather intricate issue.

Example 8 (Existence of algorithms in compressed sensing) Fix real constants $\rho \in (1/3, 1)$, $\tau > 14$, $b_A > 6$, and $b_y > 2$. Let Ω be the collection of all inputs $(A, y) \in \mathbb{R}^{m\times N} \times \mathbb{R}^m$ (with any dimensions m, N) where $\|A\|_2 \leq b_A$, $\|y\|_2 \leq b_y$, A satisfies the robust nullspace property (as in Sect. 2) of order s for any $s \in \mathbb{N}$ with parameters ρ, τ, and where $y = Ax$ for any x that is s-sparse.

(i) There exists a constant $C > 0$ independent of ρ, τ, b_A, and b_y such that if we define $\Omega' \subset \Omega$ to be the collection of $(A, y) \in \mathbb{R}^{m\times N} \times \mathbb{R}^m$, where A satisfies the robust nullspace property of any order $s \in N$ with any dimensions $N \geq 10s$ and $m \geq Cs\log(eN/s)$, then we have the following. For $\eta \in (0, 1]$ and

$$K \geq \lceil |\log_{10}(\eta/2)| \rceil,$$

there does not exist any algorithm (even randomised) that can compute the BP problem (22) for all inputs in Ω' (with probability greater than $p > 1/2$ in the randomised case). The statement is true even if one restricts to well-conditioned and bounded inputs such that $\|A\|_2 \leq c$, $\text{Cond}(AA^*) \leq c$ and $\|y\|_2 \leq c$, for some constant $c > 0$.

(iii) There exist an algorithm and a polynomial $P : \mathbb{R}^2 \to \mathbb{R}$, such that for $\eta \in [0, 1]$ and $K \in \mathbb{N}$ satisfying

$$K \leq \left\lfloor \left| \log_{10} \left(\frac{2(3+\rho)\tau\eta}{1-\rho} \wedge 1 \right) \right| \right\rfloor, \tag{23}$$

the algorithm can compute the BP problem (22) for all inputs in Ω where the runtime is bounded by $P(m + N, K)$. In particular, for $\delta \in (0, 1]$ and for fixed K satisfying (23), the BP problem (22) is in P, meaning solvable in polynomial time in $m + N$. Moreover, for $\delta = 0$, the BP problem (22) is in P, meaning solvable in polynomial time in $m + N$ and K.

(iv) For sufficiently large $m, N \in \mathbb{N}$, there exist inputs $\iota = (A, y) \in \Omega$ such that A is a subsampled Hadamard matrix or Bernoulli matrix and

$$C_{\mathrm{RCC}}(\iota) = \infty.$$

In particular, given (iii) with K satisfying (23) and (iv), there exist inputs in Ω with infinite RCC condition number, yet the problem (22) is in P.

The impossibility results in (i) and (ii) and the 'in P' statements in (iii) are valid in both Turing and BSS models.

Example 8 may be viewed as the practical cousin of Example 4. Indeed, Example 8 demonstrates the facets of Example 4 in actual applications. However, it is important to emphasise that while Examples 8 and 4 explain why things fail, they also explain why things often work in practice. In fact Example 8 (iii) and (iv) explain the success of compressed sensing in practice despite the paradoxical results of Example 8 (i) and (ii). The key is that, although one may not be able to get five digits, say, of accuracy in certain cases, for sufficiently small values of η, one may be able to get four digits, and that can be done quickly. Moreover, as discussed in Example 5, four digits of accuracy is more than enough for any application involving imaging.

6.3 Connections to Other Work in Optimisation

Note that to establish (iii) in Example 8, one needs to link approximations of the objective function of BP to approximations of the set of minimisers when $A \in \mathbb{R}^{m\times N}$ satisfies the robust nullspace property. This link was first considered by Ben-Tal & Nemirovski in [56, Sec. 1.3.1].

The reader may recognise that there are results in optimisation that are based on inexact inputs, often referred to as 'robust optimisation'. The comprehensive book by A. Ben-Tal, L. El Ghaoui and A. Nemirovski [57] presents an excellent overview of this field. However, classical robust optimisation is mostly concerned with computing the objective function rather than the minimisers, as is the main issue in many aspects of mathematics of information. Moreover, the phenomena we discuss in Examples 4, 6, and 8 will typically only happen for the problem of computing minimisers. In some sense, the results presented here can be viewed as a robust optimisation theory for computing minimisers.

Also, the reader may consult [58, 59] that discuss computations of optimisation problems in compressed sensing; however, these results are not about the extended model with inexact input suggested by Smale.

References

1. E.J. Candès, J. Romberg, T. Tao, Robust uncertainty principles: exact signal reconstruction from highly incomplete frequency information. IEEE T Inf. Theory **52**(2), 489–509 (2006)
2. D.L. Donoho, Compressed sensing. IEEE T Inf. Theory **52**(4), 1289–1306 (2006)
3. Y.C. Eldar, G. Kutyniok (eds.), *Compressed Sensing: Theory and Applications* (Cambridge University Press, Cambridge/New York, 2012)
4. S. Foucart, H. Rauhut, *A Mathematical Introduction to Compressive Sensing* (Birkhäuser, New York, 2013)
5. E.J. Candès, Y. Plan, A probabilistic and RIPless theory of compressed sensing. IEEE T Inf. Theory **57**(11), 7235–7254 (2011)
6. B. Adcock, A.C. Hansen, C. Poon, B. Roman, Breaking the coherence barrier: a new theory for compressed sensing. Forum Math. Sigma **5**, 1–84, 001 (2017)
7. B. Roman, B. Adcock, A. Hansen, On asymptotic structure in compressed sensing. arXiv:1406.4178 (2014)
8. F. Krahmer, H. Rauhut, Structured random measurements in signal processing. arXiv:1401.1106v2 (2014)
9. F. Krahmer, R. Ward, Stable and robust recovery from variable density frequency samples. IEEE Trans. Image Proc. (to appear) **23**(2), 612–22 (2014)
10. M. Lustig, D.L. Donoho, J.M. Pauly, Sparse MRI: the application of compressed sensing for rapid MRI imaging. Magn. Reson. Imaging **58**(6), 1182–1195 (2007)
11. G. Puy, P. Vandergheynst, Y. Wiaux, On variable density compressive sampling. IEEE Signal Process. Lett. **18**, 595–598 (2011)
12. A.F. Stalder, M. Schmidt, H.H. Quick, M. Schlamann, S. Maderwald, P. Schmitt, Q. Wang, M.S. Nadar, M.O. Zenge, Highly undersampled contrast-enhanced MRA with iterative reconstruction: integration in a clinical setting. Magn. Reson. Med. **74**(6), 1652–1660 (2015)
13. Q. Wang, M. Zenge, H.E. Cetingul, E. Mueller, M.S. Nadar, Novel sampling strategies for sparse mr image reconstruction. In: *Proceedings of the International Society for Magnetic Resonance in Medicine, ISMRM'14*, vol 22, pp 1549 (2014)
14. A. Cohen, W. Dahmen, R. DeVore, Compressed sensing and best k-term approximation. J. Am. Math. Soc. **22**(1), 211–231 (2009)
15. E. Candès, J. Romberg, Robust signal recovery from incomplete observations. In: *IEEE International Conference on Image Processing*, pp 1281–1284 (2006)
16. L. Gan, T.T. Do, T.D. Tran, Fast compressive imaging using scrambled hadamard ensemble. In: *Proceedings of European Signal Processing Conference*, pp 139–154 (2008)
17. M. Duarte, R. Baraniuk, Kronecker compressive sensing. IEEE T Image Process. **21**(2), 494–504 (2012)
18. T. Goldstein, L. Xu, K.F. Kelly, R.G. Baraniuk, The stone transform: multi-resolution image enhancement and real-time compressive video. arXiv:1311.3405 (2013)
19. H. Rauhut, R. Ward, Interpolation via weighted l^1 minimization. arXiv:1401.1106v2 (2014)

20. V. Studer, J. Bobin, M. Chahid, H. Moussavi, E. Candès, M. Dahan, Compressive fluorescence microscopy for biological and hyperspectral imaging. Natl. Acad. Sci. USA **109**(26), 1679–1687 (2011)
21. B. Adcock, A.C. Hansen, B. Roman, The quest for optimal sampling: computationally efficient, structure-exploiting measurements for compressed sensing. In: *Compressed Sensing and Its Applications (to appear)* (Springer, Cham, 2014)
22. A. Bastounis, A.C. Hansen, On the absence of uniform recovery in many real-world applications of compressed sensing and the restricted isometry property and nullspace property in levels. SIAM J. Imaging Sci. **10**(1), 335–371 (2017)
23. Y. Traonmilin, R. Gribonval, Stable recovery of low-dimensional cones in hilbert spaces: one rip to rule them all. Appl. Comput. Harmon. Anal. **45**(1), 170–205 (2018)
24. D. Takhar, J.N. Laska, M.B. Wakin, M.F. Duarte, D. Baron, S. Sarvotham, K.F. Kelly, R.G. Baraniuk, A new compressive imaging camera architecture using optical-domain compression. In: *Computational Imaging IV at SPIE Electronic Imaging*, pp 43–52 (2006)
25. G. Huang, H. Jiang, K. Matthews, P.A. Wilford, Lensless imaging by compressive sensing. In: *IEEE International Conference on Image Processing*, pp 2101–2105 (2013)
26. R.G. Baraniuk, V. Cevher, M.F. Duarte, C. Hedge, Model-based compressive sensing. IEEE T Inf. Theory **56**(4), 1982–2001 (2010)
27. S. Som, P. Schniter, Compressive imaging using approximate message passing and a markov-tree prior. IEEE T Signal Process. **60**(7), 3439–3448 (2012)
28. L. He, L. Carin, Exploiting structure in wavelet-based Bayesian compressive sensing. IEEE T Signal Process. **57**(9), 3488–3497 (2009)
29. N. Kingsbury, Image processing with complex wavelets. Phil. Trans. Royal Society London A **357**, 2543–2560 (1997)
30. E. Candès, D.L. Donoho, Recovering edges in ill-posed inverse problems: optimality of curvelet frames. Ann. Stat. **30**(3), 784–842 (2002)
31. S. Dahlke, G. Kutyniok, P. Maass, C. Sagiv, H.-G. Stark, G. Teschke, The uncertainty principle associated with the continuous shearlet transform. Int. J. Wavelets Multiresolut. Inf. Process. **6**(2), 157–181 (2008)
32. S.G. Mallat, *A Wavelet Tour of Signal Processing: The Sparse Way*, 3rd edn (Academic, Burlington, 2009)
33. S. Smale, Mathematical problems for the next century. Math. Intell. **20**, 7–15 (1998)
34. A. Bastounis, A.C. Hansen, V. Vlacic, On computational barriers and paradoxes in estimation, regularisation, learning and computer assisted proofs. Preprint (2020)
35. A.C. Hansen, On the solvability complexity index, the n-pseudospectrum and approximations of spectra of operators. J. Am. Math. Soc. **24**(1), 81–124 (2011)
36. J. Ben-Artzi, M.J. Colbrook, A.C. Hansen, O. Nevanlinna, M. Seidel, On the solvability complexity index hierarchy and towers of algorithms. Preprint (2018)
37. J. Ben-Artzi, A.C. Hansen, O. Nevanlinna, M. Seidel, Can everything be computed? – on the solvability complexity index and towers of algorithms. arXiv:1508.03280v1 (2015)
38. J. Ben-Artzi, A.C. Hansen, O. Nevanlinna, M. Seidel, New barriers in complexity theory: on the solvability complexity index and the towers of algorithms. C. R. Math. **353**(10), 931–936 (2015)
39. S. Boyd, L. Vandenberghe, *Convex Optimization* (Cambridge University Press, New York, 2004)
40. P. Bürgisser, F. Cucker, *Condition: The Geometry of Numerical Algorithms*. Grundlehren der Mathematischen Wissenschaften (Springer, Berlin/Heidelberg/New York, 2013)
41. A. Chambolle, T. Pock, A first-order primal-dual algorithm for convex problems with applications to imaging. J. Math. Imaging Vis. **40**(1), 120–145 (2011)
42. Y. Nesterov, A. Nemirovskii, *Interior-Point Polynomial Algorithms in Convex Programming*. SIAM studies in applied mathematics. Society for Industrial and Applied Mathematics, Philadelphia (1994)
43. Y. Nesterov, *Introductory Lectures on Convex Optimization: A Basic Course*. Applied Optimization (Kluwer Academic Publisher, Boston/Dordrecht/London, 2004)

44. A. Ben-Tal, A.S. Nemirovski, *Lectures on Modern Convex Optimization: Analysis, Algorithms, and Engineering Applications*. Society for Industrial and Applied Mathematics (Philadelphia, 2001)
45. A. Chambolle, An algorithm for total variation minimization and applications. J. Math. Imaging Vis. **20**(1), 89–97 (2004)
46. M. Grötschel, L. Lovász, A. Schrijver, *Geometric Algorithms and Combinatorial Optimization* (Springer, Berlin/New York, 1988)
47. A.M. Turing, On computable numbers, with an application to the entscheidungsproblem. Proc. Lond. Math. Soc. **S2–42**(1), 230 (1936)
48. L. Blum, M. Shub, S. Smale, On a theory of computation and complexity over the real numbers: NP-completeness, recursive functions and universal machines. Bull. Am. Math. Soc. (N.S.) **21**(1), 1–46 (1989)
49. L. Lovasz, *An Algorithmic Theory of Numbers, Graphs and Convexity*. CBMS-NSF Regional Conference Series in Applied Mathematics. Society for Industrial and Applied Mathematics (1987)
50. L. Blum, F. Cucker, M. Shub, S. Smale, *Complexity and Real Computation* (Springer, New York, Inc., Secaucus, 1998)
51. L. Valiant, *Probably Approximately Correct: Nature's Algorithms for Learning and Prospering in a Complex World* (Basic Books, Inc., New York, 2013)
52. J. Renegar, Linear programming, complexity theory and elementary functional analysis. Math. Program. **70**(1), 279–351 (1995)
53. J. Renegar, Incorporating condition measures into the complexity theory of linear programming. SIAM J. Optim. **5**(3), 506–524 (1995)
54. P. Bürgisser, F. Cucker, On a problem posed by Steve Smale. Ann. Math. (2) **174**(3), 1785–1836 (2011)
55. F. Cucker, A theory of complexity, condition, and roundoff. Forum Math. Sigma **3**, 002 (2015)
56. A. Ben-Tal, A. Nemirovski, *Lectures on Modern Convex Optimization: Analysis, Algorithms, and Engineering Applications*. Available online at https://www2.isye.gatech.edu/~nemirovs/ (2000)
57. A. Ben-Tal, L. El Ghaoui, A. Nemirovski, *Robust Optimization*, Princeton Series in Applied Mathematics (Princeton University Press, Princeton, 2009)
58. J. Liang, J. Fadili, G. Peyre, Activity identification and local linear convergence of forward–backward-type methods. SIAM J. Optim. **27**(1), 408–437 (2017)
59. V. Roulet, N. Boumal, A. dõAspremont, Computational complexity versus statistical performance on sparse recovery problems. Inf. Inference J. IMA **9**(1), 1–32, 01 (2019)

Reflections on a Theorem of Boas and Pollard

Christopher Heil

Abstract Inspired by an elegant theorem of Boas and Pollard (and related results by Kazarian, Price, Talalyan, Zink, and others), we discuss multiplicative completion of redundant systems in Hilbert and Banach function spaces.

1 Introduction

From the point of view of a harmonic analyst, the most important orthonormal basis is the trigonometric system $\{e^{2\pi inx}\}_{n\in\mathbb{N}}$ in $L^2(\mathbb{T})$, where $\mathbb{T} = [0, 1]$. Since this system is basis, it becomes incomplete if we remove any elements. In a short but attractive paper published in 1948, Boas and Pollard [6] proved that if we remove finitely many elements from the trigonometric system, thereby leaving an incomplete set, then we can restore completeness by a simple multiplication. Specifically, if F is any finite subset of $\mathbb{Z}$, then there exists a bounded function m such that $\{e^{2\pi inx}m(x)\}_{n\notin F}$ is complete in $L^2(\mathbb{T})$; in fact, Boas and Pollard proved that this holds for any orthonormal basis for $L^2(\mathbb{T})$. There are a number of very surprising equivalent reformulations and interesting related results. We will mention a few of these below (see Sect. 5), but refer to the important paper by Kazarian and Zink [21] and the references contained therein for full details.

Completeness is a fairly weak condition; in many situations, we would like to know if we have a Schauder basis or other basis-like properties, such as being a frame. Kazarian has shown that if finitely many elements are removed from the trigonometric system, then the resulting set $\{e^{2\pi inx}\}_{n\notin F}$ cannot be a Schauder basis for $L^p(\mu)$, where μ is a bounded Radon measure [18, 19], and in [20] he studied systems $\{e^{2\pi inx}\}_{n\notin F}$ in $L^p(\mu)$ where F is a finite sequence of consecutive integers.

This work was partially supported by a grant from the Simons Foundation.

C. Heil (✉)
School of Mathematics, Georgia Tech, Atlanta, GA, USA
e-mail: heil@math.gatech.edu

M. Th. Rassias (ed.), *Harmonic Analysis and Applications*, Springer Optimization and Its Applications 168, https://doi.org/10.1007/978-3-030-61887-2_7

In this expository note, we will review and discuss several examples related to these issues and derive a generalization of results in the spirit of Boas and Pollard for frames and other redundant systems.

2 Preliminaries

We let $\mathbb{T} = [0, 1]$, and consider functions on $\mathbb{T}$ to be extended 1-periodically to the real line. The inner product on $L^2(\mathbb{T})$ is

$$\langle f, g \rangle = \int_0^1 f(x)\,\overline{g(x)}\,dx.$$

We use standard notations for frames, Riesz and Schauder bases, and related concepts, as found in texts such as [10, 11, 14], or [29]. We outline below some particular terminology and facts that we will need.

If X is a Banach space, then we let X^* denote its dual space. The action of a functional $\mu \in X^*$ on an element $f \in X$ will be written $\langle f, \mu \rangle$.

Let $\{f_i\}_{i\in\mathbb{N}}$ be a sequence in a Banach space X. We say that $\{f_i\}_{i\in\mathbb{N}}$ is *complete* if its finite linear span is dense in X. It is *minimal* if there exists a sequence $\{\widetilde{f_i}\}_{i\in\mathbb{N}}$ in X^* that is *biorthogonal* to $\{f_i\}_{i\in\mathbb{N}}$, i.e., $\langle f_i, \widetilde{f_j} \rangle = \delta_{ij}$ for $i, j \in \mathbb{N}$. Equivalently, $\{f_i\}_{i\in\mathbb{N}}$ is minimal if $f_j \notin \overline{\operatorname{span}}\{f_i\}_{i\neq j}$ for each $j \in \mathbb{N}$. A sequence that is both minimal and complete is called *exact*. In this case, the biorthogonal sequence is unique.

The sequence $\{f_i\}_{i\in\mathbb{N}}$ is a *Schauder basis* for X if for each $f \in X$ there exist unique scalars c_i such that $f = \sum_{i=1}^{\infty} c_i f_i$, with convergence in the norm of X. Every Schauder basis is exact, and the biorthogonal sequence $\{\widetilde{f_i}\}_{i\in\mathbb{N}}$ is a Schauder basis for its closed span in X^* (if X is reflexive, then the biorthogonal sequence is a Schauder basis for X^*). We have $f = \sum_{i=1}^{\infty} \langle f, \widetilde{f_i} \rangle f_i$ for all $f \in X$. A Schauder basis is called an *unconditional basis* if this series converges unconditionally for every $f \in X$.

We say that a Schauder basis $\{f_i\}_{i\in\mathbb{N}}$ is *bounded* if $0 < \inf \|f_i\| \le \sup \|f_i\| < \infty$. In this case, $0 < \inf \|\widetilde{f_i}\| \le \sup \|\widetilde{f_i}\| < \infty$.

A *Riesz basis* is the image of an orthonormal basis for a Hilbert space H under a continuously invertible linear mapping of H onto itself. Every Riesz basis is a bounded unconditional basis for H, and conversely.

We say $\{f_i\}_{i\in\mathbb{N}}$ is a *frame* for a Hilbert space H if there exist constants $A, B > 0$, called *frame bounds*, such that

$$A\,\|f\|^2 \le \sum_{i=1}^{\infty} |\langle f, f_i \rangle|^2 \le B\,\|f\|^2, \qquad \text{for all } f \in H. \tag{1}$$

All Riesz bases are frames, but not conversely. A *frame sequence* is a sequence $\{f_i\}_{i\in\mathbb{N}}$ that is a frame for its closed span in H.

If $\{f_i\}_{i\in\mathbb{N}}$ satisfies at least the second inequality in (1), then we say that $\{f_i\}_{i\in\mathbb{N}}$ is a *Bessel sequence* or that it *possesses an upper frame bound*, and we call B a *Bessel bound*. Likewise if at least the first inequality in (1) is satisfied, then we say that $\{f_i\}_{i\in\mathbb{N}}$ *possesses a lower frame bound*.

If $\{f_i\}_{i\in\mathbb{N}}$ is Bessel, then the *analysis operator* $Cf = \{\langle f, f_i\rangle\}_{i\in\mathbb{N}}$ is a bounded mapping $C\colon H \to \ell^2$. If $\{f_i\}_{i\in\mathbb{N}}$ is a frame, then the *frame operator* $Sf = C^*Cf = \sum\langle f, f_i\rangle f_i$ is a bounded, positive definite, invertible map of H onto itself. Every frame $\{f_i\}_{i\in\mathbb{N}}$ has a *canonical dual frame* $\{\widetilde{f_i}\}_{i\in\mathbb{N}}$ given by $\widetilde{f_i} = S^{-1}f_i$ where S is the frame operator. We have

$$f = \sum_{i=1}^{\infty} \langle f, \widetilde{f_i}\,\rangle\, f_i = \sum_{i=1}^{\infty} \langle f, f_i\rangle\, \widetilde{f_i}, \qquad \text{for all } f \in H. \tag{2}$$

Furthermore, the series in (2) converges unconditionally for every f (so any countable index set can be used to index a frame). In general, for a frame, the coefficients in (2) need not be unique. In fact, uniqueness holds for every f if and only if $\{f_i\}_{i\in\mathbb{N}}$ is a Riesz basis.

3 Examples

Consider the lattice system of weighted exponentials

$$\mathscr{E}_g = \{e^{2\pi i n x} g(x)\}_{n\in\mathbb{Z}},$$

where g is a function in $L^2(\mathbb{T})$. The basis and frame properties of this system in $L^2(\mathbb{T})$ are summarized in the following theorem. Part (c) of this theorem is a consequence of the classical theory developed by Hunt, Muckenhoupt, and Wheeden, e.g., see [17]. For the proof of part (e), see Benedetto and Li [4]. The proofs of the other parts of the theorem are straightforward, e.g., see [14]. In this result, we let Z_g denote the zero set of g:

$$Z_g = \{x \in \mathbb{T} : g(x) = 0\}.$$

Technically, Z_g is only defined up to sets of measure zero, i.e., if we choose a different representative of g, then we may get a different set Z_g, but the symmetric difference between any two such sets will have measure zero.

Theorem 1 *If $g \in L^2(\mathbb{T})$, then the following statements hold:*

(a) *$\mathscr{E}_g$ is complete in $L^2(\mathbb{T})$ if and only if $g \neq 0$ a.e.*

(b) $\mathcal{E}_g$ *is minimal in* $L^2(\mathbb{T})$ *if and only if* $1/g \in L^2(\mathbb{T})$. *Moreover, in this case, it is exact and the biorthogonal system is* $\mathcal{E}_{\widetilde{g}}$ *where* $\widetilde{g}(x) = 1/\overline{g(x)}$.
(c) *With respect to the ordering* $\mathbb{Z} = \{0, -1, 1, -2, 2, \dots\}$, $\mathcal{E}_g$ *is a Schauder basis for* $L^2(\mathbb{T})$ *if and only if* $|g|^2$ *belongs to the Muckenhoupt weight class* $\mathcal{A}_2(\mathbb{T})$.
(d) $\mathcal{E}_g$ *is a Bessel sequence in* $L^2(\mathbb{T})$ *if and only if* $g \in L^\infty[0, 1]$. *Moreover, in this case,* $|g(x)|^2 \le B$ *a.e. where* B *is a Bessel bound.*
(e) $\mathcal{E}_g$ *is a frame sequence in* $L^2(\mathbb{T})$ *if and only if there exist* $A,\ B > 0$ *such that* $A \le |g(x)|^2 \le B$ *for a.e.* $x \notin Z_g$. *In this case, the closed span of* $\mathcal{E}_g$ *is*

$$H_g = \{f \in L^2(\mathbb{T}) : f = 0 \text{ a.e. on } Z_g\},$$

and $A,\ B$ *are frame bounds for* $\mathcal{E}_g$ *as a frame for* H_g.
(f) $\mathcal{E}_g$ *is an unconditional basis for* $L^2(\mathbb{T})$ *if and only if there exist* $A,\ B > 0$ *such that* $A \le |g(x)|^2 \le B$ *for a.e.* x, *and in this case, it is a Riesz basis for* $L^2(\mathbb{T})$.
(g) $\mathcal{E}_g$ *is an orthonormal basis for* $L^2(\mathbb{T})$ *if and only if* $|g(x)| = 1$ *for a.e.* x.

If we remove a single element from the trigonometric system, say the constant function 1 (corresponding to the index $n = 0$), then we are left with an incomplete system. The next example shows what happens if we multiply the remaining elements by the function x (parts of this example are adapted from our paper [28] with Yoon, which, among other results, studies systems of the form $\{x^N e^{2\pi inx}\}_{n\in\mathbb{Z}\setminus F}$ where F is finite). In this result, we consider series with respect to the ordering $\mathbb{Z}\setminus\{0\} = \{1, -1, 2, -2, \dots\}$,

Theorem 2

(a) $\{x\, e^{2\pi inx}\}_{n\neq 0}$ *is exact in* $L^2(\mathbb{T})$, *and its biorthogonal sequence is*

$$\{\widetilde{e}_n\}_{n\neq 0} = \left\{\frac{e^{2\pi inx} - 1}{x}\right\}_{n\neq 0}.$$

(b) *The biorthogonal system is exact in* $L^2(\mathbb{T})$, *but it is not bounded above in norm.*
(c) $\{x\, e^{2\pi inx}\}_{n\neq 0}$ *is not a Schauder basis for* $L^2(\mathbb{T})$.
(d) *If the series* $f(x) = \sum_{n\neq 0} c_n\, x\, e^{2\pi inx}$ *converges in* $L^2(\mathbb{T})$ *for some scalars* c_n, *then* $c_n = \langle f, \widetilde{e}_n\rangle$ *for every* $n \neq 0$, *and* $c_n \to 0$ *as* $n \to \pm\infty$.
(e) *There are no scalars* c_n *such that the constant function can be written as*

$$1 = \sum_{n\neq 0} c_n\, x\, e^{2\pi inx}$$

with convergence of the series in the norm of $L^2(\mathbb{T})$.

Proof

(a) The biorthogonality follows directly. To show completeness, suppose that $f \in L^2(\mathbb{T})$ satisfies $\langle f(x), x\, e^{2\pi inx}\rangle = 0$ for every $n \neq 0$. Then the function $xf(x)$, which belongs to $L^2(\mathbb{T})$, is orthogonal to $e^{2\pi inx}$ for every $n \neq 0$. Therefore, $xf(x) = c$ a.e. where c is a constant. If $c \neq 0$, then $f(x) = c/x \notin L^2(\mathbb{T})$, which is a contradiction. Therefore, $c = 0$, so $f = 0$ a.e. and $\{x\, e^{2\pi inx}\}_{n\neq 0}$ is complete.

(b) Suppose that $h \in L^2(\mathbb{T})$ satisfies $\langle h, \widetilde{e}_n\rangle = 0$ for $n \neq 0$. For convenience of notation, let

$$\widetilde{e}_0(x) = \frac{e^{-2\pi i 0\cdot x} - 1}{x} = 0.$$

Then $\langle h, \widetilde{e}_n\rangle = 0$ for all $n \in \mathbb{Z}$. Since $\widetilde{e}_n$ is a bounded function, we have $g(x) = h(x)\,\frac{e^{2\pi ix}-1}{x} \in L^2(\mathbb{T})$. Yet for every $m \in \mathbb{Z}$,

$$\begin{aligned}\langle g, e_m\rangle = \int_0^1 g(x)\, e^{-2\pi imx}\, dx &= \int_0^1 h(x)\, \frac{e^{-2\pi i(m-1)x} - 1 + 1 - e^{-2\pi imx}}{x}\, dx \\ &= \langle h, \widetilde{e}_{m-1}\rangle - \langle h, \widetilde{e}_m\rangle = 0,\end{aligned}$$

so $g = 0$ a.e., and therefore $h = 0$ a.e. Hence, $\{\widetilde{e}_n\}_{n\neq 0}$ is complete in $L^2(\mathbb{T})$.

A direct computation shows that

$$\|\widetilde{e}_n\|_2^2 = \int_0^1 |\widetilde{e}_n(x)|^2\, dx = 4\pi n \int_0^{\pi n} \frac{\sin^2 u}{u^2}\, du < \infty,$$

so we do have $\widetilde{e}_n \in L^2(\mathbb{T})$. But $\int_0^{\pi n} \frac{\sin^2 u}{u}\, du \to \frac{\pi}{2}$ as $n \to \infty$, so $\{\widetilde{e}_n\}_{n\neq 0}$ is not bounded above in norm.

(c) We have $\|x\, e^{2\pi inx}\|_2 = 3^{-1/2}$ for every n, so $\{x\, e^{2\pi inx}\}_{n\neq 0}$ is bounded above and below in norm. If it was a Schauder basis, then its biorthogonal system would also be bounded above and below in norm (e.g., see [14, Sec. 5.6]). Part (b) shows that this is not the case, so $\{x\, e^{2\pi inx}\}_{n\neq 0}$ cannot be a Schauder basis for $L^2(\mathbb{T})$.

(d) For simplicity of notation, let $f_n(x) = x\, e^{2\pi inx}$. If the series $f = \sum_{n\neq 0} c_n f_n$ converges, then $c_n = \langle f, \widetilde{e}_n\rangle$ follows from biorthogonality. Let s_n denote the nth partial sum of the series. Then

$$s_{2n} = c_1 f_1 + c_{-1} f_{-1} + \cdots + c_n f_n + c_{-n} f_{-n}$$

and

$$s_{2n-1} = c_1 f_1 + c_{-1} f_{-1} + \cdots + c_n f_n.$$

Since $s_n \to f$, it follows that

$$\|s_{2n} - s_{2n-1}\|_2 \leq \|s_{2n} - f\|_2 + \|f - s_{2n-1}\|_2 \to 0 \quad \text{as } n \to \infty.$$

But

$$s_{2n} - s_{2n-1} = c_{-n} f_{-n} = \langle f, \widetilde{e}_{-n} \rangle f_{-n},$$

so

$$\|s_{2n} - s_{2n-1}\|_2 = |\langle f, \widetilde{e}_{-n} \rangle| \, \|f_{-n}\|_2 = 3^{-1/2} \, |\langle f, \widetilde{e}_{-n} \rangle| = 3^{-1/2} \, |c_{-n}|.$$

Therefore, $c_{-n} \to 0$ as $n \to \infty$. A similar argument shows that $c_n \to 0$ as $n \to \infty$.

(e) Assume that $1 = \sum_{n \neq 0} c_n f_n$, and let s_n denote the nth partial sum of this series. Applying part (d) to $f = 1$, we have $c_n = \langle 1, \widetilde{e}_n \rangle$ and $c_n \to 0$ as $n \to \pm\infty$. However,

$$\begin{aligned}
|c_{-n}| = |\langle 1, \widetilde{e}_{-n} \rangle| &= \left| \int_0^1 \frac{e^{2\pi i n x} - 1}{x} \, dx \right| \\
&= \left| \int_0^1 \frac{\cos 2\pi n x - 1}{x} \, dx \, + i \int_0^1 \frac{\sin 2\pi n x}{x} \, dx \right| \\
&\geq \left| \int_0^1 \frac{\sin 2\pi n x}{x} \, dx \right| \to \frac{\pi}{2} \text{ as } n \to \infty.
\end{aligned}$$

But this contradicts the fact that $c_{-n} \to 0$ as $n \to \infty$. □

Boas and Pollard proved (among other examples) that if we remove infinitely many elements from the trigonometric system, then we may not be able to restore completeness by multiplying the remaining elements by some function. For inclusiveness, we present their argument. We need the following elementary lemma (functions in $L^2(\mathbb{T})$ are considered to be extended 1-periodically to the entire real line):

Lemma 1 *Let N be a positive integer. If $f \in L^2(\mathbb{T})$ is $1/N$-periodic, then we have $\langle f(x), e^{2\pi i n x} \rangle = 0$ for all $n \in \mathbb{Z}$ such that N does not divide n.*

Theorem 3 ([6]) *If $S \subseteq \mathbb{Z}$ contains an arithmetic progression, then the orthonormal sequence $\{e^{2\pi i n x}\}_{n \in \mathbb{Z} \setminus S}$ cannot be completed in $L^2(\mathbb{T})$ by multiplication by a square-integrable function.*

Proof Without loss of generality, assume $S = \{nN\}_{n \in \mathbb{Z}}$ for some $N \in \mathbb{N}$. Fix any $m \in L^2(\mathbb{T})$. If m is zero on a set of positive measure, then $\{m(x) \, e^{2\pi i n x}\}_{n \notin S}$ is incomplete, so assume $m \neq 0$ a.e. Then there exists a set $E_1 \subseteq (0, 1/N)$ on which $|m|$ is bounded above and below, and then a set $E_2 \subseteq E_1 + 1/N$ on which $|m|$ is bounded above and below, and so forth. Define

$$F = E_N \cup (E_N - \tfrac{1}{N}) \cup \cdots \cup (E_N - \tfrac{N-1}{N}).$$

Then $F + 1/N = F$ (mod 1). Moreover, $F \subseteq E_N \cup \cdots \cup E_1$, so $|m|$ is bounded above and below on F. Therefore,

$$f(x) = \begin{cases} 1/\overline{m(x)}, & x \in F, \\ 0, & x \notin F, \end{cases}$$

is a nonzero element of $L^2(\mathbb{T})$. Further, $f\,\overline{m} = \chi_F$ is $1/N$-periodic, so we apply Lemma 1 and conclude that that $\langle f(x), m(x)e^{2\pi inx}\rangle = \langle f\,\overline{m(x)}, e^{2\pi inx}\rangle = 0$ for all $n \notin S$. Therefore, $\{m(x)e^{2\pi inx}\}_{n\notin S}$ is incomplete. □

4 Boas and Pollard Revisited

We make a few observations that (modestly) extend the multiplicative completion result of Boas and Pollard. Specifically, they proved that if $\{f_n\}_{n\in\mathbb{N}}$ is an orthonormal basis for $L^2(\mathbb{T})$ and F is a finite subset of $\mathbb{Z}$, then there exists a function $m \in L^\infty(\mathbb{T})$ such that $\{mf_n\}_{n\in\mathbb{Z}\setminus F}$ is complete in $L^2(\mathbb{T})$.

Observe that if we replace $L^2(\mathbb{T})$ with an arbitrary space $L^2(\mu)$, then the analogous result can fail. For example, if $\{\delta_n\}_{n\in\mathbb{N}}$ denotes the standard basis for ℓ^2, then there is no sequence x such that $\{x\,\delta_n\}_{n>1}$ is complete, where $x\,\delta_n$ denotes the componentwise product of the two sequences.

Recall that a measure μ on a measurable space (X, Σ) is *nonatomic* if for every measurable set A satisfying $\mu(A) > 0$ there exists a measurable set $B \subsetneq A$ such that $\mu(B) > 0$. In this case, it follows that there are infinitely many sets $A = A_1 \supsetneq A_2 \supsetneq \cdots$ such that

$$\mu(A) = \mu(A_1) > \mu(A_2) > \cdots > 0.$$

Taking $B_k = A_k \setminus A_{k+1}$, we obtain disjoint measurable sets, all with positive measures, such that $A = \cup B_k$.

Lemma 2 *Assume that μ is a nonatomic measure on a measure space (X, Σ). If $1 \le p < \infty$ and $f \in L^p(\mu)$ is not the zero function in $L^p(\mu)$, then there exists a function $m \in L^\infty(\mu)$ such that $m(x) \neq 0$ at every point and $f/m \notin L^p(\mu)$.*

Proof Since f is not the zero function, there is some positive number R such that $E = \{|f| < R\}$ has positive measure. Since μ is nonatomic, there exists a measurable $A \subseteq E$ such that $0 < \mu(A) < \mu(E)$. Write $A = \cup B_k$ disjointly where each B_k is measurable and has positive measure. Define

$$m(x) = \begin{cases} \mu(B_k)^{1/p}\,|f(x)|, & x \in B_k, \\ 1, & \text{otherwise}. \end{cases}$$

Then for $x \in B_k$, we have $|m(x)| \le R\,\mu(B_k)^{1/p} \le R\,\mu(A)^{1/p}$, so $m \in L^\infty(\mu)$. Note that f/m is defined a.e., and

$$\int_X \left|\frac{f(x)}{m(x)}\right|^p d\mu(x) \geq \sum_{k\in\mathbb{N}} \int_{B_k} \left|\frac{f(x)}{m(x)}\right|^p d\mu(x) = \sum_{k\in\mathbb{N}} \int_{B_k} \frac{1}{\mu(B_k)}\, d\mu(x) = \sum_{k\in\mathbb{N}} 1 = \infty,$$

so $f/m \notin L^p(\mu)$. □

Adapting the techniques of [6], we will prove that a Boas–Pollard-type result holds whenever we are in a situation similar to the one given in the conclusion of Lemma 2. First, we need the following lemma. Here, if (X, Σ, μ) is a measure space, then we say that a Banach space $\mathscr{A}$ of measurable functions on X is *solid* if given $g \in \mathscr{A}$ and a measurable function f such that $|f| \leq |g|$ a.e., we have $f \in \mathscr{A}$ and $\|f\|_{\mathscr{A}} \leq \|g\|_{\mathscr{A}}$ (compare [25]).

Lemma 3 *Let (X, Σ, μ) be a measure space, and let $\mathscr{A}$ be a solid Banach space of measurable complex-valued functions on X. Assume that for any measurable set $E \subseteq X$ there exists a measurable function ψ on X such that $\{\psi \neq 0\} \subseteq E$ and $\psi \notin \mathscr{A}$. Then, given any $f_1, \dots, f_N \in \mathscr{A}$, there exists a function $g \in L^\infty(\mu)$ such that*

(a) *$g(x) \neq 0$ for all $x \in X$, and*
(b) *$f/g \notin \mathscr{A}$ for every $f \in \operatorname{span}\{f_1, \dots, f_N\} \setminus \{0\}$.*

Proof Without loss of generality, we assume $f_n \neq 0$ for all n. We proceed by induction.

Base step. Set $N = 1$, and let $f = cf_1$, where $c \neq 0$. Since f is not the zero function, there is some positive integer n such that $E = \{\frac{1}{n} < |f| < n\}$ has positive measure. By hypothesis, there exists a function $\psi \notin \mathscr{A}$ such that $\{\psi \neq 0\} \subseteq E$. Let $\varphi(x) = \max\{|\psi(x)|, 1\}$, and set $g = 1/\varphi$. Since $1 \leq \varphi(x) < \infty$ at all points, we have $0 < g \leq 1$. Moreover, if $x \in E$, then

$$\frac{|f(x)|}{|g(x)|} \geq \frac{\varphi(x)}{n} \geq \frac{|\psi(x)|}{n}.$$

This also holds for $x \notin E$ since $\{\psi \neq 0\} \subseteq E$. Since $\psi \notin \mathscr{A}$ and $\mathscr{A}$ is solid, it follows that $f/g \notin \mathscr{A}$.

Inductive step. Assume that the conclusions of the lemma hold for some $N \geq 1$, and let $f_1, \dots, f_{N+1} \in \mathscr{A}$ be fixed. Then, by hypothesis, there exists a function $g \in L^\infty(\mu)$, nonzero at every point, such that

$$S_N = \big\{f \in \operatorname{span}\{f_1, \dots, f_N\} \setminus \{0\} : f/g \in \mathscr{A}\big\} = \varnothing.$$

Define

$$S_{N+1} = \big\{f \in \operatorname{span}\{f_1, \dots, f_{N+1}\} \setminus \{0\} : f/g \in \mathscr{A}\big\}.$$

If $S_{N+1} = \varnothing$, then the proof is complete, so assume $F = \sum_{n=1}^{N+1} c_n f_n \in S_{N+1}$. Note that $c_{N+1} \neq 0$, for otherwise $F \in S_N = \varnothing$. Assume also that $G =$

$\sum_{n=1}^{N+1} b_n f_n \in S_{N+1}$; then $b_{N+1} \neq 0$ for the same reason. Clearly,

$$H = \frac{1}{c_{N+1}} F - \frac{1}{b_{N+1}} G \in \operatorname{span}\{f_1, \ldots, f_N\}.$$

Moreover, $H/g \in \mathscr{A}$ as both F/g and G/g are in $\mathscr{A}$. Since $S_N = \varnothing$, it follows that $H = 0$. Thus, G is a multiple of F, so $S_N \subseteq \{cF : c \neq 0\}$. Now, $F \neq 0$ since $F \in S_{N+1}$. Therefore, there is some $\varepsilon > 0$ and some $E \subseteq X$ with positive measure such that $|F(x)| \geq \varepsilon$ for $x \in E$. By hypothesis, there exists a function $\psi \notin \mathscr{A}$ with $\{\psi \neq 0\} \subseteq E$. Set $\varphi(x) = \max\{|\psi(x)|, 1/|g(x)|\}$ and define $h = 1/\varphi$. Then $h \leq |g|$ so $h \in L^\infty(\mu)$. Moreover, if $x \in E$, then $|F(x)/h(x)| \geq \varepsilon\, \varphi(x) \geq \varepsilon\, |\psi(x)|$. This also holds for $x \notin E$ since $\{\psi \neq 0\} \subseteq E$. As $\psi \notin \mathscr{A}$ and $\mathscr{A}$ is solid, it follows that $F/h \notin \mathscr{A}$.

Finally, to finish the proof, assume that $f \in \operatorname{span}\{f_1, \ldots, f_{N+1}\} \setminus \{0\}$ is given. If $f/h \in \mathscr{A}$, then $f/g \in \mathscr{A}$ since $h \leq |g|$. Therefore, $f \in S_{N+1}$, so $f = cF$ for some $c \neq 0$. However, $F/h \notin \mathscr{A}$, a contradiction. Therefore, $f/h \notin \mathscr{A}$. □

Theorem 4 *Let (X, Σ, μ) be a measure space, and let $\mathscr{B}$ be a solid Banach space $\mathscr{B}$ of measurable complex-valued functions on X. Assume that $\mathscr{B}^*$ is also a solid Banach function space on X that satisfies the hypotheses of Lemma 3. Given $S \subseteq \mathscr{B}$, define*

$$S^\perp = \{g \in \mathscr{B}^* : \langle f, g\rangle = 0 \text{ for all } f \in S\}.$$

Suppose that $\{f_n\}_{n\in\mathbb{N}} \subseteq \mathscr{B}$ and $g_1, \ldots, g_N \in \mathscr{B}^*$ satisfy

$$\{f_n\}^\perp \subseteq \operatorname{span}\{g_1, \ldots, g_N\}.$$

Then there exists a function $m \in L^\infty(\mu)$ with $m(x) \neq 0$ for every x such that $\{mf_n\}_{n>N}$ is complete in $\mathscr{B}$.

Proof By Lemma 3, there exists a function $m \in L^\infty(\mu)$ that is nonzero at every point such that

$$g \in \operatorname{span}\{g_1, \ldots, g_N\} \setminus \{0\} \quad \Longrightarrow \quad g/\overline{m} \notin \mathscr{B}^*. \tag{3}$$

Assume that $h \in \mathscr{B}^*$ satisfies $\langle mf_n, h\rangle = 0$ for all $n > N$. Since $m \in L^\infty(\mu)$ we have $h\overline{m} \in \mathscr{B}^*$. Since $\langle f_n, h\overline{m}\rangle = 0$ for all n, we also have $h\overline{m} \in \{f_n\}^\perp \subseteq \operatorname{span}\{g_1, \ldots, g_N\}$. If $h\overline{m} \neq 0$ then $h = (h\overline{m})/\overline{m} \notin \mathscr{B}^*$, which is a contradiction. Therefore, $h\overline{m} = 0$, so $h = 0$ a.e. since m is everywhere nonzero. Hence, $\{mf_n\}_{n>N}$ is complete in $\mathscr{B}$. □

Example 1

(a) Assume that $\{f_n\} \subseteq \mathscr{B}$ and $\{g_n\} \subseteq \mathscr{B}^*$ satisfy $g = \sum \langle g, f_n \rangle g_n$ for $g \in \mathscr{B}^*$ (not necessarily uniquely; such a system is called a *quasibasis* for $\mathscr{B}$). Fix $N > 0$. If $g \in \{f_n\}_{n>N}^{\perp}$, then $g = \sum_{n=1}^{N} \langle g, f_n \rangle g_n \in \operatorname{span}\{g_1, \dots, g_N\}$.
(b) If $\{f_n\}_{n\in\mathbb{N}}$ is a Schauder basis for $\mathscr{B}$ and $\mathscr{B}$ is reflexive, then there exists a *dual basis* $\{g_n\}_{n\in\mathbb{N}} \subseteq \mathscr{B}^*$, i.e., $g = \sum \langle g, f_n \rangle g_n$, uniquely, for all $g \in \mathscr{B}^*$ (e.g., see [14] for details). Therefore, by part (a) and Theorem 4, given any $N > 0$, there exists a function $m \in L^\infty(\mu)$ such that $\{mf_n\}_{n>N}$ is complete in $\mathscr{B}$.
(c) If $\{g_n\}_{n\in\mathbb{N}}$ is a frame for $\mathscr{B} = L^2(\mu)$ and $\{f_n\}_{n\in\mathbb{N}}$ is its dual frame, then $g = \sum \langle g, f_n \rangle g_n$ for all $g \in L^2(\mu)$. Therefore, by part (a) and Theorem 4, given any $N > 0$, there exists a function $m \in L^\infty(\mu)$ such that $\{mf_n\}_{n>N}$ is complete in $L^2(\mu)$.
(d) Let X be a finite set and let μ be a counting measure on X. Given $\varnothing \neq E \subseteq X$ and any finite function ψ on X with $\{\psi \neq 0\} \subseteq E$,

$$\|\psi\|_{L^p(\mu)}^p = \sum_{t\in E} |\psi(t)|^p < \infty,$$

since X is finite. Thus, $\mathscr{A} = L^p(\mu)$ does not satisfy the hypotheses of Lemma 3.

5 Related Results

As we mentioned in the introduction, there are many surprising results related to Boas–Pollard-type phenomena. We list just a few of these, but refer to the references for additional results, including those by Kazarian, Price, Talalyan, and Zink.

Given a sequence $\{f_n\}_{n\in\mathbb{N}} \subseteq L^2(\mu)$, where (X, μ) is a finite separable measure space with $\mu(\mu) = 1$, Talalyan proved that the following statements are equivalent (see [27]):

(a) Given $\varepsilon > 0$ there exists $S_\varepsilon \subseteq X$ such that $\mu(S_\varepsilon) > 1 - \varepsilon$ and $\{f_n \chi_{S_\varepsilon}\}$ is complete in $L^2(S_\varepsilon)$.
(b) For every function f on X which is finite a.e. and every $\varepsilon > 0$, there exists $S_\varepsilon \subseteq X$ and $g \in \operatorname{span}\{f_n\}$ such that $\mu(S_\varepsilon) > 1 - \varepsilon$ and $|f - g| < \varepsilon$ on S_ε.

Price and Zink proved that (a) and (b) are also equivalent to the following, seemingly unrelated, Boas–Pollard-type property (see [22, 23]):

(c) There exists a bounded, nonnegative function m such that $\{mf_n\}$ is complete in $L^2(\mu)$.

In [7–9] Byrnes and Newman consider a problem similar to the one addressed by Boas and Pollard. Instead of deleting elements from a sequence and then multiplying the remaining elements by a function, they retain all elements of the sequence and multiply only a portion of the sequence by a function. In particular, they show in [9] that if $\{f_n\}_{n\in\mathbb{Z}}$ is an orthonormal basis for $L^2(\mathbb{T})$ and $S \subseteq \mathbb{Z}$, then

$\{f_n\}_{n\in S} \cup \{mf_n\}_{n\notin S}$ is complete in $L^2(\mathbb{T})$ if and only if there exists an $\alpha \in \mathbb{C}$ such that $\mathrm{Re}(\alpha m) \geq 0$ a.e. and either $\mathrm{Im}(\alpha m) > 0$ a.e. or $\mathrm{Im}(\alpha m) < 0$ a.e. on the zero set of $\mathrm{Re}(\alpha m)$.

Finally, we remark that the basis and frame properties of "irregular" systems of weighted exponentials $\mathscr{E}(g, \Lambda) = \{e^{2\pi i\lambda x}g(x)\}_{\lambda\in\Lambda}$, where Λ is an arbitrary countable sequence in $\mathbb{R}$, are also very interesting (and usually difficult). For background and references on this subject, we refer to [16]. Similarly difficult are the cases of "irregular" Gabor systems $\{e^{2\pi i\beta x}g(x-\alpha)\}_{(\alpha,\beta)\in\Gamma}$ and wavelet systems $\{a^{1/2}\psi(ax - b)\}_{(a,b)\in\Gamma}$ in $L^2(\mathbb{R})$. Typically, in each of these cases, necessary conditions for the system to be a frame can be formulated in terms of the Beurling densities of the index set. For example, weighted exponentials and Gabor systems both exhibit a Nyquist density cutoff density [1–3, 12, 13, 24], whereas the situation for wavelet systems is much more subtle [5, 15, 26].

References

1. A. Aldroubi, K. Gröchenig, Nonuniform sampling and reconstruction in shift-invariant spaces. SIAM Rev. **43**, 585–620 (2001)
2. R. Balan, P.G. Casazza, C. Heil, Z. Landau, Density, overcompleteness, and localization of frames, I. Theory. J. Fourier Anal. Appl. **12**, 105–143 (2006)
3. R. Balan, P.G. Casazza, C. Heil, Z. Landau, Density, overcompleteness, and localization of frames, II. Gabor frames. J. Fourier Anal. Appl. **12**, 307–344 (2006)
4. J.J. Benedetto, S. Li, The theory of multiresolution analysis frames and applications to filter banks. Appl. Comput. Harmon. Anal. **5**, 389–427 (1998)
5. S. Bishop, Comparison theorems for separable wavelet frames. J. Approx. Theory **161**, 432–447 (2009)
6. R.P. Boas, H. Pollard, The multiplicative completion of sets of functions. Bull. Am. Math. Soc. **54**, 518–522 (1948)
7. J.S. Byrnes, Functions which multiply bases. Bull. Lond. Math. Soc. **4**, 330–332 (1972)
8. J.S. Byrnes, Complete multipliers. Trans. Am. Math. Soc. **172**, 399–403 (1972)
9. J.S. Byrnes, D.J. Newman, Completeness preserving multipliers. Proc. Am. Math. Soc. **21**, 445–450 (1969)
10. O. Christensen, *An Introduction to Frames and Riesz Bases* (Birkhäuser, Boston, 2003)
11. I. Daubechies, *Ten Lectures on Wavelets* (SIAM, Philadelphia, 1992)
12. K. Gröchenig, H. Razafinjatovo, On Landau's necessary density conditions for sampling and interpolation of band-limited functions. J. Lond. Math. Soc. **54**(2), 557–565 (1996)
13. C. Heil, Wiener amalgam spaces in generalized harmonic analysis and wavelet theory, Ph.D. thesis, University of Maryland, College Park, 1990
14. C. Heil, *A Basis Theory Primer*, Expanded Edition (Birkhäuser, Boston, 2011)
15. C. Heil, G. Kutyniok, The homogeneous approximation property for wavelet frames. J. Approx. Theory **147**, 28–46 (2007)
16. C. Heil, G. Kutyniok, Density of frames and Schauder bases of windowed exponentials. Houston J. Math. **34**, 565–600 (2008)
17. R. Hunt, B. Muckenhoupt, R. Wheeden, Weighted norm inequalities for the conjugate function and Hilbert transform. Trans. Am. Math. Soc. **176**, 227–251 (1973)
18. K.S. Kazarian, The multiplicative completion of basic sequences in L^p, $1 \leq p < \infty$, to bases in L^p (Russian). Akad. Nauk Armjan. SSR Dokl. **62**, 203–209 (1976)

19. K.S. Kazarian, The multiplicative complementation of some incomplete orthonormal systems to bases in L^p, $1 \leq p < \infty$ (Russian). Anal. Math. **4**, 37–52 (1978)
20. K.S. Kazarjan, Summability of generalized Fourier series and Dirichlet's problem in $L^p(d\mu)$ and weighted H^p-spaces ($p > 1$). Anal. Math. **13**, 173–197 (1987)
21. K.S. Kazarian, R.E. Zink, Some ramifications of a theorem of Boas and Pollard concerning the completion of a set of functions in L^2. Trans. Am. Math. Soc. **349**, 4367–4383 (1997)
22. J.J. Price, Topics in orthogonal functions. Am. Math. Monthly **82**, 594–609 (1975)
23. J.J. Price, R.E. Zink, On sets of functions that can be multiplicatively completed. Ann. Math. **82**, 139–145 (1965)
24. J. Ramanathan, T. Steger, Incompleteness of sparse coherent states. Appl. Comput. Harmon. Anal. **2**, 148–153 (1995)
25. H. Reiter, *Classical Harmonic Analysis and Locally Compact Groups* (Oxford University Press, Oxford, 1968)
26. W. Sun, X. Zhou, Density of irregular wavelet frames. Proc. Am. Math. Soc. **132**, 2377–2387 (2004)
27. A.A. Talalyan, On the convergence almost everywhere of subsequences of partial sums of general orthogonal series. Izv. Akad. Nauk Armyan SSR Ser. Fiz.-Mat. **10**, 17–34 (1957)
28. G.J. Yoon, C. Heil, Duals of weighted exponentials. Acta Appl. Math. **119**, 97–112 (2012)
29. R. Young, *An Introduction to Nonharmonic Fourier Series*, Revised First Edition (Academic, San Diego, 2001)

The Andoni–Krauthgamer–Razenshteyn Characterization of Sketchable Norms Fails for Sketchable Metrics

Subhash Khot and Assaf Naor

Abstract Andoni, Krauthgamer, and Razenshteyn (AKR) proved (STOC 2015) that a finite-dimensional normed space $(X, \|\cdot\|_X)$ admits a $O(1)$ sketching algorithm (namely, with $O(1)$ sketch size and $O(1)$ approximation) if and only if for every $\varepsilon \in (0,1)$, there exist $\alpha \geqslant 1$ and an embedding $f : X \to \ell_{1-\varepsilon}$ such that $\|x-y\|_X \leqslant \|f(x)-f(y)\|_{1-\varepsilon} \leqslant \alpha\|x-y\|_X$ for all $x, y \in X$. The "if part" of this theorem follows from a sketching algorithm of Indyk (FOCS 2000). The contribution of AKR is therefore to demonstrate that the mere availability of a sketching algorithm implies the existence of the aforementioned geometric realization. Indyk's algorithm shows that the "if part" of the AKR characterization holds true for any metric space whatsoever, i.e., the existence of an embedding as above implies sketchability even when X is not a normed space. Due to this, a natural question that AKR posed was whether the assumption that the underlying space is a normed space is needed for their characterization of sketchability. We resolve this question by proving that for arbitrarily large $n \in \mathbb{N}$, there is an n-point metric space $(M(n), d_{M(n)})$ which is $O(1)$-sketchable yet for every $\varepsilon \in (0, \frac{1}{2})$, if $\alpha(n) \geqslant 1$ and $f_n : M(n) \to \ell_{1-\varepsilon}$ are such that $d_{M(n)}(x,y) \leqslant \|f_n(x) - f_n(y)\|_{1-\varepsilon} \leqslant \alpha(n) d_{M(n)}(x,y)$ for all $x, y \in M(n)$, then necessarily $\lim_{n\to\infty} \alpha(n) = \infty$.

An extended abstract announcing this work appeared in the Thirtieth Annual ACM-SIAM Symposium on Discrete Algorithms.

S.K. was supported by NSF CCF-1422159 and the Simons Foundation. A.N. was supported by NSF CCF-1412958, the Packard Foundation and the Simons Foundation. This work was carried out under the auspices of the Simons Algorithms and Geometry (A&G) Think Tank.

S. Khot
Department of Computer Science, Courant Institute of Mathematical Sciences, New York University, New York, NY, USA
e-mail: khot@cs.nyu.edu

A. Naor (✉)
Mathematics Department, Princeton University, Princeton, NJ, USA
e-mail: naor@math.princeton.edu

M. Th. Rassias (ed.), *Harmonic Analysis and Applications*, Springer Optimization and Its Applications 168, https://doi.org/10.1007/978-3-030-61887-2_8

1 Introduction

We shall start by recalling the notion of sketchability; it is implicit in the seminal work [2] of Alon, Matias, and Szegedy, though the formal definition that is described below was put forth by Saks and Sun [42]. This is a crucial and well-studied algorithmic primitive for analyzing massive data sets, with several powerful applications; surveying them here would be needlessly repetitive, so we refer instead to, e.g., [4, 20] and the references therein.

Given a set X, a function $K : X \times X \to \mathbb{R}$ is called a nonnegative kernel if $K(x, y) \geqslant 0$ and $K(x, y) = K(y, x)$ for every $x, y \in X$. In what follows, we will be mainly interested in the geometric setting when the kernel $K = d_X$ is in fact a metric on X, but even for that purpose, we will also need to consider nonnegative kernels that are not metrics.

Fix $D \geqslant 1$ and $s \in \mathbb{N}$. Say that a nonnegative kernel $K : X \times X \to [0, \infty)$ is (s, D)-sketchable if for every $r > 0$ there is a mapping $\mathsf{R} = \mathsf{R}_r : \{0, 1\}^s \times \{0, 1\}^s \to \{0, 1\}$ and a probability distribution over mappings $\mathsf{Sk} = \mathsf{Sk}_r : X \to \{0, 1\}^s$ such that

$$\inf_{\substack{x,y\in X\\ K(x,y)\leqslant r}} \mathbf{Prob}\Big[\mathsf{R}\big(\mathsf{Sk}(x), \mathsf{Sk}(y)\big) = 0\Big] \geqslant \tfrac{3}{5} \quad \text{and}$$

$$\inf_{\substack{x,y\in X\\ K(x,y)> Dr}} \mathbf{Prob}\Big[\mathsf{R}\big(\mathsf{Sk}(x), \mathsf{Sk}(y)\big) = 1\Big] \geqslant \tfrac{3}{5}. \tag{1}$$

The value $\frac{3}{5}$ in (1) can be replaced throughout by any constant that is strictly bigger than $\frac{1}{2}$; we chose to fix an arbitrary value here in order to avoid the need for the notation to indicate dependence on a further parameter. A kernel (or, more formally, a family of kernels) is said to be sketchable if it is (s, D)-sketchable for some $s = O(1)$ and $D = O(1)$.

The way to interpret the above definition is to think of Sk as a randomized method to assign one of the 2^s labels $\{0, 1\}^s$ to each point in X and to think of R as a reconstruction algorithm that takes as input two such labels in $\{0, 1\}^s$ and outputs either 0 or 1, which stand for "small" or "large," respectively. The meaning of (1) becomes that for every pair $x, y \in X$, if one applies the reconstruction algorithm to the random labels $\mathsf{Sk}(x)$ and $\mathsf{Sk}(y)$, then with substantially high probability, its output is consistent with the value of the kernel $K(x, y)$ at scale r and approximation D, namely, the algorithm declares "small" if $K(x, y)$ is at most r, and it declares "large" if $K(x, y)$ is greater than Dr.

Suppose that $\alpha, \beta, \theta > 0$ and that $K : X\times X \to [0, \infty)$ and $L : Y\times Y \to [0, \infty)$ are nonnegative kernels on the sets X and Y, respectively. Suppose also that there is $f : Y \to X$ such that $\alpha L(x, y)^\theta \leqslant K(f(x), f(y)) \leqslant \beta L(x, y)^\theta$ for all $x, y \in Y$. It follows formally from this assumption and the above definition that if K is (s, D)-sketchable for some $s \in \mathbb{N}$ and $D \geqslant 1$, then L is $(s, (\beta D/\alpha)^{1/\theta})$-sketchable. Such an "embedding approach" to deduce sketchability is used frequently in the literature. As an example of its many consequences, since ℓ_2 is sketchable by the works of

Indyk and Motwani [21] and Kushilevitz, Ostrovsky, and Rabani [30], so is any metric space of negative type, where we recall that a metric space (X, d) is said to be of negative type (see, e.g., [16]) if the metric space (X, ρ) with $\rho = \sqrt{d}$ is isometric to a subset of ℓ_2.

1.1 *The Andoni–Krauthgamer–Razenshteyn Characterization of Sketchable Norms*

The following theorem from [4] is a remarkable result of Andoni, Krauthgamer, and Razenshteyn (AKR) that characterizes those norms that are sketchable[1] in terms of their geometric embeddability into a classical kernel (which is not a metric).

Theorem 1 (AKR characterization of sketchability) *Fix $s \in \mathbb{N}$ and $D \geqslant 1$. A finite-dimensional normed space $(X, \|\cdot\|_X)$ is (s, D)-sketchable if and only if for any $\varepsilon \in (0, 1)$, there exists $\alpha = \alpha(s, D, \varepsilon) > 0$ and an embedding $f : X \to \ell_{1-\varepsilon}$ such that*

$$\forall x, y \in X, \qquad \|x - y\|_X \leqslant \|f(x) - f(y)\|_{1-\varepsilon} \leqslant \alpha \|x - y\|_X.$$

Thus, a finite-dimensional normed space is sketchable if and only if it can be realized as a subset of a the classical sequence space $\ell_{1-\varepsilon}$ so that the kernel $\|\cdot\|_{1-\varepsilon}$ reproduces faithfully (namely, up to factor α) all the pairwise distances in X. See [4, Theorem 1.2] for an explicit dependence in Theorem 1 of $\alpha(s, D, \varepsilon)$ on the parameters s, D, ε.

L_p space notation In Theorem 1 and below, we use the following standard notation for L_p spaces. If $p \in (0, \infty)$ and (Ω, μ) is a measure space, then $L_p(\mu)$ is the set of (equivalence classes up to measure 0 of) measurable functions $\varphi : \Omega \to \mathbb{R}$ with $\int_\Omega |\varphi(\omega)|^p \, d\mu(\omega) < \infty$. When μ is the counting measure on $\mathbb{N}$, write $L_p(\mu) = \ell_p$. When μ is the counting measure on $\{1, \ldots, n\}$ for some $n \in \mathbb{N}$, write $L_p(\mu) = \ell_p^n$. When μ is the Lebesgue measure on $[0, 1]$, write $L_p(\mu) = L_p$. When the underlying measure is clear from the context (e.g., counting measure or Lebesgue measure), one sometimes writes $L_p(\mu) = L_p(\Omega)$. The $L_p(\mu)$ (quasi)norm is defined by setting $\|\varphi\|_p^p = \int_\Omega |\varphi(\omega)|^p \, d\mu(\omega)$ for $\varphi \in L_p(\mu)$. While if $p \geqslant 1$, then $(\varphi, \psi) \mapsto \|\varphi - \psi\|_p$ is a metric on $L_p(\mu)$, if $p = 1 - \varepsilon$ for some $\varepsilon \in (0, 1)$, then $\|\cdot\|_{1-\varepsilon}$ is not a metric; if $L_{1-\varepsilon}(\mu)$ is infinite dimensional, then $\|\cdot\|_{1-\varepsilon}$ is not even equivalent to a metric in the sense that there do not exist

[1] In [4], the conclusion of Theorem 1 is proven under a formally weaker assumption, namely, it uses a less stringent notion of sketchability which allows for the random sketches of the points $x, y \in X$ to be different from each other and for the reconstruction algorithm to depend on the underlying randomness that was used to produce those sketches. Since our main result, namely, Theorem 2, is an impossibility statement, it becomes only stronger if we use the simpler and stronger notion of sketchability that we stated above.

any $c, C \in (0, \infty)$ and a metric $d : L_{1-\varepsilon}(\mu) \times L_{1-\varepsilon}(\mu) \to [0, \infty)$ such that $cd(\varphi, \psi) \leqslant \|\varphi - \psi\|_{1-\varepsilon} \leqslant Cd(\varphi, \psi)$ for all $\varphi, \psi \in L_{1-\varepsilon}(\mu)$; see, e.g., [24]. But, $\|\cdot\|_{1-\varepsilon}$ is a nonnegative kernel on $L_{1-\varepsilon}(\mu)$, and there is a canonical metric $\mathfrak{d}_{1-\varepsilon}$ on $L_{1-\varepsilon}(\mu)$, which is given by

$$\forall \varphi, \psi \in L_{1-\varepsilon}(\mu), \qquad \mathfrak{d}_{1-\varepsilon}(\varphi, \psi) \stackrel{\text{def}}{=} \|\varphi-\psi\|_{1-\varepsilon}^{1-\varepsilon} = \int_{\Omega} |\varphi(\omega)-\psi(\omega)|^{1-\varepsilon}\, d\mu(\omega). \tag{2}$$

See the books [33, 34] and [24] for much more on the structure for $L_p(\mu)$ spaces when $p \geqslant 1$ and $0 < p < 1$, respectively.

1.1.1 Beyond Norms?

Fix $\varepsilon \in (0, 1)$. The sketchability of the nonnegative kernel on $\ell_{1-\varepsilon}$ that is given by $\|\varphi - \psi\|_{1-\varepsilon}$ for $\varphi, \psi \in \ell_{1-\varepsilon}$ was proved by Indyk [20] (formally, using the above terminology, it is sketchable provided ε is bounded away from 1; when $\varepsilon \to 1^+$ the space $s = s(\varepsilon)$ of Indyk's algorithm becomes unbounded); alternatively, one could combine the sketchability of ℓ_2 that was established in [21, 30] with the embedding of [10], through the above embedding approach (the proof in [20] is different, and it has further useful algorithmic features that we will not discuss here).

Thus, any metric space (M, d_M) for which there exist $\alpha \in [1, \infty)$ and an embedding $f : M \to \ell_{1-\varepsilon}$ that satisfies

$$\forall x, y \in M, \qquad d_M(x, y) \leqslant \|f(x) - f(y)\|_{1-\varepsilon} \leqslant \alpha d_M(x, y) \tag{3}$$

is sketchable with sketch size $O_\varepsilon(1)$ and approximation $O(\alpha)$. Therefore, the "if part" of Theorem 1 holds for any metric space whatsoever, not only for norms. The "only if" part of Theorem 1, namely, showing that the mere availability of a sketching algorithm for a normed space implies that it can be realized faithfully as a subset of $\ell_{1-\varepsilon}$, is the main result of [4]. This major achievement demonstrates that a fundamental algorithmic primitive *coincides* with a geometric/analytic property that has been studied long before sketchability was introduced (other phenomena of this nature were discovered in the literature, but they are rare). The underlying reason for Theorem 1 is deep, as the proof in [4] relies on a combination of major results from the literature on functional analysis and communication complexity.

A natural question that Theorem 1 leaves open is whether one could obtain the same result for $\varepsilon = 0$, namely, for embeddings into ℓ_1. As discussed in [4], this is equivalent to an old question [31] of Kwapień; a positive result in this direction (for a certain class of norms) is derived in [4] using classical partial progress of Kalton [23] on Kwapień's problem, but fully answering this long-standing question seems difficult (and it may very well have a negative answer).

Another natural question that Theorem 1 leaves open is whether its assumption that the underlying metric space is a norm is needed. Given that the "if part" of Theorem 1 holds for any metric space, this amounts to understanding whether a sketchable metric space (M, d_M) admits for every $\varepsilon \in (0, 1)$ an embedding $f : M \to \ell_{1-\varepsilon}$ that satisfies (3). This was a central open question of [4]. Theorem 2 resolves this question. It should be noted that the authors of [4] formulated their question while hinting that they suspect that the answer is negative, namely, in [4, page 893], they wrote: *we are not aware of any counter-example to the generalization of Theorem 1.2 to general metrics* (Theorem 1.2 in [4] corresponds to Theorem 1 here). One could therefore view Theorem 2 as a confirmation of a prediction of [4].

Theorem 2 (failure of the AKR characterization for general metrics) *For arbitrarily large $n \in \mathbb{N}$, there exists an n-point metric space $(M(n), d_{M(n)})$ which is $\big(O(1), O(1)\big)$-sketchable, yet for every $\varepsilon \in \big(0, \frac{1}{2}\big)$ and $\alpha \geqslant 1$, if there were a mapping $f : M(n) \to \ell_{1-\varepsilon}$ that satisfies $d_{M(n)}(x, y) \leqslant \|f(x) - f(y)\|_{1-\varepsilon} \leqslant \alpha d_{M(n)}(x, y)$ for all $x, y \in M(n)$, then necessarily*

$$\alpha \gtrsim (\log\log n)^{\frac{1-2\varepsilon}{2(1-\varepsilon)}}. \tag{4}$$

Asymptotic notation In addition to the usual "$O(\cdot), o(\cdot), \Omega(\cdot), \Theta(\cdot)$" notation, it will be convenient to use throughout this article (as we already did in (4)) the following (also standard) asymptotic notation. Given two quantities $Q, Q' > 0$, the notations $Q \lesssim Q'$ and $Q' \gtrsim Q$ mean that $Q \leqslant CQ'$ for some universal constant $C > 0$. The notation $Q \asymp Q'$ stands for $(Q \lesssim Q') \wedge (Q' \lesssim Q)$. If we need to allow for dependence on parameters, we indicate this by subscripts. For example, in the presence of auxiliary objects (e.g., numbers or spaces) $\phi, \mathfrak{Z}$, the notation $Q \lesssim_{\phi,\mathfrak{Z}} Q'$ means that $Q \leqslant C(\phi, \mathfrak{Z})Q'$, where $C(\phi, \mathfrak{Z}) > 0$ is allowed to depend only on $\phi, \mathfrak{Z}$; similarly for the notations $Q \gtrsim_{\phi,\mathfrak{Z}} Q'$ and $Q \asymp_{\phi,\mathfrak{Z}} Q'$.

We will see that the metric spaces $\{(M(n), d_{M(n)})\}_{n=1}^{\infty}$ of Theorem 2 are of negative type, so by the above discussion, their sketchability follows from the sketchability of Hilbert space [21, 30]. In fact, these metric spaces are (subsets of) the metric spaces of negative type that were considered by Devanur, Khot, Saket, and Vishnoi in [15] as integrality gap examples for the Goemans–Linial semidefinite relaxation of the Sparsest Cut problem with uniform demands. Hence, our contribution is the geometric aspect of Theorem 2, namely, demonstrating the non-embeddability into $\ell_{1-\varepsilon}$, rather than its algorithmic component (sketchability). This is a special case of the more general geometric phenomenon of Theorem 7, which is our main result. It amounts to strengthening our work [26] which investigated the ℓ_1 non-embeddability of quotients of metric spaces using Fourier-analytic techniques. Here, we derive the (formally stronger) non-embeddability into ℓ_1 of snowflakes of such quotients (the relevant terminology is recalled in Section 1.2). It suffices to mention at this juncture (with further discussion in Section 1.2.4 below) that on a conceptual level, the strategy of [26] (as well as that of [15, 29]) for proving non-embeddability using the classical theorem [22] of Kahn, Kalai, and Linial (KKL) on influences of variables does not imply the required ℓ_1 non-embeddability of snowflakes of quotients. Instead, we revisit the use of Bourgain's noise sensitivity theorem [8], which was applied for other (non-embeddability) purposes in [26, 27], but subsequent work [15, 29] realized that one could use the much simpler KKL theorem in those contexts (even yielding quantitative improvements). Thus, prior to the present work, it seemed that, after all, Bourgain's theorem does not have a decisive use in metric embedding theory, but here we see that in fact it has a qualitative advantage over the KKL theorem in some geometric applications.

The present work also shows that the Khot–Vishnoi approach [27] to the Sparsest Cut integrality gap has a further qualitative advantage (beyond its relevance to the case of uniform demands) over the use of the Heisenberg group for this purpose [32], which yields a better [13] (essentially sharp [39]) lower bound. Indeed, the Heisenberg group is a $O(1)$-doubling metric space (see, e.g., [18]), and by Assouad's embedding theorem [6], any such space admits for any $\varepsilon \in (0,1)$ an embedding into $\ell_{1-\varepsilon}$ which satisfies (3) with $\alpha \lesssim_\varepsilon 1$ (for the connection to Assouad's theorem, which may not be apparent at this point, see Fact 6 below). Thus, despite its quantitative superiority as an integrality gap example for Sparsest Cut with general demands, the Heisenberg group cannot yield Theorem 2, while the Khot–Vishnoi spaces do (strictly speaking, we work here with a simpler different construction than that of [27], but an inspection of the ensuing proof reveals that one could have also used the metric spaces of [27] to answer the question of [4]).

Question 3 The obvious question that is left open by Theorem 2 is to understand what happens when $\varepsilon \in \left[\frac{1}{2}, 1\right)$. While we established a marked qualitative gap vis-à-vis sketchability between the behaviors of general normed spaces and general metric spaces, the possibility remains that there exists some $\varepsilon_0 \in \left[\frac{1}{2}, 1\right)$ such that any sketchable metric space (M, d_M) admits an embedding into $\ell_{1-\varepsilon_0}$ that satisfies (3) with $\alpha = O(1)$; perhaps one could even take $\varepsilon_0 = \frac{1}{2}$ here. This possibility is of course tantalizing, as it would be a complete characterization of sketchable metric spaces that is nevertheless qualitatively different from its counterpart for general normed spaces. At present, there is insufficient evidence to speculate that this is so, and it seems more likely that other counterexamples could yield a statement that is analogous to Theorem 2 also in the range $\varepsilon \in \left[\frac{1}{2}, 1\right)$, though a new idea would be needed for that.

Question 4 Even in the range $\varepsilon \in \left(0, \frac{1}{2}\right)$ of Theorem 2, it would be interesting to determine if one could improve (4) to $\alpha \gtrsim (\log n)^{c(\varepsilon)}$ for some $c(\varepsilon) > 0$ (see Remark 5 for a technical enhancement that yields an asymptotic improvement of (4) but does not achieve such a bound). For the corresponding question when $\varepsilon = 0$, namely, embeddings into ℓ_1, it follows from [39] that one could improve (4) to $\alpha \gtrsim \sqrt{\log n}$. However, the example that exhibits this stronger lower bound for $\varepsilon = 0$ is a doubling metric space, and hence by Assouad's theorem [6] for every $\varepsilon > 0$, it does admit an embedding into $\ell_{1-\varepsilon}$ that satisfies (4) with $\alpha \lesssim_\varepsilon 1$. Note that by [5, 38] we see that if an n-point metric space (M, d_M) is sketchable for the reason that for some $\theta \in (0,1]$ the metric space (M, d_M^θ) is bi-Lipschitz to a subset of ℓ_2, then (3) holds for $\varepsilon = 0$ and $\alpha \lesssim (\log n)^{\frac{1}{2}+o(1)}$. It would be worthwhile to determine if this upper bound on α (for $\varepsilon = 0$) holds for any sketchable metric space whatsoever, i.e., not only for those whose sketchability is due to the fact that some power of the metric is Hilbertian. It seems plausible that the latter question is accessible using available methods.

Remark 5 The lower bound (4) can be improved by incorporating the "enhanced short code argument" of Kane and Meka [25] (which is in essence a derandomization step) into the ensuing reasoning. This yields a more complicated construction for which (4) can be improved to $\alpha \geqslant \exp\big(c(1-2\varepsilon)\sqrt{\log\log n}\big)$ for some universal constant $c > 0$. Because it becomes a significantly more intricate case-specific argument that does not pertain to the more general geometric phenomenon that we study in Theorem 7, the details of this quantitative enhancement of Theorem 2 are omitted.

1.2 Metric Embeddings

The distortion of a metric space (U, d_U) in a metric space (V, d_V) is a numerical invariant that is denoted $\mathsf{c}_{(V,d_V)}(U, d_U)$ and defined to be the infimum over those $\alpha \in [1, \infty]$ for which there exist an embedding $f : U \to V$ and a scaling factor $\lambda \in (0, \infty)$ such that $\lambda d_U(x, y) \leqslant d_V\big(f(x), f(y)\big) \leqslant \alpha\lambda d_U(x, y)$ for all distinct $x, y \in U$. Given $p \geqslant 1$, the infimum of $\mathsf{c}_{(V,d_V)}(U, d_U)$ over all possible[2] $L_p(\mu)$ spaces (V, d_V) is denoted $\mathsf{c}_p(U, d_U)$.

1.2.1 Snowflakes

Because for every $\varepsilon \in (0, 1)$ the quasi-norm $\|\cdot\|_{1-\varepsilon}$ does not induce a metric on $\ell_{1-\varepsilon}$, the embedding requirement (3) does not fit into the above standard metric embedding framework. However, as we explain in Fact 6, it is possible to situate (3) within this framework (even without mentioning $\ell_{1-\varepsilon}$ at all) by considering embeddings of the $(1-\varepsilon)$-snowflake of a finite metric space into ℓ_1. Recall the commonly used terminology (see, e.g., [14]) that the $(1-\varepsilon)$-snowflake of a metric space (M, d_M) is the metric space $(M, d_M^{1-\varepsilon})$.

[2]When (U, d_U) is a finite metric space, it suffices to consider embeddings into ℓ_p rather than a general $L_p(\mu)$ space, as follows via a straightforward approximation by simple functions. We warn that this is not so for general (infinite) separable metric spaces, in which case one must consider embeddings into L_p; by [12, Corollary 1.5] there is even a doubling subset of L_1 that does not admit a bi-Lipschitz embedding into ℓ_1.

Fact 6 *Let (M, d_M) be a finite*[3] *metric space and fix $\varepsilon \in (0,1)$. The quantity $\mathsf{c}_1(M, d_M^{1-\varepsilon})^{\frac{1}{1-\varepsilon}}$ is equal to the infimum over those $\alpha \geqslant 1$ for which there exists an embedding $f : M \to \ell_{1-\varepsilon}$ that satisfies* (3).

Proof Suppose that $f : M \to \ell_{1-\varepsilon}$ satisfies (3). Then, recalling the notation (2) for the metric $\eth_{1-\varepsilon}$ on $\ell_{1-\varepsilon}$, we have $d_M(x,y)^{1-\varepsilon} \leqslant \eth_{1-\varepsilon}(f(x), f(y)) \leqslant \alpha^{1-\varepsilon} d_M(x,y)^{1-\varepsilon}$ for all $x, y \in M$. It follows from general principles [10, 43] that the metric space $(\ell_{1-\varepsilon}, \eth_{1-\varepsilon})$ admits an isometric embedding into an $L_1(\mu)$ space (an explicit formula for such an embedding into $L_1(\mathbb{R}^2)$ can be found in [35, Remark 5.10]). Hence, $\mathsf{c}_1(M, d_M^{1-\varepsilon}) \leqslant \alpha^{1-\varepsilon}$. Conversely, there is an explicit embedding (see equation (2) in [36]) $T : \ell_1 \to L_{1-\varepsilon}(\mathbb{N} \times \mathbb{R})$ which is an isometry when one takes the metric $\eth_{1-\varepsilon}$ on $L_{1-\varepsilon}(\mathbb{N} \times \mathbb{R})$. Hence, if $\beta > \mathsf{c}_1(M, d_M^{1-\varepsilon})$, then take an embedding $g : M \to \ell_1$ such that $d_M(x,y)^{1-\varepsilon} \leqslant \|g(x) - g(y)\|_1 \leqslant \beta d_M(x,y)^{1-\varepsilon}$ for all $x, y \in M$ and consider the embedding $T \circ g$ which satisfies (3) with $\alpha = \beta^{1/(1-\varepsilon)}$, except that the target space is $L_{1-\varepsilon}(\mathbb{N} \times \mathbb{R})$ rather than $\ell_{1-\varepsilon}$. By an approximation by simple functions, we obtain the desired embedding into $\ell_{1-\varepsilon}$. □

Standard examples of metric spaces (M, d_M) such that $\mathsf{c}_1(M, d_M^{1-\varepsilon})$ is large for any $\varepsilon \in (0,1)$ are $O(1)$-expanders [19], namely, M is the vertex set of a large $O(1)$-regular graph such that the second-largest eigenvalue λ_2 of its adjacency matrix satisfies $1/(1-\lambda_2) = O(1)$, and d_M is the associated shortest-path metric; see [17] for this snowflake non-embeddability statement (even coarse non-embeddability) and [37, Lemma 48] for the best-known distortion bound here. However, it is known that no expander is $O(1)$-sketchable, due to the unpublished manuscript [3], so this natural route (in light of Fact 6) toward a potential resolution of the aforementioned question of [4] cannot work.

1.2.2 Quotients

Suppose that G is a group that acts on a metric space (X, d_X) by isometries. The quotient space $X/G = \{Gx\}_{x \in X}$ of all the orbits of G can be equipped with the following quotient metric $d_{X/G} : (X/G) \times (X/G) \to [0, \infty)$:

$$\forall\, x, y \in X, \qquad d_{X/G}(Gx, Gy) \stackrel{\mathrm{def}}{=} \inf_{(u,v) \in (Gx) \times (Gy)} d_X(u,v) = \inf_{g \in G} d_X(gx, y). \tag{5}$$

See [11, Section 5.19] for more on this basic construction (in particular, for a verification that (5) indeed gives a metric).

[3] The only reason for the finiteness assumption here (the present article deals only with finite metric space) is to ensure that the embedding is into $\ell_{1-\varepsilon}$ rather than a more general $L_{1-\varepsilon}(\mu)$ space. For embeddings of finite-dimensional normed spaces, i.e., the setting of [4], a similar reduction to embeddings into $\ell_{1-\varepsilon}$ is possible using tools from [1, 7, 40].

Given $k \in \mathbb{N}$, we will consider the Hamming cube to be the vector space $\mathbb{F}_2^k$ over the field of two elements $\mathbb{F}_2$, equipped with the Hamming metric $d_{\mathbb{F}_2^k} : \mathbb{F}_2^k \times \mathbb{F}_2^k \to \mathbb{N} \cup \{0\}$ that is given by

$$\forall\, x=(x_1,\ldots,x_k),\ y=(y_1,\ldots,y_k)\in\mathbb{F}_2^k,\quad d_{\mathbb{F}_2^k}(x,y)=|\{j \in \{1,\ldots,k\} :\ x_j \neq y_j\}|.$$

Below, $\mathbb{F}_2^k$ will always be assumed to be equipped with the metric $d_{\mathbb{F}_2^k}$. The standard basis of $\mathbb{F}_2^k$ is denoted $e_1,\ldots,e_k$.

If G is a group acting on $\mathbb{F}_2^k$ by isometries, and if it isn't too large, say, $|G| \leqslant 2^{k/2}$, then all but an exponentially small fraction of the pairs $(x,y) \in \mathbb{F}_2^k \times \mathbb{F}_2^k$ satisfy $d_{\mathbb{F}_2^k}(Gx, Gy) \gtrsim k$. Specifically, there is a universal constant $\eta > 0$ such that

$$|G| \leqslant 2^{\frac{k}{2}} \implies \left|\left\{(x,y) \in \mathbb{F}_2^k \times \mathbb{F}_2^k :\ d_{\mathbb{F}_2^k/G}(x,y) \leqslant \eta k\right\}\right| \leqslant 2^{\frac{5}{3}k}. \qquad (6)$$

A simple counting argument which verifies (6) appears in the proof of [26, Lemma 3.2].

The symmetric group S_k acts isometrically on $\mathbb{F}_2^k$ by permuting the coordinates, namely, for each permutation g of $\{1,\ldots,k\}$ and $x \in \mathbb{F}_2^k$, we write $gx = (x_{g^{-1}(1)}, x_{g^{-1}(2)},\ldots,x_{g^{-1}(k)})$. A subgroup $G \leqslant S_k$ of S_k therefore acts by isometries on $\mathbb{F}_2^k$; below we will only consider quotients of the form $(\mathbb{F}_2^k/G, d_{\mathbb{F}_2^K/G})$ when G is a transitive subgroup of S_k (namely, for any $i, j \in \{1,\ldots,k\}$, there exists a permutation $g \in G$ such that $g(i) = j$).

1.2.3 ℓ_1 Non-embeddability of Snowflakes of (Subsets of) Hypercube Quotients

In [26] we studied the ℓ_1 embeddability of quotients of $\mathbb{F}_2^k$. In particular, [26, Corollary 3] states that if G is a transitive subgroup of S_k with $|G| \leqslant 2^{k/2}$, then

$$c_1\left(\mathbb{F}_2^k/G, d_{\mathbb{F}_2^k/G}\right) \gtrsim \log k. \qquad (7)$$

In Remark 4 of [26], we (implicitly) asked about the sketchability of $\mathbb{F}_2^k/G$, by inquiring whether its $(1/2)$-snowflake embeds into a Hilbert space with $O(1)$ distortion, as a possible alternative approach for obtaining integrality gaps (quantitatively stronger than what was known at the time) for the Goemans–Linial semidefinite relaxation of the Sparsest Cut problem. This hope was realized in [15] for the special case when $G = \langle \mathfrak{S}_k \rangle \leqslant S_k$ is the cyclic group that is generated by the cyclic shift $\mathfrak{S}_k = (1, 2, \ldots, k) \in S_k$. Specifically, it follows from [15] that there exists a large subset $M \subseteq \mathbb{F}_2^k$, namely, $|\mathbb{F}_2^k \setminus M| \lesssim 2^k/k^2$, and a metric ρ on $M/\langle \mathfrak{S}_k \rangle$ satisfying $\rho(\mathcal{O}, \mathcal{O}') \asymp d_{\mathbb{F}_2^k/\langle \mathfrak{S}_k \rangle}(\mathcal{O}, \mathcal{O}')$ for all pairs of orbits $\mathcal{O}, \mathcal{O}' \in M/\langle \mathfrak{S}_k \rangle$, and such that the metric space $(M/\langle \mathfrak{S}_k \rangle, \sqrt{\rho})$ embeds isometrically into ℓ_2. Strictly speaking, a stronger statement than this was obtained in [15] for a larger metric space (namely, for the quotient of $\mathbb{F}_2^k \times \mathbb{F}_2^k$ by the group $\langle \mathfrak{S}_k \rangle \times \langle \mathfrak{S}_k \rangle$), but here it suffices to consider the above smaller metric space which inherits the stated properties.

Recalling Fact 6, this discussion leads naturally, as a strategy toward proving Theorem 2, to investigating whether a lower bound as (7) holds for the $(1-\varepsilon)$-snowflake of the hypercube quotient $\mathbb{F}_2^k/G$ rather than that quotient itself. We will see that the method of [26] does not yield any such lower bound that tends to ∞ as $k \to \infty$ for fixed $\varepsilon > 0$, but we do obtain the desired statement here, albeit with an asymptotically weaker lower bound than the $\log k$ of (7). Note that an application of Theorem 7 to the above subset $M \subseteq \mathbb{F}_2^k$ from [15] yields Theorem 2, because of Fact 6.

Theorem 7 (non-embeddability of snowflakes of quotients of large subsets of the hypercube) *Fix $k \in \mathbb{N}$ and $\varepsilon \in (0, \frac{1}{2})$. Let G be a transitive subgroup of S_k with $|G| \leqslant 2^{k/2}$. Then, every $M \subseteq \mathbb{F}_2^k$ with $|\mathbb{F}_2^k \setminus M| \leqslant 2^k/\sqrt{\log k}$ satisfies*

$$\mathsf{c}_1\Big(M/G, d_{\mathbb{F}_2^k/G}^{1-\varepsilon}\Big) \gtrsim (\log k)^{\frac{1}{2}-\varepsilon}. \tag{8}$$

It would be interesting to determine the asymptotically sharp behavior (up to universal constant factors) in (8) for $M = \mathbb{F}_2^k$, though understanding the dependence on the transitive subgroup $G \leqslant S_k$ may be challenging; see [9] for investigations along these lines. Even in the special case $G = \langle \mathfrak{S}_k \rangle$, we do not know the sharp bound and in particular how it transitions from the $(\log k)^{1/2-\varepsilon}$ of (8) to the $\log k$ of (7) as $\varepsilon \to 0$ (it could be that neither bound is tight).

1.2.4 Bourgain's Fourier Tails Versus the Kahn–Kalai–Linial Influence of Variables

In [26, Theorem 3.8], we applied the important theorem [22] of Kahn, Kalai, and Linial on the influence of variables on Boolean functions to show that if G is a transitive subgroup of S_k, then every $f : \mathbb{F}_2^k/G \to \ell_1$ satisfies the following Cheeger/Poincaré inequality:

$$\frac{1}{4^k} \sum_{(x,y)\in \mathbb{F}_2^k\times\mathbb{F}_2^k} \big\|f(Gx)-f(Gy)\big\|_1 \lesssim \frac{1}{\log k}\sum_{j=1}^{k}\frac{1}{2^k}\sum_{x\in\mathbb{F}_2^k}\big\|f\big(G(x+e_j)\big)-f(Gx)\big\|_1. \tag{9}$$

Fix $(\varepsilon, \alpha) \in (0, 1) \times [1, \infty)$. If $|G| \leqslant 2^{k/2}$ and $d_{\mathbb{F}_2^k/G}(Gx, Gy)^{1-\varepsilon} \leqslant \|f(Gx) - f(Gy)\|_1 \leqslant \alpha d_{\mathbb{F}_2^k/G}(Gx, Gy)^{1-\varepsilon}$ for $x, y \in \mathbb{F}_2^k$, then

$$
\begin{aligned}
k^{1-\varepsilon} &\overset{(6)}{\lesssim} \frac{1}{4^k} \sum_{(x,y)\in\mathbb{F}_2^k\times\mathbb{F}_2^k} d_{\mathbb{F}_2^k/G}(Gx,Gy)^{1-\varepsilon} \leqslant \frac{1}{4^k} \sum_{(x,y)\in\mathbb{F}_2^k\times\mathbb{F}_2^k} \|f(Gx)-f(Gy)\|_1\\
&\overset{(9)}{\lesssim} \frac{1}{\log k}\sum_{j=1}^k \frac{1}{2^k}\sum_{x\in\mathbb{F}_2^k} \big\|f\big(G(x+e_j)\big)-f(Gx)\big\|_1\\
&\leqslant \frac{\alpha}{\log k}\sum_{j=1}^k \frac{1}{2^k}\sum_{x\in\mathbb{F}_2^k} d_{\mathbb{F}_2^k/G}\big(G(x+e_j),Gx\big)^{1-\varepsilon}\\
&\overset{(5)}{\leqslant} \frac{\alpha}{\log k}\sum_{j=1}^k \frac{1}{2^k}\sum_{x\in\mathbb{F}_2^k} d_{\mathbb{F}_2^k}(x+e_j,x)^{1-\varepsilon} = \frac{\alpha k}{\log k}.
\end{aligned}
$$

It follows that

$$
\mathsf{c}_1\Big(\mathbb{F}_2^k/G, d_{\mathbb{F}_2^k/G}^{1-\varepsilon}\Big) \gtrsim \frac{\log k}{k^{\varepsilon}}. \tag{10}
$$

This is how (7) was derived in [26], but the right-hand side of (10) tends to ∞ as $k\to\infty$ only if $\varepsilon = o((\log\log k)/\log k)$.

Following the above use of the KKL theorem [26], it was used elsewhere in place of applications [26, 27] of a more substantial theorem of Bourgain [8] on the Fourier tails of Boolean functions that are not close to juntas; notably this was first done by Krauthgamer and Rabani [29] to obtain an asymptotically improved analysis of the Khot–Vishnoi integrality gap [27] for Sparsest Cut. We have seen above that the KKL-based approach does not yield Theorem 7 (though, of course, one cannot rule out the availability of a more sophisticated application of KKL that does), but our use of Bourgain's theorem in the ensuing proof of Theorem 7 shows that this theorem does sometime provide qualitatively stronger geometric information. One should note here that (8) follows from an application of a sharp form of Bourgain's theorem that was more recently obtained by Kindler, Kirshner, and O'Donnell [28]; an application of Bourgain's original formulation yields a bound that is asymptotically weaker by a lower-order factor.

2 Proof of Theorem 7

Here we will prove Theorem 7, thereby completing the justification of Theorem 2 as well.

2.1 Fourier-Analytic Preliminaries

We will include here some basic facts and notation related to Fourier analysis on the hypercube $\mathbb{F}_2^k$; an extensive treatment of this topic can be found in, e.g., the monograph [41]. Fix $k \in \mathbb{N}$. From now on, let $\mu = \mu_k$ denote the normalized counting measure on $\mathbb{F}_2^k$. Given $A \subseteq \{1, \ldots, k\}$, the Walsh function $\mathsf{W}_A : \mathbb{F}_2^k \to \{-1, 1\}$ and Fourier coefficient $\widehat{\varphi}(A) \in \mathbb{R}$ of a function $\varphi : \mathbb{F}_2^k \to \mathbb{R}$ are defined by

$$\forall x \in \mathbb{F}_2^k, \qquad \mathsf{W}_A(x) = (-1)^{\sum_{j=1}^n x_j} \qquad \text{and} \qquad \widehat{\varphi}(A) = \int_{\mathbb{F}_2^k} \varphi(x)\mathsf{W}_A(x)\,\mathrm{d}\mu(x).$$

The convolution $\varphi * \psi : \mathbb{F}_2^k \to \mathbb{R}$ of two functions $\varphi, \psi : \mathbb{F}_2^k \to \mathbb{R}$ is defined by

$$\forall x \in \mathbb{F}_2^k, \qquad (\varphi*\psi)(x) = \int_{\mathbb{F}_2^k} \varphi(y)\psi(x+y)\,\mathrm{d}\mu(y) = \sum_{A\subseteq\{1,\ldots,k\}} \widehat{\varphi}(A)\widehat{\psi}(A)\mathsf{W}_A(x),$$

where the last equality is valid because the 2^k Walsh functions $\{\mathsf{W}_A\}_{A\subseteq\{1,\ldots,k\}}$ consist of all of the characters of the additive group $\mathbb{F}_2^k$, hence forming an orthonormal basis of $L_2(\mu)$. Suppose that $g \in \mathsf{GL}(\mathbb{F}_2^k)$ is an automorphism of $\mathbb{F}_2^k$. If $\varphi : \mathbb{F}_2^k \to \mathbb{R}$ is a g-invariant function, i.e., $\varphi(gy) = \varphi(y)$ for all $y \in \mathbb{F}_2^k$, then for every $\psi : \mathbb{F}_2^k \to \mathbb{R}$ and $x \in \mathbb{F}_2^k$,

$$(\varphi * \psi)(x) = \int_{\mathbb{F}_2^k} \varphi(y)\psi(x+y)\,\mathrm{d}\mu(y) = \int_{\mathbb{F}_2^k} \varphi(gy)\psi(x+y)\,\mathrm{d}\mu(y)$$
$$= \int_{\mathbb{F}_2^k} \varphi(z)\psi\big(x+g^{-1}z\big)\,\mathrm{d}\mu(z) = \int_{\mathbb{F}_2^k} \varphi(z)\psi\big(g^{-1}(gx+z)\big)\,\mathrm{d}\mu(z) = s\Big(\varphi*\big(\psi\circ g^{-1}\big)\Big)(gx).$$

In particular, under the above invariance assumption, we have the identity

$$\|\varphi * \psi\|_{L_2(\mu)} = \left\|\varphi * \left(\psi \circ g^{-1}\right)\right\|_{L_2(\mu)}. \tag{11}$$

Given $\mathfrak{p} \in [0, 1]$, let $\vartheta^{\mathfrak{p}} : 2^{\mathbb{F}_2^k \times \mathbb{F}_2^k} \to [0, 1]$ be the probability measure that is defined by setting for each $(x, y) \in \mathbb{F}_2^k \times \mathbb{F}_2^k$,

$$\vartheta^{\mathfrak{p}}(x, y) \stackrel{\mathrm{def}}{=} \frac{\mathfrak{p}^{d_{\mathbb{F}_2^k}(x,y)}(1-\mathfrak{p})^{k-d_{\mathbb{F}_2^k}(x,y)}}{2^k} = \frac{1}{4^k}\prod_{j=1}^k \left(1 + (1-2\mathfrak{p})(-1)^{x_j+y_j}\right)$$
$$= \frac{1}{4^k} \sum_{A\subseteq\{1,\ldots,k\}} (1-2\mathfrak{p})^{|A|}\mathsf{W}_A(x+y). \tag{12}$$

In other words, $\vartheta^{\mathfrak{p}}(x, y)$ is equal to the probability that the ordered pair (x, y) is the outcome of the following randomized selection procedure: The first element $x \in \mathbb{F}_2^k$ is chosen uniformly at random, and the second element $y \in \mathbb{F}_2^k$ is obtained by changing the sign of each entry of x independently with probability $\mathfrak{p}$. Note in passing that both marginals of $\vartheta^{\mathfrak{p}}$ are equal to μ, i.e., $\vartheta^{\mathfrak{p}}(\Omega \times \mathbb{F}_2^k) = \vartheta^{\mathfrak{p}}(\mathbb{F}_2^k \times \Omega) = \mu(\Omega)$ for every $\Omega \subseteq \mathbb{F}_2^k$. Also, for every $\Omega \subseteq \mathbb{F}_2^k$, we have

$$
\begin{aligned}
\vartheta^{\mathfrak{p}}\big(\Omega \times (\mathbb{F}_2^k \smallsetminus \Omega)\big) &= \frac{1}{8}\int_{\mathbb{F}_2^k \times \mathbb{F}_2^k} \Big((-1)^{\mathbf{1}_\Omega(x)} - (-1)^{\mathbf{1}_\Omega(y)}\Big)^2 \, \mathrm{d}\vartheta^{\mathfrak{p}}(x, y) \\
&= \frac{1}{4}\bigg(1 - \int_{\mathbb{F}_2^k \times \mathbb{F}_2^k} (-1)^{\mathbf{1}_\Omega}(x)(-1)^{\mathbf{1}_\Omega}(y) \, \mathrm{d}\vartheta^{\mathfrak{p}}(x, y)\bigg) \qquad (13)\\
&= \frac{1}{4}\sum_{A \subseteq \{1,\ldots,k\}} \Big(1 - (1 - 2\mathfrak{p})^{|A|}\Big)\Big(\widehat{(-1)^{\mathbf{1}_\Omega}}(A)\Big)^2,
\end{aligned}
$$

where the last equality in (13) is a direct consequence of Parseval's identity and the final expression in (12) for $\vartheta^{\mathfrak{p}}(\cdot, \cdot)$.

For $\varphi : \mathbb{F}_2^k \to \mathbb{R}$ and $j, m \in \{1, \ldots, k\}$, the level-$m$ influence of the jth variable on φ, denoted $\mathsf{Inf}_j^{\leqslant m}[\varphi]$, is the quantity

$$
\mathsf{Inf}_j^{\leqslant m}[\varphi] = \sum_{\substack{A \subseteq \{1,\ldots,k\} \smallsetminus \{j\} \\ |A| \leqslant m-1}} \widehat{\varphi}(A \cup \{j\})^2 = \Big\| \varphi * \mathcal{R}_j^{\leqslant m} \Big\|_{L_2(\mu)}^2, \qquad (14)
$$

where the last equality is a consequence of Parseval's identity, using the notation

$$
\mathcal{R}_j^{\leqslant m} \stackrel{\mathrm{def}}{=} \sum_{\substack{A \subseteq \{1,\ldots,k\} \smallsetminus \{j\} \\ |A| \leqslant m-1}} \mathsf{W}_{A \cup \{j\}}. \qquad (15)
$$

It follows from the first equation in (14) that

$$
\begin{aligned}
\sum_{j=1}^{k} \mathsf{Inf}_j^{\leqslant m}[\varphi] &= \sum_{\substack{B \subseteq \{1,\ldots,k\} \\ |B| \leqslant m}} |B| \widehat{\varphi}(B)^2 \leqslant m \sum_{\substack{B \subseteq \{1,\ldots,k\} \\ B \neq \varnothing}} \widehat{\varphi}(B)^2 \\
&= m\bigg(\int_{\mathbb{F}_2^k} \varphi^2 \, \mathrm{d}\mu - \widehat{\varphi}(\varnothing)^2\bigg) = m\mathsf{Var}_\mu[\varphi], \qquad (16)
\end{aligned}
$$

where $\mathsf{Var}_\mu[\cdot]$ denotes the variance with respect to the probability measure μ. By considering the symmetric group S_k as a subgroup of $\mathsf{GL}(\mathbb{F}_2^k)$, where the action is permutation of coordinates, an inspection of definition (15) reveals that $\mathcal{R}_j^{\leqslant m} \circ g =$

$\mathcal{R}_{gj}^{\leqslant m}$ for $g \in S_k$ and $j, m \in \{1, \dots, k\}$. By (11) and the second equality in (14), if $\varphi : \mathbb{F}_2^k \to \mathbb{R}$ is g-invariant, then

$$\forall\, j, m \in \{1, \dots, k\}, \qquad \mathsf{Inf}_j^{\leqslant m}[\varphi] = \mathsf{Inf}_{g^{-1}j}^{\leqslant m}[\varphi].$$

A combination of this observation with (16) yields the following statement, which we record for ease of later reference:

Fact 8 *Fix $k \in \mathbb{N}$. Let G be a subgroup of S_k that acts transitively on the coordinates $\{1, \dots, k\}$. Suppose that $\varphi : \mathbb{F}_2^k \to \mathbb{R}$ is a G-invariant function, i.e., $f(gx) = f(x)$ for every $g \in G$ and $x \in \mathbb{F}_2^n$. Then, for every $m \in \{1, \dots, k\}$, we have*

$$\max_{j \in \{1,\dots,k\}} \mathsf{Inf}_j^{\leqslant m}[\varphi] \leqslant \frac{m}{k} \mathsf{Var}_\mu[\varphi].$$

Throughout what follows, given a subgroup $G \leqslant S_k$, we denote by $\pi_G : \mathbb{F}_2^k \to \mathbb{F}_2^k/G$ its associated quotient mapping, i.e., $\pi_G(x) = Gx$ for all $x \in \mathbb{F}_2^k$. We denote by $\mu_{\mathbb{F}_2^k/G} = \mu \circ \pi_G^{-1}$ the probability measure on $\mathbb{F}_2^k/G$ that is given by

$$\forall\, \mathcal{O} \in \mathbb{F}_2^k/G, \qquad \mu_{\mathbb{F}_2^k/G}(\mathcal{O}) = \mu(\mathcal{O}).$$

In a similar vein, for every $\mathfrak{p} \in [0, 1]$, the probability measure $\vartheta^{\mathfrak{p}}$ on $\mathbb{F}_2^k \times \mathbb{F}_2^k$ that is given in (12) descends to a probability measure $\vartheta^{\mathfrak{p}}_{\mathbb{F}_2^k/G} = \vartheta^{\mathfrak{p}} \circ (\pi_G \times \pi_G)^{-1}$ on $(\mathbb{F}_2^k/G) \times (\mathbb{F}_2^k/G)$ by setting

$$\forall\, \mathcal{O}, \mathcal{O}' \subseteq \mathbb{F}_2^k/G, \qquad \vartheta^{\mathfrak{p}}_{\mathbb{F}_2^k/G}(\mathcal{O}, \mathcal{O}') = \vartheta^{\mathfrak{p}}(\mathcal{O} \times \mathcal{O}').$$

2.2 A Cheeger/Poincaré Inequality for Transitive Quotients

Our main technical result is the following inequality.

Lemma 9 *There is a universal constant $\beta \in (0, 1)$ with the following property. Fix an integer $k \geqslant 55$ and a transitive subgroup G of S_k. Suppose that $X \subseteq \mathbb{F}_2^k/G$ is a sufficiently large subset in the following sense:*

$$\mu_{\mathbb{F}_2^k/G}(X) \geqslant 1 - \frac{1}{\sqrt{\log k}}. \tag{17}$$

Then there is a further subset $Y \subseteq X$ with $\mu_{\mathbb{F}_2^k/G}(Y) \geqslant \frac{3}{4}\mu_{\mathbb{F}_2^k/G}(X)$ such that every function $f : Y \to \ell_1$ satisfies

$$\iint_{Y\times Y} \|f(\mathcal{O})-f(\mathcal{O}')\|_1 \,\mathrm{d}\mu_{\mathbb{F}_2^k/G}(\mathcal{O})\,\mathrm{d}\mu_{\mathbb{F}_2^k/G}(\mathcal{O}')$$
$$\lesssim \sqrt{\log k}\iint_{Y\times Y} \|f(\mathcal{O})-f(\mathcal{O}')\|_1 \,\mathrm{d}\vartheta_{\mathbb{F}_2^n/G}^{\frac{1}{\beta\log k}}(\mathcal{O},\mathcal{O}'). \tag{18}$$

Prior to proving Lemma 9, we shall assume its validity for the moment and proceed to prove Theorem 7.

Proof of Theorem 7 assuming Lemma 9 Fix $\alpha \geqslant 1$ and suppose that $f: M/G \to \ell_1$ satisfies

$$\forall x,y\in M,\qquad d_{\mathbb{F}_2^k/G}(Gx,Gy)^{1-\varepsilon} \leqslant \|f(Gx)-f(Gy)\|_1 \leqslant \alpha d_{\mathbb{F}_2^k/G}(Gx,Gy)^{1-\varepsilon}. \tag{19}$$

Our task is to bound α from below by the right-hand side of (8).

An application of Lemma 9 to $X = M/G$, which satisfies the requirement (17) by the assumption of Theorem 7, produces a subset Y with $\mu(\pi_G^{-1}(Y)) \geqslant \frac{1}{2}$ for which (18) holds true. It follows that

$$\iint_{\pi_G^{-1}(Y)\times\pi_G^{-1}(Y)} d_{\mathbb{F}_2^k/G}(Gx,Gy)^{1-\varepsilon}\,\mathrm{d}\mu(x)\,\mathrm{d}\mu(y)$$
$$\overset{(18)\wedge(19)}{\lesssim} \alpha\sqrt{\log k}\iint_{(\mathbb{F}_2^k/G)\times(\mathbb{F}_2^k/G)} d_{\mathbb{F}_2^k/G}(\mathcal{O},\mathcal{O}')^{1-\varepsilon}\,\mathrm{d}\vartheta_{\mathbb{F}_2^n/G}^{\frac{1}{\beta\log k}}(\mathcal{O},\mathcal{O}')$$
$$\overset{(5)}{\leqslant} \alpha\sqrt{\log k}\int_{\mathbb{F}_2^k\times\mathbb{F}_2^k} d_{\mathbb{F}_2^k}(x,y)^{1-\varepsilon}\,\mathrm{d}\vartheta^{\frac{1}{\beta\log k}}(x,y)$$
$$\overset{(12)}{=} \alpha\sqrt{\log k}\sum_{\ell=0}^{k}\ell^{1-\varepsilon}\binom{k}{\ell}\left(\frac{\beta}{\log k}\right)^{\ell}\left(1-\frac{\beta}{\log k}\right)^{k-\ell} \leqslant \alpha\sqrt{\log k}\left(\frac{\beta k}{\log k}\right)^{1-\varepsilon}. \tag{20}$$

Since $|G|\leqslant 2^{k/2}$, by (6), there exists $\eta\gtrsim 1$ such that, since $\mu(\pi_G^{-1}(Y))\gtrsim 1$, we have

$$\mu\times\mu\Big(\big\{(x,y)\in\pi_G^{-1}(Y)\times\pi_G^{-1}(Y):\ d_{\mathbb{F}_2^k/G}(Gx,Gy)>\eta k\big\}\Big)$$
$$\geqslant \mu\big(\pi_G^{-1}(Y)\big)^2-\mu\times\mu\Big(\big\{(x,y)\in\mathbb{F}_2^k\times\mathbb{F}_2^k:\ d_{\mathbb{F}_2^k/G}(Gx,Gy)\leqslant\eta k\big\}\Big)\geqslant$$
$$\mu\big(\pi_G^{-1}(Y)\big)^2-2^{-\frac{k}{3}}\gtrsim 1.$$

So, the first quantity in (20) is at least a constant multiple of $k^{1-\varepsilon}$, and the desired lower bound on α follows. □

***Proof of Lemma* 9** Suppose that $Z \subseteq X$ satisfies

$$\frac{1}{4} \leqslant \frac{\mu_{\mathbb{F}_2^k/G}(Z)}{\mu_{\mathbb{F}_2^k/G}(X)} \leqslant \frac{2}{3}. \tag{21}$$

Writing $\mathfrak{q} = \mu_{\mathbb{F}^k/G}(Z)$, the function $(-1)^{\mathbf{1}_{\pi_G^{-1}(Z)}} : \mathbb{F}_2^k \to \{-1, 1\}$ is G-invariant, and its variance is equal to $4\mathfrak{q}(1-\mathfrak{q}) \asymp 1$. Let $\beta \in (2/\log k, 1)$ be a small enough universal constant that will be determined later. Also, let $C \in (1, \infty)$ be a large enough universal constant; specifically take C to be the universal constant that appears in the statement of [28, Theorem 3.1]. If we denote $m = \lceil \beta \log k \rceil$, then it follows from Fact 8 that, provided β is a sufficiently small constant, we have

$$\max_{j\in\{1,\ldots,k\}} \mathsf{Inf}_j^{\leqslant m}\Big[(-1)^{\mathbf{1}_{\pi_G^{-1}(Z)}}\Big] \leqslant \frac{m}{k}\mathsf{Var}\Big[(-1)^{\mathbf{1}_{\pi_G^{-1}(Z)}}\Big] \leqslant \frac{\mathsf{Var}\Big[(-1)^{\mathbf{1}_{\pi_G^{-1}(Z)}}\Big]^4}{C^m}.$$

This is precisely the assumption of [28, Theorem 3.1], from which we deduce the following Fourier tail bound:

$$\begin{aligned}
&\sum_{\substack{A\subseteq\{1,\ldots,k\}\\ |A|>\lceil\beta\log k\rceil}} \Big(\widehat{(-1)^{\mathbf{1}_{\pi_G^{-1}(Z)}}}(A)\Big)^2 \\
&= \sum_{\substack{A\subseteq\{1,\ldots,k\}\\ |A|>m}} \Big(\widehat{(-1)^{\mathbf{1}_{\pi_G^{-1}(Z)}}}(A)\Big)^2 \gtrsim \frac{\mathsf{Var}\Big[(-1)^{\mathbf{1}_{\pi_G^{-1}(Z)}}\Big]}{\sqrt{m}} \asymp \frac{1}{\sqrt{\beta\log k}}.
\end{aligned} \tag{22}$$

Next, by the identity (13), we have

$$\begin{aligned}
&\vartheta_{\frac{1}{\beta\log k}}\Big(\pi_G^{-1}(Z)\times\big(\mathbb{F}_2^k\smallsetminus\pi_G^{-1}(Z)\big)\Big) \\
&= \frac{1}{4}\sum_{A\subseteq\{1,\ldots,k\}}\bigg(1-\Big(1-\frac{2}{\beta\log k}\Big)^{|A|}\bigg)\Big(\widehat{(-1)^{\mathbf{1}_{\pi_G^{-1}(Z)}}}(A)\Big)^2 \\
&\geqslant \frac{1}{4}\bigg(1-\Big(1-\frac{2}{\beta\log k}\Big)^{\lceil\beta\log k\rceil+1}\bigg)\sum_{\substack{A\subseteq\{1,\ldots,k\}\\ |A|>\lceil\beta\log k\rceil}}\Big(\widehat{(-1)^{\mathbf{1}_{\pi_G^{-1}(Z)}}}(A)\Big)^2 \overset{(22)}{\geqslant} \frac{\gamma}{\sqrt{\beta\log k}},
\end{aligned} \tag{23}$$

for some universal constant $\gamma \in (0, 1)$. Therefore,

$$
\begin{aligned}
&\vartheta^{\frac{1}{\beta\log k}}\Big(\pi_G^{-1}(Z)\times\big(\pi_G^{-1}(X)\smallsetminus\pi_G^{-1}(Z)\big)\Big)\\
&\geqslant \vartheta^{\frac{1}{\beta\log k}}\Big(\pi_G^{-1}(Z)\times\big(\mathbb{F}_2^k\smallsetminus\pi_G^{-1}(Z)\big)\Big)-\vartheta^{\frac{1}{\beta\log k}}\Big(\mathbb{F}_2^k\times\big((\mathbb{F}_2^k\smallsetminus\pi_G^{-1}(X)\big)\Big)\\
&=\vartheta^{\frac{1}{\beta\log k}}\Big(\pi_G^{-1}(Z)\times\big(\mathbb{F}_2^k\smallsetminus\pi_G^{-1}(Z)\big)\Big)-\mu\big(\mathbb{F}_2^k\smallsetminus\pi_G^{-1}(X)\big)\\
&\overset{(17)\wedge(23)}{\geqslant}\frac{\gamma}{\sqrt{\beta\log k}}-\frac{1}{\sqrt{\log k}}\overset{(21)}{\asymp}\frac{1}{\sqrt{\log k}}\cdot\frac{\mu_{\mathbb{F}^k/G}(Z)}{\mu_{\mathbb{F}^k/G}(X)},
\end{aligned}
\tag{24}
$$

where the final step of (24) holds provided $1\asymp\beta\leqslant\gamma^2/4$, which is our final requirement from the universal constant β.

Observe that

$$
\begin{aligned}
&\vartheta_{\mathbb{F}_2^k/G}^{\frac{1}{\beta\log k}}(X\times X)\geqslant\vartheta^{\frac{1}{\beta\log k}}\big(\mathbb{F}_2^k\times\mathbb{F}_2^k\big)\\
&-\vartheta^{\frac{1}{\beta\log k}}\Big(\big(\mathbb{F}_2^k\smallsetminus\pi_G^{-1}(X)\big)\times\mathbb{F}_2^k\Big)-\vartheta^{\frac{1}{\beta\log k}}\Big(\mathbb{F}_2^k\times\big(\mathbb{F}_2^k\smallsetminus\pi_G^{-1}(X)\big)\Big)\\
&=1-2\mu\big(\mathbb{F}_2^k\smallsetminus\pi_G^{-1}(X)\big)=1-2\big(1-\mu_{\mathbb{F}_2^k/G}(X)\big)\overset{(17)}{\geqslant}1-\frac{2}{\sqrt{\log k}}\asymp 1.
\end{aligned}
\tag{25}
$$

Hence,

$$
\begin{aligned}
&\frac{\vartheta_{\mathbb{F}_2^k/G}^{\frac{1}{\beta\log k}}\Big(\big(Z\times(X\smallsetminus Z)\big)\cup\big((X\smallsetminus Z)\times Z\big)\Big)}{\vartheta_{\mathbb{F}_2^k/G}^{\frac{1}{\beta\log k}}(X\times X)}\\
&=\frac{2\vartheta^{\frac{1}{\beta\log k}}\Big(\pi_G^{-1}(Z)\times\big(\pi_G^{-1}(X)\smallsetminus\pi_G^{-1}(Z)\big)\Big)}{\vartheta_{\mathbb{F}_2^k/G}^{\frac{1}{\beta\log k}}(X\times X)}\overset{(24)}{\gtrsim}\frac{1}{\sqrt{\log k}}\cdot\frac{\mu_{\mathbb{F}^k/G}(Z)}{\mu_{\mathbb{F}^k/G}(X)}.
\end{aligned}
\tag{26}
$$

We are now in position to apply [26, Lemma 6] with the parameters $\delta=\frac{1}{4}$, $\alpha\asymp 1/\sqrt{\log k}$, and the probability measures:

$$
\sigma\overset{\mathrm{def}}{=}\frac{\mu_{\mathbb{F}_2^k/G}}{\mu_{\mathbb{F}_2^k/G}(X)}:2^X\to[0,1]\qquad\text{and}\qquad \tau\overset{\mathrm{def}}{=}\frac{\vartheta_{\mathbb{F}_2^k/G}^{\frac{1}{\beta\log k}}}{\vartheta_{\mathbb{F}_2^k/G}^{\frac{1}{\beta\log k}}(X\times X)}:2^{X\times X}\to[0,1].
\tag{27}
$$

Due to (26), by the proof of [26, Lemma 6] (specifically, equation (7) in [26]), there exists a subset $Y\subseteq\mathbb{F}_2^k/G$ with $\sigma(Y)\geqslant 3/4$, i.e., $\mu_{\mathbb{F}_2^k/G}(Y)\geqslant 3\mu_{\mathbb{F}_2^k/G}(X)/4$, such that every $f : Y \to L_1$ satisfies

$$\iint_{Y\times Y} \|f(\mathcal{O}) - f(\mathcal{O}')\|_1 \,\mathrm{d}\mu_{\mathbb{F}_2^k/G}(\mathcal{O})\,\mathrm{d}\mu_{\mathbb{F}_2^k/G}(\mathcal{O}') \overset{(21)\wedge(27)}{\asymp} \iint_{Y\times Y} \|f(\mathcal{O}) - f(\mathcal{O}')\|_1 \,\mathrm{d}\sigma(\mathcal{O})\,\mathrm{d}\sigma(\mathcal{O}')$$
$$\lesssim \sqrt{\log k} \iint_{Y\times Y} \|f(\mathcal{O}) - f(\mathcal{O}')\|_1 \,\mathrm{d}\tau(\mathcal{O},\mathcal{O}') \overset{(25)\wedge(27)}{\asymp} \sqrt{\log k} \iint_{Y\times Y} \|f(\mathcal{O}) - f(\mathcal{O}')\|_1 \,\mathrm{d}\vartheta_{\mathbb{F}_2^n/G}^{\frac{1}{\beta\log k}}(\mathcal{O},\mathcal{O}').$$

□

Acknowledgments We are grateful to the anonymous referee for helpful corrections and suggestions.

References

1. I. Aharoni, B. Maurey, B.S. Mityagin, Uniform embeddings of metric spaces and of Banach spaces into Hilbert spaces. Israel J. Math. **52**(3), 251–265 (1985)
2. N. Alon, Y. Matias, M. Szegedy, The space complexity of approximating the frequency moments. J. Comput. Syst. Sci. **58**(1, part 2), 137–147 (1999). Twenty-eighth Annual ACM Symposium on the Theory of Computing, Philadelphia, 1996
3. A. Andoni, R. Krauthgamer, Distance estimation protocols for general metrics. Unpublished manuscript (2008)
4. A. Andoni, R. Krauthgamer, I. Razenshteyn, Sketching and embedding are equivalent for norms. SIAM J. Comput. **47**(3), 890–916 (2018). Preliminary version appeared in STOC'2015
5. S. Arora, J.R. Lee, A. Naor, Euclidean distortion and the sparsest cut. J. Am. Math. Soc. **21**(1), 1–21 (2008)
6. P. Assouad, Plongements lipschitziens dans $\mathbf{R}^n$. Bull. Soc. Math. France **111**(4), 429–448 (1983)
7. Y. Benyamini, J. Lindenstrauss, *Geometric Nonlinear Functional Analysis. Vol. 1*. Volume 48 of American Mathematical Society Colloquium Publications (American Mathematical Society, Providence, 2000)
8. J. Bourgain, On the distributions of the Fourier spectrum of Boolean functions. Israel J. Math. **131**, 269–276 (2002)
9. J. Bourgain, G. Kalai, Influences of variables and threshold intervals under group symmetries. Geom. Funct. Anal. **7**(3), 438–461 (1997)
10. J. Bretagnolle, D. Dacunha-Castelle, J.-L. Krivine, Fonctions de type positif sur les espaces L^p. C. R. Acad. Sci. Paris **261**, 2153–2156 (1965)
11. M. R. Bridson, A. Haefliger, *Metric Spaces of Non-positive Curvature*. Volume 319 of Grundlehren der Mathematischen Wissenschaften [Fundamental Principles of Mathematical Sciences] (Springer, Berlin, 1999)
12. J. Cheeger, B. Kleiner, Realization of metric spaces as inverse limits, and bilipschitz embedding in L_1. Geom. Funct. Anal. **23**(1), 96–133 (2013)

13. J. Cheeger, B. Kleiner, A. Naor, Compression bounds for Lipschitz maps from the Heisenberg group to L_1. Acta Math. **207**(2), 291–373 (2011)
14. G. David, S. Semmes, *Fractured Fractals and Broken Dreams*. Volume 7 of Oxford Lecture Series in Mathematics and its Applications (The Clarendon Press/Oxford University Press, New York, 1997). Self-similar geometry through metric and measure
15. N.R. Devanur, S.A. Khot, R. Saket, N.K. Vishnoi, Integrality gaps for sparsest cut and minimum linear arrangement problems, in *STOC'06: Proceedings of the 38th Annual ACM Symposium on Theory of Computing* (ACM, New York, 2006), pp. 537–546
16. M.M. Deza, M. Laurent, *Geometry of Cuts and Metrics*. Volume 15 of Algorithms and Combinatorics (Springer, Berlin, 1997)
17. M. Gromov, Random walk in random groups. Geom. Funct. Anal. **13**(1), 73–146 (2003)
18. J. Heinonen, *Lectures on Analysis on Metric Spaces*. Universitext (Springer, New York, 2001)
19. S. Hoory, N. Linial, A. Wigderson, Expander graphs and their applications. Bull. Am. Math. Soc. (N.S.) **43**(4), 439–561 (2006)
20. P. Indyk, Stable distributions, pseudorandom generators, embeddings, and data stream computation. J. ACM **53**(3), 307–323 (2006). Preliminary version appeared in FOCS'2000
21. P. Indyk, R. Motwani, Approximate nearest neighbors: towards removing the curse of dimensionality, in *STOC'98*, Dallas (ACM, New York, 1999), pp. 604–613
22. J. Kahn, G. Kalai, N. Linial, The influence of variables on boolean functions (extended abstract), in *29th Annual Symposium on Foundations of Computer Science*, White Plains, 24–26 Oct 1988 (IEEE Computer Society, 1988), pp. 68–80
23. N.J. Kalton, Banach spaces embedding into L_0. Israel J. Math. **52**(4), 305–319 (1985)
24. N.J. Kalton, N.T. Peck, J.W. Roberts, *An F-Space Sampler*. Volume 89 of London Mathematical Society Lecture Note Series (Cambridge University Press, Cambridge, 1984)
25. D. Kane, R. Meka, A PRG for Lipschitz functions of polynomials with applications to sparsest cut, in *STOC'13—Proceedings of the 2013 ACM Symposium on Theory of Computing* (ACM, New York, 2013), pp. 1–10
26. S. Khot, A. Naor, Nonembeddability theorems via Fourier analysis. Math. Ann. **334**(4), 821–852 (2006)
27. S.A. Khot, N.K. Vishnoi, The unique games conjecture, integrability gap for cut problems and embeddability of negative-type metrics into ℓ_1. J. ACM **62**(1), Art. 8, 39 (2015)
28. G. Kindler, N. Kirshner, R. O'Donnell, Gaussian noise sensitivity and Fourier tails. Israel J. Math. **225**(1), 71–109 (2018)
29. R. Krauthgamer, Y. Rabani, Improved lower bounds for embeddings into L_1. SIAM J. Comput. **38**(6), 2487–2498 (2009)
30. E. Kushilevitz, R. Ostrovsky, Y. Rabani, Efficient search for approximate nearest neighbor in high dimensional spaces. SIAM J. Comput. **30**(2), 457–474 (2000)
31. S. Kwapień, Unsolved Problems. Stud. Math. **38**, 467–483 (1970). Problem 3, p. 469
32. J.R. Lee, A. Naor, L_p metrics on the Heisenberg group and the Goemans–Linial conjecture, in *47th Annual IEEE Symposium on Foundations of Computer Science (FOCS 2006)*, Berkeley, Proceedings, 21–24 Oct 2006 (IEEE Computer Society, 2006), pp. 99–108
33. J. Lindenstrauss, L. Tzafriri, *Classical Banach Spaces. I*. Sequence spaces, Ergebnisse der Mathematik und ihrer Grenzgebiete, vol. 92 (Springer, Berlin/New York, 1977)
34. J. Lindenstrauss, L. Tzafriri, *Classical Banach Spaces. II*. Volume 97 of Ergebnisse der Mathematik und ihrer Grenzgebiete [Results in Mathematics and Related Areas]. Function spaces (Springer, Berlin/New York, 1979)
35. M. Mendel, A. Naor, Euclidean quotients of finite metric spaces. Adv. Math. **189**(2), 451–494 (2004)
36. A. Naor, L_1 embeddings of the Heisenberg group and fast estimation of graph isoperimetry, in *Proceedings of the International Congress of Mathematicians. Volume III* (Hindustan Book Agency, New Delhi, 2010), pp. 1549–1575
37. A. Naor, An average John theorem. Preprint available at https://arxiv.org/abs/1905.01280 (2019)

38. A. Naor, Y. Rabani, A. Sinclair, Quasisymmetric embeddings, the observable diameter, and expansion properties of graphs. J. Funct. Anal. **227**(2), 273–303 (2005)
39. A. Naor, R. Young, Vertical perimeter versus horizontal perimeter. Ann. Math. (2) **188**(1), 171–279 (2018)
40. E.M. Nikišin, A resonance theorem and series in eigenfunctions of the Laplace operator. Izv. Akad. Nauk SSSR Ser. Mat. **36**, 795–813 (1972)
41. R. O'Donnell, *Analysis of Boolean Functions* (Cambridge University Press, New York, 2014)
42. M. Saks, X. Sun, Space lower bounds for distance approximation in the data stream model, in *Proceedings of the Thirty-Fourth Annual ACM Symposium on Theory of Computing* (ACM, New York, 2002), pp. 360–369
43. J.H. Wells, L.R. Williams, *Embeddings and Extensions in Analysis*. Ergebnisse der Mathematik und ihrer Grenzgebiete, Band 84 (Springer, New York/Heidelberg, 1975)

Degree of Convergence of Some Operators Associated with Hardy-Littlewood Series for Functions of Class $Lip(\alpha, p)$, $p > 1$

Manish Kumar, Benjamin A. Landon, R. N. Mohapatra, and Tusharakanta Pradhan

Abstract In this article, we study the degree of convergence of Euler, Borel, and (e, c) transforms of the Fourier series of functions of class $Lip(\alpha, p)$, for $p > 1$. When p tends to infinity, the results yield known results in the supremum norm studied by P. Sadangi (Sadangi, Degree of Convergence of functions in the Hölder metric, Ph.D. Thesis, Utkal University, 2006). The results of this chapter set the stage for further generalizations in other function spaces.

Mathematics Subject Classification: 40A05, 41A10, 42A10

1 Basic Definitions and Introduction

A series is divergent if its sum diverges to infinity or oscillates finitely. Summability methods are used to assign a sum to series which oscillates finitely. Methods used to sum such series include Cesàro, Nörlund, Riesz, Abel, Euler, Borel, (e, c), and Karamata means. See Hardy [11] for all definitions and related results. Also see references [1, 4, 7–10, 13, 14, 19–21] for works related to our investigation in this paper.

M. Kumar · T. Pradhan
Department of Mathematics, Birla Institute of Technology and Sciences-Pilani, Hyderabad, India
e-mail: manishkumar@hyderabad.bits-pilani.ac.in

B. A. Landon
School of Mathematics, Daytona State College, Daytona Beach, FL, USA
e-mail: Benjamin.Landon@daytonastate.edu

R. N. Mohapatra (✉)
Department of Mathematics, University of Central Florida, Orlando, FL, USA
e-mail: ram.mohapatra@ucf.edu

M. Th. Rassias (ed.), *Harmonic Analysis and Applications*, Springer Optimization and Its Applications 168, https://doi.org/10.1007/978-3-030-61887-2_9

1.1 Summation Methods

Given $\sum a_n$ with partial sum s_n, let $t_n = ps_n$, where p is some transform of s_n. Let c be the collection of all convergent sequences. If $p : c \to c$, c is the collections of all convergent sequences of real numbers, then p is said to be *conservative*.
If $s_n \to s$ implies that $ps_n \to s$ as $n \to \infty$, then method p is said to be *regular*.
The *degree of convergence* of a summation method to a given function f is a measure of how fast t_n converges to f. This means that we need to find λ_n such that

$$||t_n - f|| = O\left(\frac{1}{\lambda_n}\right), \tag{1}$$

where $\lambda_n \to \infty$ as $n \to \infty$ and the norm is the supremum norm. A significant application of the summation methods is to Fourier series.

1.2 Fourier Series

Let $f \in L(0, 2\pi)$ be periodic with period 2π. $C_{2\pi}$ is the collection of all continuous functions with period 2π. The Fourier series of f is given by

$$f \sim \frac{1}{2}a_0 + \sum_{n=1}^{\infty}(a_n \cos nx + b_n \sin nx), \tag{2}$$

where a_n and b_n are the Fourier coefficients. The series conjugate to (2) is given by

$$f \sim \sum_{n=1}^{\infty}(a_n \sin nx - b_n \cos nx). \tag{3}$$

Zygmund [22] showed that if $f \in C_{2\pi} \cap Lip\, \alpha, 0 < \alpha \leq 1$ and s_n is the nth partial sum of the Fourier series of f, then

$$||s_n(f; x) - f(x)|| = O\left(\frac{\log n}{n^{\alpha}}\right). \tag{4}$$

1.3 Hardy-Littlewood Series

Let

$$\sum_{n=0}^{\infty} A_n(x) = \frac{a_0}{2} + \sum_{n=1}^{\infty}(a_n \cos nx + b_n \sin nx). \tag{5}$$

If

$$S_n^{\star}(x) = \sum_{k=0}^{n-1} A_k(x) + \frac{1}{2}A_n(x), \tag{6}$$

then the Hardy-Littlewood series (HL-series) is defined as

$$\sum_{n=1}^{\infty} \frac{S_n^{\star}(x) - f(x)}{n}. \tag{7}$$

Let

$$\sum_{n=0}^{\infty} B_n(x) = \sum_{n=1}^{\infty}(a_n \sin nx - b_n \cos nx). \tag{8}$$

If we write

$$\widetilde{S}_n^{\star}(x) = \sum_{k=1}^{n-1} B_k(x) + \frac{1}{2}B_n(x), \tag{9}$$

then the associated Hardy-Littlewood series is defined as

$$\sum_{n=1}^{\infty} \frac{\widetilde{S}_n^{\star}(x) - \widetilde{f}(x)}{n}. \tag{10}$$

The convergence of the above series is addressed in a theorem due to Das et al. [5]. In this article, we shall determine the degree of convergence of certain means of the Hardy-Littlewood series of a function f to itself in $H_{\alpha,p}$.

2 Introduction to Some Summation Methods

In this section, we define some methods of summation that will be used throughout the chapter. We also give the definition of Hölder continuity.

2.1 Borel, Euler, and (e, c) Means

There are several methods of summing divergent series. We shall state several such methods:

Borel's exponential mean: Let $\sum_{n=0}^{\infty} u_n(x)$ be an infinite series with sequence of partial sums $\{t_n(x)\}$. Borel's exponential mean $B_p(t; x)$ of the sequence $\{t_n(x)\}$ is defined by

$$B_p(t; x) = e^{-p} \sum_{n=0}^{\infty} t_n(x) \frac{p^n}{n!}, \quad (p > 0). \tag{11}$$

Euler mean: Given any sequence $\{t_n(x)\}$ its (E, q), $q > 0$, mean $E_n^q(t; x)$ is defined by

$$E_n^q(t; x) = (q + 1)^{-n} \sum_{k=0}^{n} \binom{n}{k} q^{n-k} t_k(x). \tag{12}$$

(e,c) mean: Let $\sum_{n=-\infty}^{\infty} c_n(x)$ be an infinite series with the partial sums $\{t_n(x)\}$. The (e, c), $(c > 0)$ mean $e_n^c(t; x)$ of $\{t_n(x)\}$ is defined by

$$e_n^c(t; x) = \sqrt{\frac{c}{\pi n}} \sum_{k=-\infty}^{\infty} t_{n+k}(x) e^{-\frac{ck^2}{n}}, \tag{13}$$

where it is understood that $t_{n+k}(x) = 0$, when $n + k < 0$.

2.2 Fourier Series and Conjugate Series in the Hölder Metric

Let $C_{2\pi}$ be the space of all 2π periodic functions defined on $[0, 2\pi]$ and let for $0 < \alpha \leq 1$ and for all x, y

$$H_\alpha = \{f \in C_{2\pi} : |f(x) - f(y)| \leq M|x - y|^\alpha\}, \tag{14}$$

where M is a positive constant. The functions H_α are called Hölder continuous functions. Then space $H_\alpha (0 < \alpha \leq 1)$ is a Banach space [16] under the norm $||.||_\alpha$:

$$||f||_\alpha = ||f||_c + \sup_{x \neq y} \Delta^\alpha f(x, y), \quad (f \in H_\alpha), \tag{15}$$

where $||f||_c$ denotes the sup norm of f with respect to x,

$$\Delta^{\alpha} f(x, y) = \frac{|f(x) - f(y)|}{|x - y|^{\alpha}}, \quad x \neq y, \tag{16}$$

and by convention

$$\Delta^{0} f(x, y) = 0.$$

The metric induced by the norm $||.||_{\alpha}$ on H_{α} is called the Hölder metric. It can be seen that

$$||f||_{\beta} \leq (2\pi)^{\alpha-\beta} ||f||_{\alpha}, \tag{17}$$

for $0 \leq \beta < \alpha \leq 1$. Thus $\{H_{\alpha}, ||.||_{\alpha}\}$ is a Banach space which decreases as α increases, i.e.,

$$C_{2\pi} \supseteq H_{\beta} \supseteq H_{\alpha} \text{ for } 0 \leq \beta < \alpha \leq 1. \tag{18}$$

2.3 The Measure of Convergence of the Euler, Borel, and (e, c) Means of a Series Associated with the Hardy-Littlewood Series in the Hölder Metric

Let

$$g(x) = \frac{2}{\pi} \int_{0+}^{\pi} \psi_x(t) \frac{1}{2} \cot\left(\frac{1}{2}t\right) \log\left(\frac{1}{2} \csc \frac{1}{2}t\right) dt, \tag{19}$$

where

$$\psi_x(t) = \frac{1}{2} \left\{ f(x+t) - f(x-t) \right\}. \tag{20}$$

Das, Ray, and Sadangi [6] obtained the rate of convergence of the associated Hardy-Littlewood series (10) to $g(x)$ in the Hölder metric.

Theorem 2.1 *Let $\widetilde{T}_n(x)$ be the nth partial sum of the Hardy-Littlewood series* (7). *Let $0 \leq \beta < \alpha \leq 1$ and $f \in H_{\alpha}$. Then*

$$||\widetilde{T}_n - g||_{\beta} = O(1) \begin{cases} \frac{1}{n^{\alpha-\beta}} & , \alpha - \beta \neq 1 \\ \frac{\log n}{n} & , \alpha - \beta = 1. \end{cases} \tag{21}$$

Sadangi [18] *obtained the degrees of approximation of $g(x)$ in the Hölder metric using the Euler, Borel, and (e, c) means of* (7).

Theorem 2.2 *Let* $0 \leq \beta < \alpha \leq 1$ *and let* $f \in H_\alpha$. *Then*

$$||E_n^q(\widetilde{T}) - g||_\beta = O(1) \begin{cases} \frac{1}{n^{\alpha-\beta}} & , \alpha - \beta \neq 1 \\ \frac{\log n}{n} & , \alpha - \beta = 1. \end{cases} \tag{22}$$

Theorem 2.3 *Let* $0 \leq \beta < \alpha \leq 1$ *and let* $f \in H_\alpha$. *Then*

$$||B_p(\widetilde{T}) - g||_\beta = O(1) \begin{cases} \frac{1}{p^{\alpha-\beta}} & , \alpha - \beta \neq 1 \\ \frac{\log p}{p} & , \alpha - \beta = 1. \end{cases} \tag{23}$$

Theorem 2.4 *Let* $0 \leq \beta < \alpha \leq 1$ *and let* $f \in H_\alpha$. *Then*

$$||e_n(\widetilde{T}) - g||_\beta = O(1) \begin{cases} \frac{1}{n^{\alpha-\beta}} & , 0 < \alpha - \beta \leq \frac{1}{2} \\ \frac{1}{\sqrt{n}} & , \frac{1}{2} < \alpha - \beta \leq 1. \end{cases} \tag{24}$$

3 Definitions and Notations

Let $L_p[0, 2\pi]$ be the space of all 2π-periodic integrable functions and for all t

$$H_{\alpha,p} := \left\{ f \in L_p[0, 2\pi] : \left(\int_0^{2\pi} |f(x+t) - f(x)|^p dx \right)^{\frac{1}{p}} \leq K|t|^\alpha \right\}, \tag{25}$$

where K is a positive constant. The space $H_{\alpha,p}(p > 1, \alpha \leq 0 < 1)$ is a Banach space under the norm $||.||_{\alpha,p}$:

$$||f||_{\alpha,p} := ||f||_p + \sup_{t \neq o} \frac{||f(y+t) - f(y)||_p}{|t|^\alpha}. \tag{26}$$

The metric induced by norm $||.||_{\alpha,p}$ on $H_{\alpha,p}$ is called Hölder continuous with degree p. It can be seen that

$$||f||_{\beta,p} \leq (2\pi)^{\alpha-\beta} ||f||_{\alpha,p}.$$

Since $f \in H_{\alpha,p}$ if and only if $||f||_{\alpha,p} < \infty$, we have

$$L_p[0, 2\pi] \supseteq H_{\beta,p} \supseteq H_{\alpha,p}, \quad p > 1, \quad 0 \leq \beta < \alpha \leq 1. \tag{27}$$

We write

$$\varphi_x(t) = \frac{1}{2}\{f(x+t) + f(x-t) - 2f(x)\}$$

$$\psi_x(t) = \frac{1}{2}\{f(x+t) - f(x-t)\}$$

$$\chi_x(t) = \int_t^{\pi} \varphi_x(u)\frac{1}{2}\cot\frac{1}{2}udu$$

$$\theta_x(t) = -\frac{2}{\pi}\int_0^t \psi_x(u)\frac{1}{2}\cot\frac{1}{2}udu$$

$$\widetilde{f}(x) = -\frac{2}{\pi}\int_{0+}^{\pi} \psi_x(t)\frac{1}{2}\cot\frac{1}{2}tdt \tag{28}$$

$$g(x) = \frac{2}{\pi}\int_{0+}^{\pi} \psi_x(t)\frac{1}{2}\cot\left(\frac{1}{2}t\right)\log\left(\frac{1}{2}\csc\frac{1}{2}t\right)dt \tag{29}$$

$$\chi_x(0^+) = \int_{0+}^{\pi} \varphi_x(u)\frac{1}{2}\cot\frac{1}{2}udu \tag{30}$$

$$h_x(t) = \frac{\pi}{2}\theta_x(t) - \frac{t}{2}\widetilde{f}(x), \quad 0 < t \leq \pi,$$

and defined elsewhere by periodicity with period 2π.

4 Main Results

It was Prossdorf [17] who initiated the work on the degree of approximations of the H_α class in the Hölder metric by Fejér means of the Fourier series. Chandra [2] obtained a generalization of Prossdorf's work on the Nörlund mean setup. Later, Mohapatra and Chandra [15] consider the problem by matrix means. Chandra [2, 3] also studied the degree of approximation of functions of the H_α class in the Hölder metric by their Fourier series using Borel's exponential means and Euler means. Das, Ojha, and Ray [5] have studied the degree of approximation of the integral

$$\chi_x(0^+) = \int_{0+}^{\pi} \varphi_x(u)\frac{1}{2}\cot\frac{1}{2}udu,$$

by the Euler, Borel, and (e,c) transforms of the HL-series in the Hölder metric. Das, Ray, and Sadangi [6] obtained the following result on the rate of convergence of the series (10) to the integral $g(x)$ in the Hölder metric.

Let

$$\widetilde{T}(x) = \sum_{k=1}^{n} \frac{\widetilde{S}_n^*(x) - \widetilde{f}(x)}{n}, \quad n \geq 1,$$

and zero otherwise. Let $E_n^q(\widetilde{T}; x)$, $B_p(\widetilde{T}; x)$ and $e_n(\widetilde{T}; x)$ be respectively the (E, q), Borel, and (e, c) means of $\{\widetilde{T}_n(x)\}$. We prove the following theorems:

Theorem 4.1 *Let $0 \leq \beta < \alpha \leq 1$ and let $f \in H_{\alpha,p}$. Then*

$$||E_n^q(\widetilde{T}) - g||_{\beta,p} = O(1) \begin{cases} \frac{1}{n^{\alpha-\beta}} & , \alpha - \beta \neq 1 \\ \frac{\log n}{n} & , \alpha - \beta = 1. \end{cases} \tag{31}$$

Theorem 4.2 *Let $0 \leq \beta < \alpha \leq 1$ and let $f \in H_{\alpha,p}$. Then*

$$||E_p(\widetilde{T}) - g||_{\beta,p} = O(1) \begin{cases} \frac{1}{p^{\alpha-\beta}} & , \alpha - \beta \neq 1 \\ \frac{\log p}{p} & , \alpha - \beta = 1. \end{cases} \tag{32}$$

Theorem 4.3 *Let $0 \leq \beta < \alpha \leq 1$ and let $f \in H_{\alpha,p}$. Then*

$$||e_n(\widetilde{T}) - g||_{\beta,p} = O(1) \begin{cases} \frac{1}{n^{\alpha-\beta}} & , 0 < \alpha - \beta \leq \frac{1}{2} \\ \frac{1}{\sqrt{n}} & , \frac{1}{2} < \alpha - \beta \leq 1. \end{cases} \tag{33}$$

In proving these theorems, our main observation is that the kernels for Euler, Borel, and (e,c) means have some common important characteristic even though they appear to be different of each other. In what follows, we shall prove our theorems in a unified manner by taking full advantage of the common properties possessed by the kernels of Euler, Borel, and (e,c) means.

Recall the series (10). It is known from Zygmund [22] that

$$\widetilde{S}_n^\star = -\frac{2}{\pi} \int_0^\pi \psi_x(t) \frac{1 - \cos nt}{2 \tan \frac{1}{2}t} dt,$$

from which it follows that

$$\widetilde{S}_n^\star - \widetilde{f}(x) = \frac{2}{\pi} \int_0^\pi \psi_x(t) \frac{\cos nt}{2 \tan \frac{1}{2}t} dt. \tag{34}$$

For $n \geq 1$, we have, for the odd function $h_x(t)$,

$$
\begin{aligned}
c_n &= \frac{2}{\pi}\int_0^h h_x(t)\sin nt dt \\
&= \frac{2}{\pi}\int_0^\pi \left\{\frac{\pi}{2}\theta_x(t) - \frac{t}{2}\widetilde{f}(x)\right\}\sin nt dt \\
&= \frac{1}{\pi}\left[\left\{t\widetilde{f}(x) - \pi\theta_x(t)\right\}\frac{\cos nt}{n}\right]_{t=0}^{\pi} \\
&- \frac{1}{\pi}\int_0^\pi \left\{\widetilde{f}(x) + \psi_x(t)\cot\frac{1}{2}t\right\}\frac{\cos nt}{n}dt \\
&= -\frac{2}{n\pi}\int_0^\pi \psi_x(t)\frac{\cos nt}{2\tan\frac{1}{2}t}dt \\
&= -\frac{\widetilde{S}_n^\star(x) - \widetilde{f}(x)}{n}.
\end{aligned}
\tag{35}
$$

The series conjugate to $h(t) = \sum_{n=1}^{\infty} c_n \sin nt$ is $-\sum_{n=1}^{\infty} c_n \cos nt$ and hence, we have:

Proposition *The series* (10) *is the series conjugate to the Fourier series of the odd function $h_x(t)$ at $t = 0$.*
In this case,

$$
\widetilde{T}_n(x) = -\sum_{k=1}^{n} c_k = -\frac{2}{\pi}\int_0^\pi h_x(t)\widetilde{D}_n(t)dt,
\tag{36}
$$

where

$$
\widetilde{D}_n(t) = -\sum_{k=1}^{n} \sin kt \frac{\cos\frac{1}{2}t - \cos\left(n+\frac{1}{2}\right)t}{2\sin\frac{1}{2}t}.
\tag{37}
$$

At this stage, we may note that Das, Ojha, and Ray [5] *have established the Fourier character of the HL-series* (7).

5 Notations, Lemmas, and Generalized Minkowski Inequality

For generalized Minkowski inequality, see [12]:

Lemma 5.1 *If $h(y, t)$ is a function of two variables defined for $0 \le t \le \pi$, $0 \le y \le 2\pi$, then*

$$
\left\| \int h(y,t)dt \right\|_p \le \int \|h(y,t)\|_p dt, \quad p > 1.
$$

For the proof of this inequality, please see Hardy, Littlewood, and Pólya [12]. Throughout the section, we use the following additional notations:

$$\begin{aligned} G(x, y) &= g(x) - g(y), \\ \widetilde{F}(x, y) &= \widetilde{f}(x) - \widetilde{f}(y), \\ G(t) &= \theta_x(t) - \theta_y(t). \end{aligned}$$

We need the following lemmas for proof of our theorems:

Lemma 5.2 *Let* $0 \leq \beta < \alpha \leq 1$ *and let* $f \in H_\alpha$, *then for* $0 < t \leq \pi$

$$\begin{aligned} ||\psi_{y+u}(t) - \psi_y(t)||_p &= O(t^\alpha) && (38) \\ &= O(|u|^\alpha) && (39) \\ &= O(|u|^\beta t^{\alpha-\beta}) && (40) \\ \theta_{y+u}(t) &= O(t^\alpha) && (41) \\ ||G(t)||_p &= O(|u|^\beta t^{\alpha-\beta}). && (42) \end{aligned}$$

Proof The proofs of (38) and (39) are omitted as they are immediate consequences of the definition of $\psi_y(t)$ and $H_{\alpha,p}$. Writing

$$|\psi_{y+u}(t) - \psi_y(t)| = |\psi_{y+u}(t) - \psi_y(t)|^{1-\beta/\alpha} |\psi_{y+u}(t) - \psi_y(t)|^{\beta/\alpha},$$

and using the estimates (38) and (39), we obtain (40).
As

$$\theta_x(t) = -\frac{2}{\pi} \int_0^t \psi_x(u) \frac{1}{2} \cot \frac{1}{2} u du,$$

estimates (41) follows from the fact that $\psi_x(u) = O(u^\alpha)$.
As

$$||G(t)||_p = \left\| -\frac{2}{\pi} \int_0^t \frac{\psi_{y+u}(\zeta) - \psi_y(\zeta)}{2 \tan \frac{1}{2}(\zeta)} d\zeta \right\|_p,$$

estimates (42) follows by applying (40).

Lemma 5.3 *Let* $f \in H_{\alpha,p}$ *and* $0 \leq \beta < \alpha \leq 1$. *Then*

$$||G(t+h) - G(t)||_p = O(h|u|^\beta t^{\alpha-\beta-1}).$$

Proof Applying the mean value theorem and (40), we obtain for some θ with $0 < \theta < 1$

$$G(t+h) - G(t) = hG^{'}(t+\theta h)$$
$$= h\left[\frac{2}{\pi}\left\{\psi_{y+u}(t+\theta h) - \psi_y(t+\theta h)\right\}\frac{1}{2}\cot\frac{1}{2}(t+\theta h)\right]$$
$$||G(t+h) - G(t)||_p = O(h|u|^\beta(t+\theta h)^{\alpha-\beta-1})$$
$$= O(h|u|^\beta t^{\alpha-\beta-1}).$$

Lemma 5.4 *Let $0 \leq \beta < \alpha \leq 1$ and let $f \in H_{\alpha,p}$. Then*

$$(i)\ \ \widetilde{F}(x,y) = O(|u|^\beta)$$
$$(ii)\ \ G(x,y) = O(|u|^\beta).$$

Proof Since

$$\widetilde{F}(y+u,y) = \widetilde{f}(y+u) - \widetilde{f}(y) = \theta_{y+u}(\pi) - \theta(\pi) = G(\pi).$$

Lemma 5.4(i) follows from (42). Using (40), we have

$$||G(y+u,y)||_p = ||g(y+u) - g(y)||_p$$
$$\leq \frac{2}{\pi}\int_0^\pi ||\psi_{y+u}(\zeta) - \psi_y(\zeta)||_p \left|\frac{1}{2}\cot\frac{1}{2}\zeta\log\frac{1}{2}\csc\frac{1}{2}\zeta\right| d\zeta$$
$$= O(1)|u|^\beta\int_0^\pi \zeta^{\alpha-\beta-1}\log\frac{2\pi}{\zeta}d\zeta,$$

which ensures Lemma 5.4(ii) as the last integral is finite.

Lemma 5.5 (Das, Ojha, and Ray [5]) *Suppose that A and δ are both positive constants. Let β be any real number. Then as $\lambda \to \infty$,*

$$\int_{\pi/\lambda}^\delta t^\beta e^{-A\lambda t^2}dt = O(\lambda^{-\beta-1}),\ \ \beta < -1 \tag{43}$$

$$\int_{\pi/\lambda}^\delta t^\beta e^{-A\lambda t^2}dt = O(1)\log\lambda,\ \ \beta = -1 \tag{44}$$

$$\int_{\pi/\lambda}^\delta t^\beta e^{-A\lambda t^2}dt = O(1)\frac{1}{\lambda^k},\ \ 2k-1 < \beta \leq 2k,\ \ k = 0, 1, 2, \ldots \tag{45}$$

$$\int_{\pi/\lambda}^{\delta} t^{\beta} e^{-A\lambda t^2} dt = O(1)\frac{1}{\lambda^{\beta-k}}, \quad 2k < \beta \le 2k+1, \quad k = 0, 1, 2, \ldots. \tag{46}$$

Lemma 5.6 *Let $c(\lambda, t)$ be defined for all $\lambda \ge 0$ and $0 \le t \le \pi$. Suppose that*

(i) *$c(\lambda, t) \ge 0$ for all $\lambda \ge 0$ and $0 \le t \le \pi$.*
(ii) *$c(\lambda, t)$ is monotonically decreasing in t over $[0, \pi]$ for each positive constant A.*
(iii) *$c(\lambda, t) = O(e^{-A\lambda t^2})$ as $\lambda \to \infty$, $0 < t \le \pi$ for some positive constant A.*
(iv) *$c(\lambda, t) - c(\lambda, t+h) = O(tc(\lambda, t))$, $\frac{\pi}{\lambda} < t \le \pi$, where $h = \pi/\lambda$.*

Let $\theta_x(t)$ and $G(t)$ be respectively defined as in Sections 3 and 5. If $f \in H_{\alpha,p}$, $0 \le \beta < \alpha \le 1$, then for $0 < \delta \le \pi$

$$\text{(a)} \quad \int_{\frac{\pi}{\lambda}}^{\delta} \frac{G(t)}{t} c(\lambda, t) \sin \lambda t dt = O(1)|u|^{\beta} \begin{cases} \frac{1}{\lambda^{\alpha-\beta}} & , \alpha - \beta \ne 1 \\ \frac{\log \lambda}{\lambda} & , \alpha - \beta = 1 \end{cases} \tag{47}$$

$$\text{(b)} \quad \int_{\frac{\pi}{\lambda}}^{\delta} \frac{\theta_x(t)}{t} c(\lambda, t) \sin \lambda t dt = O(1)|u|^{\beta} \begin{cases} \frac{1}{\lambda^{\alpha}} & , 0 < \alpha < 1 \\ \frac{\log \lambda}{\lambda} & , \alpha = 1. \end{cases} \tag{48}$$

Proof (a) putting $h = \pi/\lambda$, we write

$$J = \int_{h}^{\delta} \frac{G(t)}{t} c(\lambda, t) \sin \lambda t dt \tag{49}$$

$$\begin{aligned} &= \left(\int_{2h}^{\delta+h} + \int_{h}^{2h} - \int_{\delta}^{\delta+h} \right) \frac{G(h)}{t} c(\lambda, t) \sin \lambda t dt \\ &- \int_{h}^{\delta} \frac{G(t+h)}{t+h} c(\lambda, t+h) \sin \lambda t dt + \int_{h}^{2h} \frac{G(t)}{t} c(\lambda, t) \sin \lambda t dt \\ &- \int_{\delta}^{\delta+h} \frac{G(t)}{t} c(\lambda, t) \sin \lambda t dt. \end{aligned} \tag{50}$$

From (49) and (50), we obtain

$$\begin{aligned} 2J &= \int_{h}^{\delta} \left[\frac{G(t)}{t} c(\lambda, t) - \frac{G(t+h)}{t+h} c(\lambda, t+h) \right] \sin \lambda t dt \\ &+ \int_{h}^{2h} \frac{G(t)}{t} c(\lambda, t) \sin \lambda t dt - \int_{\delta}^{\delta+h} \frac{G(t)}{t} c(\lambda, t) \sin \lambda t dt \\ &= P + Q - R. \end{aligned} \tag{51}$$

By Minkowski's inequality,

$$2||J||_p \leq ||P||_p + ||Q||_p + ||R||_p,$$

since

$$\begin{aligned}
P &= \int_h^\delta \left[\frac{G(t)}{t}c(\lambda,t) - \frac{G(t+h)}{t+h}c(\lambda,t+h)\right]\sin\lambda t dt \\
&= \int_\lambda^\delta \frac{G(t)-G(t+h)}{t}c(\lambda,t)\sin\lambda t dt \\
&+ \int_h^\delta \frac{G(t+h)}{t}[c(\lambda,t)-c(\lambda,t+h)]\sin\lambda t dt \\
&+ \int_\lambda^\delta G(t+h)\left[\frac{1}{t}-\frac{1}{t+h}\right]c(\lambda,t+h)\sin\lambda t dt \\
&= P_1 + P_2 + P_3. \qquad (52)
\end{aligned}$$

By Minkowski's inequality,

$$||P||_p \leq ||P_1||_p + ||P_2||_p + ||P_3||_p.$$

By using the generalized Minkowski inequality and usual method of estimation, we have

$$\begin{aligned}
||P_1|| &= O(1)|u|^\beta h \int_h^\delta t^{\alpha-\beta-2}dt, \\
&= O(1)|u|^\beta \begin{cases} h^{\alpha-\beta} & , \alpha-\beta \neq 1 \\ h\log h^{-1} & , \alpha-\beta = 1. \end{cases} \qquad (53)
\end{aligned}$$

where we have used Lemma 5.3 and (iii) of Lemma 5.6.

By Lemma 5.2 and the definition of $c(\lambda,t)$, and using the method similar to that used to obtain (53), we get

$$\begin{aligned}
||P_2||_p &= O(1)|u|^\beta \int_h^\delta t^{\alpha-\beta-1}tc(\lambda,t)dt \\
&= O(1)|u|^\beta \int_h^\delta t^{\alpha-\beta}c^{-A\lambda t^2}dt \\
&= O(1)|u|^\beta h^{\alpha-\beta}, \quad 0 \leq \beta < \alpha \leq 1, \qquad (54)
\end{aligned}$$

using Lemma 5.5.

Adopting the technique similar to those used in splitting J into P, Q, and R, we can write

$$\begin{aligned}
2P_3 &= \int_h^\delta \left[G(t+h)\left(\frac{1}{t} - \frac{1}{t+h}\right) c(\lambda, t+h) \right. \\
&\quad \left. - G(t+2h)\left(\frac{1}{t+h} - \frac{1}{t+2h}\right) c(\lambda, t+2h) \right] \sin \lambda t dt \\
&\quad + h \int_h^{2h} \frac{G(t+h)}{t(t+h)} c(\lambda, t+h) \sin \lambda t dt - \int_\delta^{\delta+h} \frac{G(t+h)}{t(t+h)} c(\lambda, t+h) \sin \lambda t dt \\
&= L + M - N. \qquad (55)
\end{aligned}$$

By Minkowski's inequality,

$$2||P_3|| \leq ||L||_p + ||M||_p + ||N||_p.$$

We have

$$\begin{aligned}
L &= \int_h^\delta \left[G(t+h)\left(\frac{1}{t} - \frac{1}{t+h}\right) c(\lambda, t+h) \right. \\
&\quad \left. - G(t+2h)\left(\frac{1}{t+h} - \frac{1}{t+2h}\right) c(\lambda t + 2h) \right] \sin \lambda t dt \\
&= h \int_h^\delta \frac{G(t+h) - G(t+2h)}{t(t+h)} c(\lambda, t+h) \sin \lambda t dt \\
&\quad + h \int_h^\delta \frac{G(t+2h)}{t(t+h)} [c(\lambda, t+h) - c(t\lambda, t+2h)] \sin \lambda t dt \\
&= h \int_h^\delta \frac{g(t+2h)}{t+h} \left(\frac{1}{t} - \frac{1}{t+2h}\right) c(\lambda, t+2h) \sin \lambda t dt \\
&= L_1 + L_2 + L_3. \qquad (56)
\end{aligned}$$

By Minkowski's inequality,

$$||L||_p \leq ||L_1||_p + ||L_2||_p + ||L_3||_p.$$

By the generalized Minkowski inequality mentioned in Section 5, and the usual method of estimation, we get by usual estimation technique

$$||L_1||_p \leq |h| \int_h^\delta \frac{||G(t+h) - G(t+2h)||_p}{|t(t+h)|} |c(\lambda, t+h) \sin \lambda t| dt$$

$$
\begin{aligned}
&= O(1)|u|^{\beta}h^{2}\int_{h}^{\delta}\frac{(t+h)^{\alpha-\beta-1}}{t(t+h)}e^{-At^{2}}dt\\
&= O(1)|u|^{\beta}h^{2}\int_{h}^{\delta}t^{\alpha-\beta-3}dt\\
&= O(1)|u|^{\beta}h^{\alpha-\beta},\ \ 0\leq\beta<\alpha\leq 1.
\end{aligned}
\tag{57}
$$

Using Lemma 5.2 and properties of $c(\lambda, t)$, we get

$$
\begin{aligned}
||L_{2}||_{p} &\leq |h|\int_{h}^{\delta}\frac{||G(t+2h)||_{p}}{|t(t+h)|}|[c(\lambda,t+h)-c(\lambda,t+2h)]\sin\lambda t|dt\\
&= O(1)|u|^{\beta}h\int_{h}^{\delta}\frac{(t+2h)^{\alpha-\beta}}{t(t+h)}(t+h)c(\lambda,t+h)dt\\
&= O(1)|u|^{\beta}h\int_{h}^{\delta}t^{\alpha-\beta-1}dt\\
&= O(1)|u|^{\beta}h,\ \ 0\leq\beta<\alpha\leq 1,
\end{aligned}
\tag{58}
$$

and

$$
\begin{aligned}
||L_{3}||_{p} &\leq |h|\int_{h}^{\delta}\frac{||G(t+2h)||_{p}}{|t+h|}\left|\left(\frac{1}{t}-\frac{1}{t+2h}\right)c(\lambda,t+2h)\sin\lambda t\right|dt\\
&= O(1)|u|^{\beta}h^{2}\int_{h}^{\delta}\frac{(t+2h)^{\alpha-\beta}}{t(t+h)(t+2h)}e^{-A\lambda t^{2}}dt\\
&= O(1)|u|^{\beta}h^{2}\int_{h}^{\delta}t^{\alpha-\beta-3}dt\\
&= O(1)|u|^{\beta}h^{\alpha-\beta},\ \ 0\leq\beta<\alpha\leq 1.
\end{aligned}
\tag{59}
$$

Using Lemma 5.2 and the boundedness of $c(\lambda, t)$, we get

$$
\begin{aligned}
||M||_{p} &\leq \int_{h}^{2h}\frac{||G(t+h)||_{p}}{|t(t+h)|}|c(\lambda,t+h)\sin\lambda|dt\\
&= O(1)|u|^{\beta}h\int_{h}^{2h}t^{\alpha-\beta-2}dt\\
&= O(1)|u|^{\beta}\begin{cases} h^{\alpha-\beta} &, \ \alpha-\beta\neq 1\\ h\log 2 &, \ \alpha-\beta=1,\end{cases}
\end{aligned}
\tag{60}
$$

and

$$\begin{aligned} ||N||_p &\le |h| \int_\delta^{\delta+h} \frac{G(t+h)}{|t(t+h)|} |c(\lambda, t+h)\sin\lambda t| dt \\ &= O(1)|u|^\beta h \int_\delta^{\delta+h} t^{\alpha-\beta-2} dt \\ &= O(1)|u|^\beta h, \ \ 0 \le \beta < \alpha \le 1. \end{aligned} \tag{61}$$

Collecting the results from (55)–(61), we obtain

$$||P_3||_p = O(1)|u|^\beta h^{\alpha-\beta}, \ \ 0 \le \beta < \alpha \le 1. \tag{62}$$

Combining the results of (52), (53), (54), and (62), we have

$$||P_3||_p = O(1)|u|^\beta \begin{cases} h^{\alpha-\beta} & , \ \alpha-\beta \ne 1 \\ h\log h^{-1} & , \ \alpha-\beta = 1. \end{cases} \tag{63}$$

By Lemma 5.2, we have, for $0 \le \beta < \alpha \le 1$,

$$||R||_p \le \int_\delta^{\delta+h} \frac{||G(t)||_p}{|t|} |c(\lambda, t)|| \sin\lambda t| dt \tag{64}$$

$$\begin{aligned} &= O(1)|u|^\beta e^{-A\lambda\delta^2} \\ &= O(1)|u|^\beta \lambda^{-\Delta}, \end{aligned} \tag{65}$$

for every positive Δ, however large. Collecting the above estimates for P, Q, and R, we obtain

$$J = O(1)|u|^\beta \begin{cases} h^{\alpha-\beta} & , \ \alpha-\beta \ne 1 \\ h\log h^{-1} & , \ \alpha-\beta = 1, \end{cases} \tag{66}$$

and this completes the proof (a). We omit the proof of (b) because it is similar to that of part (a). The case where $\sin\lambda t$ is replaced with $\cos\lambda t$ can also be dealt with in a similar manner.

Lemma 5.7 *If $f \in H_{\alpha,p}$, $0 \le \beta < \alpha \le 1$, then as $\lambda \to \infty$*

$$\text{(a)} \quad \int_0^{\pi/\lambda} \frac{||G(t)||_p}{|t|} dt = \frac{O(1)|u|^\beta}{\lambda^{\alpha-\beta}}$$

$$\text{(b)} \quad \int_0^{\pi/\lambda} \frac{||\theta_x(t)||_p}{|t|} dt = O(1)\lambda^{-\alpha}.$$

Proof The result follows from Lemma 5.2.

Lemma 5.8 *If* $f \in H_{\alpha,p}, 0 \le \beta < \alpha \le 1$, *then as* $\lambda \to \infty$

$$\text{(a)} \quad \int_{\delta}^{\pi} \frac{||G(t)||_p}{|t|} |e^{-A\lambda t^2}| dt = O(1)|u|^{\beta}\lambda^{-\Delta}$$

and

$$\text{(b)} \quad \int_{\delta}^{\pi} \frac{||\theta_x(t)||_p}{|t|} e^{-A\lambda t^2} dt = O(1)\lambda^{-\Delta}.$$

Proof By Lemma 5.2

$$\begin{aligned}\int_{\delta}^{\pi} \frac{|G(t)|}{t} e^{-A\lambda t^2} dt &= O(1)|u|^{\beta} \int_{\delta}^{\pi} t^{\alpha-\beta-1} e^{-A\lambda t^2} dt \\ &= O(1) u^{\beta} e^{-A\lambda\delta^2} \\ &= O(1)|u|^{\beta}\lambda^{-\Delta}, \quad \Delta > 0.\end{aligned}$$

Part (b) can be dealt with in a similar fashion.

6 Proof of Theorem 1

We will use the following additional notations for the proof of Theorem 1.

$$\begin{aligned} l_n^E(x) &= E_n^q(x) - g(x) \\ p_q^n(t) &= (q+1)^{-n}(1+q^2+2q\cos t)^{\frac{n}{2}} \\ \theta = \theta(t) &= \tan^{-1} \frac{\sin t}{q+\cos t} \\ P(n,t) &= (q+1)^{-1} \sum_{k=0}^{n} \binom{n}{k} q^{n-k} \cos\left(k+\frac{1}{2}\right)t, \quad q>0 \\ Q(n,t) &= (q+1)^{-1} \sum_{k=0}^{n} \binom{n}{k} q^{n-k} \sin\left(k+\frac{1}{2}\right)t, \quad q>0 \\ E(n) &= (q+1)^{-n} \sum_{k=1}^{n} \binom{n}{k} q^{n-k} \sum_{\nu=k+1}^{\infty} \frac{(-1)^{\nu-1}}{\nu} \\ \lambda &= \frac{n}{1+q} + \frac{1}{2}. \end{aligned}$$

We need the following lemmas:

Lemma 6.1 *Let* $0 < t \leq \pi$. *Then*

$$p_q^n(n) \leq e^{-Ant^2}$$

where $A = 2q[\pi(1+q)]^{-2}$.

Lemma 6.2 *For* $0 < t \leq \pi$,

$$\text{(i)}\ \ P(n,t) = p_q^n(t)\cos\left(n\Phi + \frac{1}{2}t\right) \tag{67}$$

$$\text{(ii)}\ \ P(n,t) = O(1) \tag{68}$$

$$\text{(iii)}\ \ E(n) = O(n^{-1}). \tag{69}$$

Proof By simple computation, we have

$$\begin{aligned} P(n,t) + iQ(n,t) &= (q+1)^{-1}\sum_{k=0}^{n}\binom{n}{k}q^{n-k}e^{i\left(k+\frac{1}{2}\right)t} \\ &= (q+1)^{-n}e^{i\frac{1}{2}t}(q+e^{it})^n \\ &= p_q^n(t)e^{i\frac{1}{2}t}\left[\cos\left(n\Phi + \frac{1}{2}t\right) + i\sin\left(n\Phi + \frac{1}{2}t\right)\right], \end{aligned}$$

from which (ii) follows. As $\left|\cos\left(k+\frac{1}{2}\right)t\right| \leq 1$ and $\sum_{k=0}^{n}\binom{n}{k}q^{n-k} = (1+q)^n$, estimates (iii) follows. As

$$\sum_{\nu=k+1}^{\infty}\frac{(-1)^{\nu-1}}{\nu} = O\left(\frac{1}{k+1}\right),$$

we have,

$$\begin{aligned} (q+1)^n E(n) &= \sum_{k=1}^{n}\binom{n}{k}q^{n-k}\sum_{\nu=k+1}^{\infty}\frac{(-1)^{\nu-1}}{\nu} \\ &= O(1)\sum_{k=1}^{n}\binom{n}{k}q^{n-k}\frac{1}{k+1}. \end{aligned} \tag{70}$$

Now

$$\sum_{k=1}^{n}\binom{n}{k}q^{n-k}\frac{1}{k+1} = \frac{1}{n+1}\sum_{k=2}^{n+1}\binom{n+1}{k}q^{n+1-k}$$

$$< \frac{1}{n+1}\sum_{k=0}^{n+1}\binom{n+1}{k}q^{n+1-k}$$

$$= \frac{(q+1)^{n+1}}{n+1}. \tag{71}$$

Using (71) in (70), we obtain (69).

Lemma 6.3 *Let* $\lambda = \frac{n}{1+q} + \frac{1}{2}, h = \frac{\pi}{\lambda}$, *and* $0 < t < \frac{\pi}{4}$. *Then for* $h \le t < \delta$

$$p_q^n(t+h) - p_q^n(t) = O(1)tp_q^n(t). \tag{72}$$

Proof By the mean value theorem, we have for some ζ with $0 < \zeta < 1$

$$\begin{aligned} p_q^n(t+h) - p_q^n(t) &= h\left[\frac{d}{dx}p_q^n(x)\right] \\ &= \frac{-nhp_q^n(t+\zeta h)}{1+q^2+2q\cos(1+\zeta h)}\sin(t+\zeta h) \\ &= O(1)tp_q^n(t). \end{aligned}$$

Lemma 6.4 *Let* $\lambda = \frac{n}{1+q} + \frac{1}{2}$ *and* $0 < \delta < \frac{\pi}{4}$. *Then for* $0 < t < \delta$

$$\cos\left(n\Phi + \frac{1}{2}t\right) - \cos\lambda t = O(nt^3). \tag{73}$$

Proof *We have*

$$\begin{aligned} \left|\cos\left(n\Phi + \frac{1}{2}t\right) - \cos\lambda t\right| &= \left|2\sin\frac{1}{2}\left(n\Phi + \frac{1}{2}t + \lambda t\right)\sin\frac{1}{2}\left(\lambda t - n\Phi - \frac{1}{2}t\right)\right| \\ &\le \left|\lambda t - n\Phi - \frac{1}{2}t\right| \\ &= n\left|\Phi - \frac{t}{1+q}\right| \\ &\le \left[\left|tan^{-1}\frac{\sin t}{q+\cos t} - \frac{\sin t}{q+\cos t}\right| + \left|\frac{\sin t}{q+\cos t} - \frac{t}{1+q}\right|\right] \\ &= n\left[O\left(\left(\frac{\sin t}{q+\cos t}\right)^3\right) + O(t^3)\right] \\ &= O(nt^3). \end{aligned}$$

Proof Proof of Theorem 1. Using (34), we have

$$
\begin{aligned}
\widetilde{T}_n(x) &= \sum_{k=1}^{n} \frac{\widetilde{S}_k^{\star}(x) - \widetilde{f}(x)}{k} \\
&= -\frac{2}{\pi} \int_0^{\pi} h_x(x) \widetilde{D}_x(t) dt \\
&= -\frac{2}{\pi} \int_0^{\pi} \left[\frac{t}{2} \widetilde{f}(x) - \frac{\pi}{2} \theta_x(t) \right] \widetilde{D}_n(t) dt \\
&= -\frac{1}{\pi} \widetilde{f}(x) \int_0^{\pi} t \widetilde{D}_n(t) dt - \int_0^{\pi} \theta_x(t) \widetilde{D}_n(t) dt \\
&= -\widetilde{f}(x) \sum_{k=1}^{n} \frac{\cos k\pi}{k} - \int_0^{\pi} \theta_x(t) \frac{\cos \frac{1}{2} t - \cos \left(n + \frac{1}{2}\right) t}{2 \sin \frac{t}{2}} dt
\end{aligned}
$$

$$
= \widetilde{f}(x) \left[\log 2 - \sum_{\nu=n+1}^{\infty} \frac{(-1)^{\nu-1}}{\nu} \right] - \int_0^{\pi} \theta_x(t) \frac{1}{2} \cot \frac{1}{2} t dt
$$

$$
+ \int_0^{\pi} \theta_x(t) \frac{\cos \left(n + \frac{1}{2}\right) t}{2 \sin \frac{t}{2}} dt. \tag{74}
$$

Now,

$$
\begin{aligned}
\int_0^{\pi} \theta_x(t) \frac{1}{2} \cot \frac{1}{2} t dt &= -\frac{2}{\pi} \int_0^{\pi} \frac{1}{2} \cot \frac{1}{2} t dt \left(\int_0^{t} \psi_x(u) \frac{1}{2} \cot \frac{1}{2} u du \right) \\
&= -\frac{2}{\pi} \int_0^{\pi} \psi_x(u) \frac{1}{2} \cot \frac{1}{2} u du \int_u^{\pi} \frac{1}{2} \cot \frac{1}{2} t dt \\
&= -\frac{2}{\pi} \int_0^{\pi} \psi_x(u) \frac{1}{2} \cot \frac{1}{2} u \left[\log \frac{1}{2} \csc \frac{1}{2} u + \log 2 \right] du \\
&= -g(x) + \widetilde{f}(x) \log 2.
\end{aligned} \tag{75}
$$

From (29) and (30), it follows that, for $n \geq 1$,

$$
\widetilde{T}_n(x) = g(x) + \int_0^{\pi} \theta_x(t) \frac{\cos \left(n + \frac{1}{2}\right) t}{2 \sin \frac{t}{2}} dt - \widetilde{f}(x) \sum_{\nu=n+1}^{\infty} \frac{(-1)^{\nu-1}}{\nu}. \tag{76}
$$

Using (30) and (29), we obtain

$$
E_n^q(\widetilde{T}; x) = (q+1)^{-n} \sum_{k=0}^{n} \binom{n}{k} q^{n-k} \widetilde{T}_k(x)
$$

$$= g(x)(q+1)^{-n}\sum_{k=1}^{n}\binom{n}{k}q^{n-k}$$

$$+\int_0^{\pi}\frac{\theta_x(t)}{2\sin\frac{1}{2}t}\left[(q+1)^{-n}\sum_{k=1}^{n}\binom{n}{k}q^{n-k}\cos\left(k+\frac{1}{2}\right)t\right]dt$$

$$-\widetilde{f}(x)(q+1)^{-n}\sum_{k=1}^{n}\binom{n}{k}q^{n-k}\sum_{\nu=k+1}^{\infty}\frac{(-1)^{\nu-1}}{\nu}$$

$$= g(x)\left(1-\left(\frac{q}{1+q}\right)^n\right) \tag{77}$$

$$+\int_0^{\pi}\frac{\theta_x(t)}{2\sin\frac{1}{2}t}\left[P(n,t)-\left(\frac{q}{q+1}\right)^n\cos\frac{1}{2}t\right]dt-\widetilde{f}(x)E(n),$$

which ensures that

$$l_n^E(x) = E_n^q(\widetilde{T};x)-g(x)$$

$$= -\left(\frac{q}{1+q}\right)^n g(x)+\int_0^{\pi}\frac{\theta_x(t)}{2\sin\frac{1}{2}t}P(n,t)dt$$

$$-\int_0^{\pi}\theta_x(t)\left(\frac{q}{1+q}\right)^n\frac{1}{2}\cot\frac{1}{2}tdt-\widetilde{f}(x)E(n). \tag{78}$$

Hence,

$$l_n^E(y+u)-l_n^E(y) = -\left(\frac{q}{q+1}\right)^n G(y+u,y)+\int_0^{\pi}\frac{G(t)}{2\sin\frac{t}{2}}P(n,t)dt$$

$$-\left(\frac{q}{q+1}\right)^n\int_0^{\pi}\frac{G(t)}{2\tan\frac{t}{2}}dt-E(n)\widetilde{F}(x,y)$$

$$= -P(E)+Q(E)-R(E)+S(E). \tag{79}$$

By Minkowski's inequality,

$$||l_n^E(y+u)-l_n^E(y)|| \le ||P(E)||_p+||Q(E)||_p+||R(E)||_p+||S(E)||_p.$$

By Lemma 5.4,

$$||P(E)||_p = \left(\frac{q}{1+q}\right)^n ||G(y+u,y)||_p = O(|u|^{\beta})\left(\frac{q}{1+q}\right)^n. \tag{80}$$

By Lemmas 5.4 and 6.2(iii),

$$||G(y+u,y)||_p = |-\widetilde{F}(x,y)E(n)| = O(|u|^{\beta}n^{-1}). \tag{81}$$

Using Lemma 5.2, we get

$$\begin{aligned} ||G(y+u,y)||_p &\le \left(\frac{q}{1+q}\right)^n \int_0^{\pi} \frac{||G(t)||_p}{|2t\tan\frac{1}{2}|}dt \\ &= O(1)|u|^{\beta}\left(\frac{q}{1+q}\right)^n \int_0^{\pi} t^{\alpha-\beta-1}dt \\ &= O(1)|u|^{\beta}\left(\frac{q}{1+q}\right)^n. \end{aligned} \tag{82}$$

We put $\lambda = \frac{n}{1+q} + \frac{1}{2}$. Now for fixed δ with $0 < \delta < \pi/4$, we split the integral $Q(E)$ as follows:

$$\begin{aligned} ||Q(E)||_p &\le \left[\int_0^{\pi/\lambda} + \int_{\pi/\lambda}^{\delta} + \int_{\delta}^{\pi}\right] \frac{||Q(E)||_p}{|2\sin\frac{1}{2}t|}|P(n,t)|dt \\ &\le ||I(E)||_p + ||J(E)||_p + ||K(E)||_p, \end{aligned} \tag{83}$$

by Minkowski's inequality, Lemmas 6.2(ii), and 5.7, we have

$$\begin{aligned} ||I(E)||_p &\le \int_0^{\frac{\pi}{\lambda}} \frac{||G(t)||_p}{|2\sin\frac{1}{2}t|}|P(n,t)|dt \\ &= O(1)|u|^{\beta} \int_0^{\frac{\pi}{\lambda}} \frac{|G(t)|}{t}dt \\ &= \frac{O(1)|u|^{\beta}}{n^{\alpha-\beta}}. \end{aligned} \tag{84}$$

By Lemmas 6.1, 6.2, and 5.8, we obtain

$$\begin{aligned} ||K(E)||_p &\le \int_{\delta}^{\pi} \frac{||G(t)||_p}{2|\sin\frac{1}{2}t|}\left|p_q^n(t)\cos\left(n\Phi + \frac{1}{2}t\right)\right|dt \\ &= O(1)|u|^{\beta} \int_{\delta}^{\pi} \frac{|G(t)|}{|t|}e^{-Ant^2}dt \\ &= \frac{O(1)|u|^{\beta}}{n^{\Delta}}, \quad \Delta > 0 \text{ however large.} \end{aligned} \tag{85}$$

We write

$$
\begin{aligned}
||J(E)||_p &\le \int_{\frac{\pi}{\lambda}}^{\delta} \frac{||G(t)||_p}{|2\sin\frac{1}{2}t|}|P(n,t)|dt \\
&= \int_{\frac{\pi}{\lambda}}^{\delta} ||G(t)||_p \left(\left|\frac{1}{2\sin\frac{1}{2}t} - \frac{1}{t}\right|\right) |p_q^n(t)| \left|\cos\left(n\Phi + \frac{1}{2}t\right)\right| dt \\
&+ \int_{\frac{\pi}{\lambda}}^{\delta} \frac{G(t)}{t} p_q^n(t) \cos\lambda t dt \\
&+ \int_{\frac{\pi}{\lambda}}^{\delta} \frac{G(t)}{t} p_q^n(t) \left[\cos\left(n\Phi + \frac{1}{2}t\right) - \cos\lambda t\right] dt \\
&= J_1(E) + J_2(E) + J_3(E).
\end{aligned} \tag{86}
$$

Using Lemmas 5.2, 6.1, and 5.5 and the fact that $\frac{1}{2\sin\frac{1}{2}t} - \frac{1}{t} = O(t)$, we obtain

$$
\begin{aligned}
||J_1(E)||_p &\le \int_{\frac{\pi}{\lambda}}^{\delta} ||G(t)||_p \left(\left|\frac{1}{2\sin\frac{1}{2}t} - \frac{1}{t}\right|\right) \left|p_q^n(t)\cos\left(n\Phi + \frac{1}{2}t\right)\right| dt \\
&= O(1)|u|^{\beta} \int_{\pi/\lambda}^{\delta} t^{\alpha-\beta+1} e^{-Ant^2} dt \\
&= O(1)|u|^{\beta} \int_{\pi/\lambda}^{\delta} t^{\alpha-\beta+1} e^{-Ant^2} dt \\
&= \frac{O(1)|u|^{\beta}}{n^{-1}}.
\end{aligned} \tag{87}
$$

Using Lemmas 5.2, 6.1, 6.4, and 5.5, we have

$$
\begin{aligned}
||J_3(E)||_p &\le \int_{\pi/\lambda}^{\delta} \frac{|G(t)|}{|t|} |p_q^n(t)| \left|\cos\left(n\Phi + \frac{1}{2}t\right) - \cos\lambda t\right| dt \\
&= O(1)|u|^{\beta} n \int_{\pi/\lambda}^{\delta} t^{\alpha-\beta+2} e^{-Ant^2} dt \\
&= O(1)|u|^{\beta} n \int_{\pi/\lambda}^{\delta} t^{\alpha-\beta+2} e^{-Ant^2} dt \\
&= \frac{O(1)|u|^{\beta}}{n^{\alpha-\beta}}.
\end{aligned} \tag{88}
$$

Collecting the estimates for $P(E)$, $S(E)$, $R(E)$, $I(E)$, $K(E)$, $J_1(E)$, and $J_3(E)$ from (80), (81), (82), (84), (85), (87), and (88), we obtain

$$||l_n^E(y+u)-l_n^E(y)||_p \le \frac{O(1)|u|^\beta}{n^{\alpha-\beta}} + J_2(x). \tag{89}$$

For $\lambda = \frac{n}{1+q} + \frac{1}{2}$,

$$p_q^n(t) = p_q^{\left(\lambda-\frac{1}{2}\right)(1+q)}(t) = c(\lambda, t).$$

Therefore, we may write

$$||l_n^E(y+u)-l_n^E(y)||_p \le \int_{\frac{\pi}{\lambda}}^{\delta} \frac{||l_n^E(y+u)-l_n^E(y)||_p}{|t|}|p_q^n(t)\cos\lambda t|dt$$

$$= \int_{\frac{\pi}{\lambda}}^{\delta} \frac{||G(t)||_p}{|t|}|c(\lambda,t)\cos\lambda t|dt. \tag{90}$$

Note that $c(\lambda, t)$ satisfies (i),(ii),(iii), and (iv) of Lemma 5.6. Therefore,

$$||J_2(E)||_p = O(1)|u|^\beta \begin{cases} \frac{1}{\lambda^{\alpha-\beta}} &, \alpha-\beta \neq 1 \\ \frac{\log\lambda}{\lambda} &, \alpha-\beta = 1 \end{cases}$$

$$= O(1)|u|^\beta \begin{cases} \frac{1}{n^{\alpha-\beta}} &, \alpha-\beta \neq 1 \\ \frac{\log n}{n} &, \alpha-\beta = 1, \end{cases} \tag{91}$$

which in conjunction with (89) gives us

$$\sup_{u\neq 0} |\Delta^\beta l_n^E(y+u,y)| = \sup_{u\neq 0} \frac{||l_n^E(y+u)-l_n^E(y)||_p}{|u|^\beta}$$

$$= O(1) \begin{cases} \frac{1}{n^{\alpha-\beta}} &, \alpha-\beta \neq 1 \\ \frac{\log n}{n} &, \alpha-\beta = 1. \end{cases} \tag{92}$$

Again $f \in H_{\alpha,p} \implies ||\theta_x(t)||_p = O(t^\alpha)$ and so using Lemma 5.6(b), 5.7(b), and 5.8(b), and proceeding as above, we obtain

$$||l_n^E(.)||_p = O(1) \begin{cases} \frac{1}{n^{\alpha}} &, 0<\alpha<1 \\ \frac{\log n}{n} &, \alpha = 1. \end{cases} \tag{93}$$

Combining (92) and (93), we get (34) and this completes the proof of Theorem 1.

7 Proof of Theorem 2

We will use the following additional notations and lemmas for the proof of Theorem 2:

$$l_p^\beta(x) = B_p(\widetilde{T}; x) - g(x)$$

$$H_p(t) = e^{-p} \sum_{n=0}^{\infty} \frac{p^n}{n!} \cos\left(n + \frac{1}{2}\right) t$$

$$G_p(t) = e^{-p} \sum_{n=0}^{\infty} \frac{p^n}{n!} \sin\left(n + \frac{1}{2}\right) t$$

$$\lambda = p + \frac{1}{2}$$

$$B(p) = e^{-p} \sum_{n=1}^{\infty} \frac{p^n}{n!} \sum_{k=n+1}^{\infty} \frac{(-1)^{k-1}}{k}.$$

Lemma 7.1 *Let* $0 < \delta < \pi/4$ *and let* $A = 2/\pi^2$. *Then*

$$\text{(i)}\ e^{-p(1-\cos t)} = O(e^{-Apt^2}) \tag{94}$$

$$\text{(ii)}\ e^{-p(1-\cos t)} - e^{-p(1-\cos(t+\pi/p))} = O(1)te^{-Apt^2}. \tag{95}$$

Lemma 7.2 *For* $0 < t \le \pi$

$$\text{(i)}\ H_p(t) = e^{-p(1-\cos t)} \cos\left(p \sin t + \frac{1}{2} t\right) \tag{96}$$

$$\text{(ii)}\ H_p(t) = O(1) \tag{97}$$

$$\text{(iii)}\ B(p) = O\left(\frac{1}{p}\right). \tag{98}$$

Proof By simple computation, we have

$$\begin{aligned} H_p(t) + iG(t) &= e^{-p} \sum_{n=0}^{\infty} \frac{p^n}{n!} e^{i\left(n+\frac{1}{2}\right)t} \\ &= e^{-p} e^{i\frac{1}{2}t} \sum_{n=0}^{\infty} \frac{(pe^{it})^n}{n!} \\ &= e^{-p} e^{i\frac{1}{2}t} e^{pe^{it}} \\ &= e^{-p(1-\cos t)} e^{i\left(p \sin t + \frac{1}{2}t\right)}, \end{aligned}$$

which ensures (ii) and (iii) follow from (i). Now,

$$\begin{aligned} B(p) &= e^{-p}\sum_{n=1}^{\infty}\frac{p^n}{n!}\sum_{k=n+1}^{\infty}\frac{(-1)^{k-1}}{k} \\ &= O(1)e^{-p}\sum_{n=1}^{\infty}\frac{p^n}{n!(n+1)} \\ &= O(1)\frac{e^{-p}}{p}\sum_{n=1}^{\infty}\frac{p^{n+1}}{(n+1)!} \\ &= O(1)\frac{e^{-p}}{p}e^{p} \\ &= O\left(\frac{1}{p}\right). \end{aligned}$$

Lemma 7.3 *Let $\lambda = p + \frac{1}{2}$ and $0 < \delta < \pi/4$. Then for $0 < t < \delta$.*

$$\cos\left(p\sin t + \frac{1}{2}t\right) - \cos\lambda t = O(pt^3). \tag{99}$$

Proof Expressing the difference $\cos\left(p\sin t + \frac{1}{2}t\right) - \cos\lambda t$ as a product and making use of the fact that $\sin t - t = O(t^3)$, the estimates (99) can be established.

Proof Proof of Theorem 2. From (76), we get for $n \geq 1$

$$\widetilde{T}_n(x) - g(x) + \int_0^{\pi}\theta_x(t)\frac{\cos\left(n+\frac{1}{2}t\right)}{2\sin\frac{1}{2}t}dt - \widetilde{f}(x)\sum_{\nu=n+1}^{\infty}\frac{(-1)^{\nu-1}}{\nu}. \tag{100}$$

Hence, Borel's exponential mean $B_p(\widetilde{T}_n; x)$ of $\{\widetilde{T}_n(x)\}$ is given by

$$\begin{aligned} B_p(\widetilde{T}_n; x) &= e^{-p}\sum_{n=1}^{\infty}\frac{p^n}{n!}\left[g(x) + \int_0^{\pi}\theta_x(t)\frac{\cos\left(n+\frac{1}{2}t\right)}{2\sin\frac{1}{2}t}dt - \widetilde{f}(x)\sum_{\nu=n+1}^{\infty}\frac{(-1)^{\nu-1}}{\nu}\right] \\ &= (1-e^{-p})g(x) + \int_0^{\pi}\frac{\theta_x(t)}{2\sin\left(\frac{1}{2}t\right)}\left(H_p(t) - e^{-p}\cos\left(\frac{1}{2}t\right)\right)dt - \widetilde{f}(x)B(p), \end{aligned}$$

which ensure that

$$l_p^{\beta}(x) = B_p(\widetilde{T}, x) - g(x)$$

$$= e^{-p}g(x) + \int_0^{\pi} \frac{\theta_x(t)}{2\sin\frac{1}{2}t} H_p(t)dt - e^{-p}\int_0^{\pi} \frac{\theta_x(t)}{2\tan\frac{1}{2}t}dt - \widetilde{f}(x)B(p). \quad (101)$$

Therefore,

$$||l_n^E(y+u) - l_n^E(y)||_p = \left\| e^{-p}G(x,y) + \int_0^{\pi} \frac{G(t)}{2\sin\frac{1}{2}t} H_p(t)dt \right.$$
$$\left. - e^{-p}\int_0^{\pi} \frac{G(t)}{2\tan\frac{1}{2}t}dt - \widetilde{F}(x,y)B(p) \right\|_p$$

$$= ||P(B) + Q(B) - R(B) - S(B)||_p. \quad (102)$$

By Minkowski's inequality,

$$||l_n^{\beta}(y+u) - l_n^{\beta}(y)||_p \le ||P(B)||_p + ||Q(B)||_p + ||R(B)||_p + ||S(B)||_p.$$

Using the estimates for $G(y+u, y), G(t), \widetilde{F}(x, y)$, and $B(p)$ and adopting the technique employed for deriving the estimates for $P(E), R(E)$, and $S(E)$ in the proof of Theorem 1, it can be shown that

$$||P(B)||_p = O(1)|u|^{\beta}e^{-p}, \quad (103)$$

$$||R(B)||_p = O(1)|u|^{\beta}e^{-p}, \quad (104)$$

and

$$||S(B)||_p = O(1)|u|^{\beta}p^{-1}, \quad (105)$$

we put $\lambda = p + \frac{1}{2}$. Now for fixed δ with $0 < \delta < \pi/4$, we write

$$||Q(B)||_p \le \left\| \left[\int_0^{\frac{\pi}{\lambda}} + \int_{\frac{\pi}{\lambda}}^{\delta} + \int_{\delta}^{\pi} \right] \frac{G(t)}{2\sin\frac{1}{2}t} H_p(t)dt \right\|_p$$

$$\le ||I(B)||_p + ||J(B)||_p + ||K(B)||_p. \quad (106)$$

By Lemmas 7.2(ii), 5.7, and the generalized Minkowski inequality given in Section 5, we have, by usual estimation technique,

$$||I(B)||_p \le O(1)\int_0^{\frac{\pi}{\lambda}} \frac{||G(t)||_p}{|t|}dt = \frac{O(1)|u|^{\beta}}{p^{\alpha-\beta}}. \quad (107)$$

By Lemmas 7.1, 7.2(i), and 5.8, and the generalized Minkowski inequality given in Section 5, we have, by usual estimation technique,

$$||K(B)||_p \leq \int_\delta^\pi \frac{||G(t)||_p}{|2\sin\frac{1}{2}t|}|e^{-p(1-\cos t)}| \left|\cos\left(p\sin t + \frac{1}{2}t\right)\right| dt$$

$$= O(1)\int_\delta^\pi \frac{||G(t)||_p}{|t|} e^{-Apt^2} dt$$

$$= O(1)\frac{|u|^\beta}{p^\Delta}, \quad \Delta \text{ positive however large.} \tag{108}$$

We write

$$||J(B)||_p = \left\| \int_{\frac{\pi}{\lambda}}^{\delta} \frac{G(t)}{|2\sin\frac{1}{2}t|} e^{-p(1-\cos t)} \cos\left(p\sin t + \frac{1}{2}t\right) dt \right\|_p$$

$$= \left\| \int_{\pi/\lambda}^{\delta} G(t)\left(\frac{1}{2\sin\frac{1}{2}t} - \frac{1}{t}\right) e^{-p(1-\cos t)} \cos\left(p\sin t + \frac{1}{2}t\right) dt \right.$$

$$+ \int_{\frac{\pi}{\lambda}}^{\delta} \frac{G(t)}{t} e^{-p(1-\cos t)} \cos \lambda t dt$$

$$\left. + \int_{\frac{\pi}{\lambda}}^{\delta} \frac{G(t)}{t} e^{-p(1-\cos t)} \left\{\cos\left(p\sin t + \frac{1}{2}\right) - \cos\lambda t\right\} dt \right\|_p$$

$$\leq ||J_1(B)||_p + ||J_2(B)||_p + ||J_3(B)||_p, \tag{109}$$

by Minkowski's inequality. Using Lemmas 5.2, 7.2(i), 7.3, and 5.5 and proceeding as in the proof of $J_1(E)$ and $J_3(E)$, it can be shown that

$$||J_1(B)||_p = O(1)|u|^\beta p^{-1}, \tag{110}$$

and

$$||J_3(B)||_p = \frac{O(1)|u|^\beta}{p^{\alpha-\beta}}. \tag{111}$$

Collecting the estimates for $P(B)$, $S(B)$, $R(B)$, $I(B)$, $K(B)$, $J_1(B)$, and $J_3(B)$ from (103), (104), (105), (107), (108), (110), and (111), we obtain

$$||l_n^\beta(y+u) - l_n^\beta(y)||_p \le ||J_2(B)||_p + O(1)\frac{|u|^\beta}{p^{\alpha-\beta}}. \tag{112}$$

For $\lambda = p + \frac{1}{2}$ $\left(i.e., p = \lambda - \frac{1}{2}\right)$, the expression $e^{-p(1-\cos t)}$ reduces to $e^{-\left(\lambda-\frac{1}{2}\right)(1-\cos t)} = c(\lambda, t)$.

In view of Lemma 7.1, the function $c(\lambda, t)$ satisfies all the requirements of Lemma 5.6 and hence

$$||J_2(B)||_p = \left\| \int_{\frac{\pi}{\lambda}}^{\delta} \frac{G(t)}{t} e^{-p(1-\cos t)} \cos \lambda t dt \right\|_p$$

$$= \left\| \int_{\frac{\pi}{\lambda}}^{\delta} \frac{G(t)}{t} c(\lambda, t) \cos \lambda t dt \right\|_p$$

$$= O(1)|u|^\beta \begin{cases} \frac{1}{p^{\alpha-\beta}} & , \alpha - \beta \neq 1 \\ \frac{\log p}{p} & , \alpha - \beta = 1, \end{cases} \tag{113}$$

by the method used previously. From (112) and (113), we obtain

$$||l_n^\beta(y+u) - l_n^\beta(y)||_p \le O(1)|u|^\beta \begin{cases} \frac{1}{p^{\alpha-\beta}} & , \alpha - \beta \neq 1 \\ \frac{\log p}{p} & , \alpha - \beta = 1, \end{cases}$$

which ensures that

$$\sup_{u\neq 0} |\Delta^\beta l_p^\beta(y+u, y)| = \sup_{u\neq 0} \frac{|l_p^\beta(y+u) - l_p^\beta(y)|}{|u|^\beta}$$

$$= O(1) \begin{cases} \frac{1}{p^{\alpha-\beta}} & , \alpha - \beta \neq 1 \\ \frac{\log p}{p} & , \alpha - \beta = 1, \end{cases} \tag{114}$$

when $f \in H_{\alpha,p}$ it can be shown that

$$||l_n^\beta(.)|| = \sup_{-\pi\le x\le\pi} \left|l_p^\beta(x)\right| = O(1) \begin{cases} \frac{1}{p^{\alpha}} & , 0 < \alpha < 1 \\ \frac{\log p}{p} & , \alpha = 1. \end{cases} \tag{115}$$

From (114) and (115), we obtain (32) and this completes the proof of Theorem 2.

8 Proof of Theorem 3

We will use the following notations and lemmas for Theorem 3:

$$\theta(n) = \sqrt{\frac{c}{\pi n}} \sum_{k=-(n-1)}^{\infty} e^{\frac{-ck^2}{n}}$$

$$K_n(t) = \sqrt{\frac{c}{\pi n}} \left(1 + 2\sum_{k=1}^{n-1} e^{\frac{-ck^2}{n}} \cos kt\right)$$

$$L_n(t) = \sqrt{\frac{c}{\pi n}} \sum_{k=1}^{\infty} e^{\frac{-ck^2}{n}} \cos\left(n + k + \frac{1}{2}\right) t$$

$$l_n^e(x) = e_n(\widetilde{T}; x) - g(x)$$

$$e(n) = \sqrt{\frac{c}{\pi n}} \sum_{k=-(n-1)}^{\infty} e^{\frac{-ck^2}{n}} \sum_{\nu=n+k+1}^{\infty} \frac{(-1)^{\nu-1}}{\nu}$$

$$\lambda = n + \frac{1}{2}, \quad A = \frac{1}{4c}, \quad h = \pi/\lambda.$$

We need the following lemmas:

Lemma 8.1 *Let $c > d > 0$. Then*

$$K_n(t) = e^{-nAt^2} + \psi(n), \tag{116}$$

where

$$\psi(n) = O(e^{-dn}).$$

Lemma 8.2 *For $c > 0$*

$$\text{(i)}\ L_n(t) = O(1)\frac{e^{-cn}}{\sqrt{n}}, \tag{117}$$

$$\text{(ii)}\ e(n) = O\left(n^{-\frac{1}{2}}\right). \tag{118}$$

Proof

(i) We have

$$L_n(t) = \sqrt{\frac{c}{\pi n}} \sum_{k=n}^{\infty} e^{\frac{-ck^2}{n}} \cos\left(n + k + \frac{1}{2}\right) t$$

$$
\begin{aligned}
&= O\left(n^{-\frac{1}{2}}\right)\sum_{k=n}^{\infty} e^{\frac{-ck^2}{n}} \\
&= O\left(n^{-\frac{1}{2}}\right)\int_{n}^{\infty} \frac{n}{2cx}\frac{d}{dx}\left(-e^{\left(-\frac{cx^2}{n}\right)}\right) dx \\
&= O\left(n^{-\frac{1}{2}}\right) e^{-cn}.
\end{aligned}
$$

(ii) Clearly,

$$
\sum_{\nu=n+k+1}^{\infty} \frac{(-1)^{\nu-1}}{\nu} = O(1)\frac{1}{n+k+1}, \quad \text{whenever } n+k+1>0,
$$

and so

$$
\begin{aligned}
\sqrt{\frac{\pi n}{c}} e(n) &= \sum_{k=-(n-1)}^{\infty} e^{\frac{-ck^2}{n}} \sum_{\nu=n+k+1}^{\infty} \frac{(-1)^{\nu-1}}{\nu} \\
&= O(1) \sum_{k=-(n-1)}^{\infty} \frac{e^{\frac{-ck^2}{n}}}{n+k+1} \\
&= O(1)\left[\frac{1}{n+1} + \sum_{k=1}^{n-1} \frac{e^{\frac{-ck^2}{n}}}{n-k+1} + \sum_{k=1}^{\infty} \frac{e^{\frac{-ck^2}{n}}}{n+k+1}\right] \\
&= O(1)\left[\frac{1}{n+1} + S_1 + S_2\right]. \qquad (119)
\end{aligned}
$$

As $e^{-\frac{ck^2}{n}} \le n/ck^2$, we have

$$
S_2 \le \frac{n}{c}\sum_{k=1}^{\infty} \frac{1}{(n+k+1)k^2} = O(1), \qquad (120)
$$

lastly,

$$
\begin{aligned}
S_1 &= \sum_{k=1}^{M} \frac{e^{\frac{-ck^2}{n}}}{n-k+1} + \sum_{k=M+1}^{n-1} \frac{e^{\frac{-ck^2}{n}}}{n-k+1}, \quad M = \left[\frac{n}{2}\right] \\
&= O(1)\frac{1}{n-M+1}\sum_{k=1}^{M} e^{\frac{-ck^2}{n}} + O(1)e^{\frac{-c(M+1)^2}{n}} \sum_{k=1}^{n} \frac{1}{n-k+1}
\end{aligned}
$$

$$= O(1)\frac{M}{n-M+1} + O(1)e^{-\frac{cn}{4}}\log n$$
$$= O(1). \tag{121}$$

From (119), (120), and (121), the second part of the lemma follows.

Lemma 8.3 *For the functions $e_n(t)$, $K_n(t)$ and $L_n(t)$, we have*

$$e_n(t) = K_n(t)\cos\left(n+\frac{1}{2}\right)t + L_n(t). \tag{122}$$

Proof We have

$$e_n(t) = \sqrt{\frac{c}{\pi n}} \sum_{k=-(n-1)}^{\infty} e^{\frac{-ck^2}{n}} \cos\left(n+k+\frac{1}{2}\right)t$$
$$= \sqrt{\frac{c}{\pi n}} \sum_{k=-(n-1)}^{n-1} e^{\frac{-ck^2}{n}} \cos\left(n+k+\frac{1}{2}\right)t + \sqrt{\frac{c}{\pi n}} \sum_{k=n}^{\infty} e^{\frac{-ck^2}{n}} \cos\left(n+k+\frac{1}{2}\right)t$$
$$= \sqrt{\frac{c}{\pi n}} \left[\sum_{k=1}^{n-1} e^{\frac{-ck^2}{n}} \left(\cos\left(n+k+\frac{1}{2}\right)t + \cos\left(n-k+\frac{1}{2}\right)t\right)\right.$$
$$\left. + \cos\left(n+\frac{1}{2}\right)t\right] + L_n(t),$$

which ensures that

$$\sqrt{\frac{\pi n}{c}}(e_n(t) - L_n(t)) = \sum_{k=1}^{n-1} e^{\frac{-ck^2}{n}} \left(\cos\left(n+k+\frac{1}{2}\right)t + \cos\left(n-k+\frac{1}{2}\right)t\right)$$
$$+\cos\left(n+\frac{1}{2}\right)t$$
$$= 2\left[\sum_{k=1}^{n-1} e^{\frac{-ck^2}{n}} \cos kt + 1\right] \cos\left(n+\frac{1}{2}\right)t$$
$$= \sqrt{\frac{\pi n}{c}} K_n(t) \cos\left(n+\frac{1}{2}\right)t,$$

from which (122) follows.

Proof Proof of Theorem 3: Collecting the expression for $\widetilde{T}$ from (81), we have

$$e_n(\widetilde{T};x) = \sqrt{\frac{c}{\pi n}}\sum_{-\infty}^{\infty} e^{\frac{-ck^2}{n}}\widetilde{T}_{n+k}(x)$$

$$= \sqrt{\frac{c}{\pi n}}\sum_{-n+1}^{\infty} e^{\frac{-ck^2}{n}}\left[g(x) + \int_0^{\pi}\theta_x(t)\frac{\cos\left(n+k+\frac{1}{2}\right)t,}{2\sin\frac{1}{2}t}dt\right.$$

$$\left. -\widetilde{f}(x)\sum_{\nu=n+k+1}^{\infty}\frac{(-1)^{\nu-1}}{\nu}\right]$$

$$= \theta(n)g(x) + \int_0^{\pi}\frac{\theta_x(t)}{2\sin\frac{1}{2}t}e_n(t)dt - \widetilde{f}(x)e(n). \tag{123}$$

Thus,

$$l_n^e(x) = e_n(\widetilde{T};x) - g(x)$$
$$= (\theta(n)-1)g(x) + \int_0^{\pi}\frac{\theta_x(t)}{2\sin\frac{1}{2}t}e_n(t)dt - \widetilde{f}(x)e(n),$$

which further ensures that

$$||l_n^e(y+u) - l_n^e(y)||_p \tag{124}$$
$$= \left\|(\theta(n)-1)G(x,y) + \int_0^{\pi}\frac{\theta_x(t)}{2\sin\frac{1}{2}t}e_n(t)dt - \widetilde{F}(x,y)e(n)\right\|_p .$$

Using Lemmas 8.3 and 8.1, we can rewrite (124) as follows:

$$||l_n^e(y+u) - l_n^e(y)||_p$$
$$\leq ||(\theta(n)-1)G(y+u,y)||_p + \int_0^{\pi}\frac{||G(t)||_p}{|2\sin\frac{1}{2}t|}|K_n(t)|\left|\cos\left(n+\frac{1}{2}\right)t\right|dt$$
$$+ \int_0^{\pi}\frac{||G(t)||_p}{|2\sin\frac{1}{2}t|}|L_n(t)|dt + |\widetilde{F}(y+u,y)e(n)|$$
$$\leq ||(\theta(n)-1)G(y+u,y)||_p + \int_0^{\pi}\frac{||G(t)||_p}{|2\sin\frac{1}{2}t|}e^{-Ant^2}\left|\cos\left(n+\frac{1}{2}\right)t\right|dt$$
$$+ \left\|\psi(n)\int_0^{\pi}G(t)\frac{\cos\left(n+\frac{1}{2}\right)t}{2\sin\frac{1}{2}t}dt\right\|_p + ||\widetilde{F}(x,y)e(n)||_p + \int_0^{\pi}\frac{||G(t)||_p}{|2\sin\frac{1}{2}t|}|L_n(t)|dt$$

$$= ||P(e)||_p + ||Q(e)||_p + ||R(e)||_p + ||S(e)||_p + ||T(e)||_p. \tag{125}$$

As $\theta(n) - 1 = O\left(n^{-\frac{1}{2}}\right)$, using Lemma 5.4(ii), we have

$$||P(e)||_p = ||(\theta(n) - 1)G(y+u, y)||_p = O(1)|u|^{\beta} n^{-\frac{1}{2}}. \tag{126}$$

By Lemmas 5.2 and 8.1,

$$\begin{aligned} ||R(e)||_p &= \left\| \psi(n) \int_0^{\pi} G(t) \frac{\cos\left(n+\frac{1}{2}\right)t}{2\sin\frac{1}{2}t} dt \right\|_p \\ &= O(1)|u|^{\beta} e^{-dn} \int_0^{\pi} t^{\alpha-\beta-1} dt \\ &= O(1)|u|^{\beta} e^{-dn}. \end{aligned} \tag{127}$$

By Lemmas 5.2 and 8.2(i),

$$\begin{aligned} ||T(e)||_p &\leq \int_0^{\pi} \frac{||T(e)||_p}{|2\sin\frac{1}{2}t|} |L_n(t)| dt \\ &= O(1)|u|^{\beta} \frac{e^{-cn}}{\sqrt{n}} \int_0^{\pi} t^{\alpha-\beta-1} dt \\ &= O(1)|u|^{\beta} \frac{e^{-c}}{\sqrt{n}}. \end{aligned} \tag{128}$$

Using Lemmas 5.4(i) and 8.2(ii), we obtain

$$||S(e)||_p = ||\widetilde{F}(x, y)e(n)||_p = O(1)|u|^{\beta} n^{-\frac{1}{2}}. \tag{129}$$

Collecting the result from (125)–(129), we get

$$||l_n^e(y+u) - l_n^e(y)||_p \leq ||Q(e)||_p + O(1)|u|^{\beta} n^{-\frac{1}{2}} + O(1)|u|^{\beta} n^{\beta-\alpha}. \tag{130}$$

We put $\lambda = n + \frac{1}{2}$. Now for fixed δ with $0 < \delta < \pi/4$, we split the integral as follows:

$$\begin{aligned} ||Q(e)||_p &\leq \left[\int_0^{\pi/4} + \int_{\pi/4}^{\delta} + \int_{\delta}^{\pi}\right] \frac{||G(t)||_p}{|2\sin\frac{1}{2}t|} \left|e^{-Ant^2}\right| \left|\cos\left(n+\frac{1}{2}\right)t\right| dt \\ &= ||I(e)||_p + ||J(e)||_p + ||K(e)||_p. \end{aligned} \tag{131}$$

Following the same lines of argument used in obtaining estimates for $I(B)$ and $K(B)$ as in the proof of Theorem 2, it can be shown for $0 \le \beta < \alpha \le 1$,

$$||I(e)||_p = O(1)\frac{|u|^\beta}{n^{\alpha-\beta}} \tag{132}$$

$$||K(e)||_p = O(1)\frac{|u|^\beta}{n^{\Delta}}, \quad \Delta > 0. \tag{133}$$

Next, we write

$$\begin{aligned}||J(e)||_p &\le \int_{\frac{\pi}{\lambda}}^{\delta} ||G(t)||_p \left(\left|\frac{1}{2\sin\frac{1}{2}t} - \frac{1}{t}\right|\right) |e^{-Ant^2}| \left|\cos\left(n+\frac{1}{2}\right)t\right| dt \\ &+ \int_{\frac{\pi}{\lambda}}^{\delta} \frac{||G(t)||_p}{|t|} e^{-Ant^2} \left|\cos\left(n+\frac{1}{2}\right)t\right| dt \\ &= ||J_1(e)||_p + ||J_2(e)||_p.\end{aligned} \tag{134}$$

Using Lemmas 5.2 and 5.5, and proceeding as in the proof of $J_1(E)$, it can be shown that

$$||J_1(e)||_p = O(1)|u|^\beta n^{-1}. \tag{135}$$

From (130)–(135), it follows that

$$||l_n^e(y+u) - l_n^e(y)||_p \le ||J_2(e)||_p + O(1)|u|^\beta n^{-\frac{1}{2}} + O(1)|u|^\beta n^{\beta-\alpha}. \tag{136}$$

For $\lambda = n + \frac{1}{2}$, $e^{-Ant^2} = e^{-\left(\lambda - \frac{1}{2}\right)t^2} = c(\lambda, t)$. Clearly, $c(\lambda, t)$ satisfies the conditions of Lemma 5.6, and hence,

$$\begin{aligned}||J_2(e)||_p &\le \int_{\frac{\pi}{\lambda}}^{\delta} \frac{||G(t)||_p}{|t|} e^{-A\left(\lambda-\frac{1}{2}\right)t^2} |\cos\lambda t| dt \\ &= \int_{\frac{\pi}{\lambda}}^{\delta} \frac{||G(t)||_p}{|t|} |c(\lambda, t)\cos\lambda t| dt \\ &= O(1)|u|^\beta \begin{cases} \frac{1}{n^{\alpha-\beta}} &, \alpha - \beta \ne 1 \\ \frac{\log n}{n} &, \alpha - \beta = 1. \end{cases}\end{aligned} \tag{137}$$

From (136) and (137), it follows that

$$||\Delta^\beta l_n^e(y+u, y)||_p = \left\| \frac{l_n^e(y+u) - l_n^e(y)}{u^\beta} \right\|_p$$

$$= O(1) \begin{cases} \frac{1}{n^{\alpha-\beta}} & , 0 < \alpha - \beta \le \frac{1}{2} \\ \frac{1}{\sqrt{n}} & , \frac{1}{2} \le \alpha - \beta \le 1. \end{cases} \tag{138}$$

Again, $f \in H_{\alpha,p} \implies ||\theta_x(t)||_p = O(t^\alpha)$, and so proceeding as above, we obtain

$$||l_n^e(.)||_p = O(1) \begin{cases} \frac{1}{n^{\alpha}} & , 0 < \alpha \le \frac{1}{2} \\ \frac{1}{\sqrt{n}} & , \frac{1}{2} \le \alpha \le 1. \end{cases} \tag{139}$$

Now (33) follows from (138) and (139) and this completes the proof of Theorem 3.

9 Concluding Remarks

The findings of this paper show that if p tends to infinity, we get the corresponding results in the supremum norm as found in Sadangi [18]. It will be interesting to see how to extend these results for functions in Orlicz spaces and Besov spaces.

References

1. G. Alexits, *Convergence Problems of Orthogonal Series* (Pergamon Elmsford, New York, 1961)
2. P. Chandra, On the generalized Fejér means in the metric of the Hölder space. Math. Nachr. **109**, 39–45 (1982)
3. P. Chandra, Degree of approximation of functions in the Hölder metric by Borel's means. J. Math Anal. Appl. **149**, 236–246 (1990)
4. R. Cooke, *Infinite Matricies and Sequence Spaces* (Macmillan, London, 1950)
5. G. Das, A.K. Ojha, B.K. Ray, Degree of approximation of functions associated with Hardy-Littlewood series in the Hölder metric by Borel Means. J. Math Anal. Appl. **219**, 279–293 (1998)
6. G. Das, B.K. Ray, P. Sadangi, Rate of convergence of a series associated with Hardy-Littlewood series. J. Orissa Math. Soc. **17–20**, 127–141 (1998–2001)
7. G. Das, T. Ghosh, B.K. Ray, Degree of approximation of functions in the Hölder metric by (e,c) Means. Proc. Indian Acad. Sci. (Math. Sci.) **105**, 315–327 (1995)
8. J.H. Freilich, J.C. Mason, Best and Near-best L1 approximations by Fourier Series and Chebyshev Series. J. Approx. Theory **4**, 183–193 (1971)
9. L. Gogoladze, On a Problem of L. Leindler concerning strong approximation by Fourier Series and Lipschitz classes. Anal. Math. **9**(3), 169–175 (1983)
10. G.H. Hardy, J.E. Littlewood, The allied series of a Fourier series. Proc. Lond. Math. Soc. **24**, 211–216 (1925)
11. G.H. Hardy, *Divergent Series* (Clarendon, Oxford, 1949)
12. G.H. Hardy, J.E. Littlewood, G. Pólya, *Inequalities* (Cambridge University Press, Oxford, UK, 1952)
13. R.A. Lasuriya, Approximation of functions on the real axis by Féjer-type operators in the generalized Hölder metric. Math. Z. **81**(4), 547–552 (2007)

14. A.V. Lototsky, On a linear transformation of sequences and series. Ped. Inst. Uch. Zap. Fix-Mat. Nauki **4**, 61–91 (1953)(Russian)
15. R.N. Mohapatra, P. Chandra, Degree of approximation of functions in the Hölder metric. Acta. Math. Hung. **41**(1–2), 67–76 (1983)
16. F. Móricz, J. Németh, Generalized zygmund classes of functions and strong approximation by Fourier series. Acta Sci. Math. **73**(3–4), 637–647 (2007)
17. S. Prössdorf, Zur konvergenz der Fourier reihen Hölder steliger Funktionen. Math. Nachr. **69**, 7–14 (1975)
18. P. Sadangi, Degree of Convergence of functions in the Hölder metric, Ph.D. Thesis, Utkal University, 2006
19. T. Singh, Degree of approximation of functions in a normed space. Publ. Math. Debr. **40**(3–4), 261–267 (1992)
20. T. Singh, Approximation to functions in the Hölder metric. Proc. Nat. Acad. Sci. India **62**(A), 224–233 (1992)
21. V. Vuckovic, The summability of Fourier series by Karamata methods. Math. Z. **89**, 192–195 (1965)
22. A. Zygmund, *Trigonometric Series*, vol. I (Cambridge University Press, New York, 1959)

Real Variable Methods in Harmonic Analysis and Navier–Stokes Equations

Pierre Gilles Lemarié-Rieusset

Abstract Real variable methods in harmonic analysis were developed throughout the works of E.M. Stein. They turn out to be a powerful tool for the study of nonlinear PDEs. We illustrate this point by discussing various points of the modern theory of Navier–Stokes equations.

1 Introduction

Among the seven Millennium problems proposed by the Clay Mathematics Institute, I shall consider the question of existence and smoothness of solutions to the Navier–Stokes equations. Let us first recall the question raised by the Clay Mathematics Institute as it has been presented by Ch. Fefferman in his 2000 talk at the Collège de France [34] :

> Let $\mathbf{u}^0$ be any smooth, divergence-free vector field in the Schwartz class $\mathscr{S}(\mathbb{R}^3)$ Do there exist smooth functions $p(t,x)$, $u_i(t,x) = (u_1(t,x), u_2(t,x), u_3(t,x))$ on $\mathbb{R}^3 \times [0,\infty)$ that satisfy
>
> $$\partial_t \mathbf{u} + (\mathbf{u} \cdot \nabla)\mathbf{u} = \Delta \mathbf{u} - \nabla p,$$
>
> $$\nabla \cdot \mathbf{u} = 0,$$
>
> $$\mathbf{u}(0,.) = \mathbf{u}^0,$$
>
> $$u, p \in \mathscr{C}^\infty(\mathbb{R}^3 \times (0,+\infty))$$
>
> and

P. G. Lemarié-Rieusset (✉)
LaMME, Univ Evry, CNRS, Université Paris-Saclay, Evry, France
e-mail: pierregilles.lemarierieusset@univ-evry.fr

M. Th. Rassias (ed.), *Harmonic Analysis and Applications*, Springer Optimization and Its Applications 168, https://doi.org/10.1007/978-3-030-61887-2_10

$$\sup_{t>0} \|\mathbf{u}(t,.)\|_2 \leq \|\mathbf{u}^0\|_2?$$

Commenting on the Clay Millennium Prize on Navier–Stokes equations, L. Tartar writes in 2006 [92]:

> Reading the text of the conjectures to be solved for winning that particular prize leaves the impression that the subject was not chosen by people interested in continuum mechanics, as the selected questions have almost no physical content. /..../ The problems seem to have been chosen in the hope that they will be solved by specialists of harmonic analysis, and it has given the occasion to some of these specialists to help others in showing the techniques that they use, as in a recent book by Pierre Gilles LEMARIÉ-RIEUSSET[1] /..../.

The question I'd like to discuss here is to which extent harmonic analysis is used or should be used to study the Clay question on Navier–Stokes equations? Or, more generally, to discuss the Navier–Stokes equations on the whole space $\mathbb{R}^3$ in various functional settings while using tools from real-variable methods in harmonic analysis.

The very first point is, of course, to define correctly what is called here harmonic analysis. While classical harmonic analysis has been devoted in the nineteenth century to the spectral analysis of the Laplace operator and of the heat equation, I shall focus on the theory that has been developed in the second half of the twentieth century, mainly in the works of E. M. Stein [90]. (For a short account of this history, see the recent paper of G. B. Folland [38]). As a matter of fact, the two Fields medalists who play influential roles on the Clay problem on Navier–Stokes equations, namely, Ch. Fefferman and T. Tao, are both former students of E. M. Stein. A good account of this harmonic analysis theory is to be found in the books by Grafakos [50, 51].

The work of E. M. Stein is well illustrated by the titles of two of his books: *Singular Integrals and Differentiability Properties of Functions* [89] and *Harmonic Analysis : Real-Variable Methods, Orthogonality, and Oscillatory Integrals* [90]. A major topic was the extension of Littlewood–Paley theory from the disc to $\mathbb{R}^n$. This is closely related to the study of Sobolev spaces and of Besov spaces, a class of spaces he studied thoroughly in his book on singular integrals [89].

Littlewood–Paley–Stein decomposition of distributions and Besov spaces turned to be a fundamental tool for the modern approach of the Navier–Stokes equations and are the center of many books devoted to harmonic analysis and Navier–Stokes equations, such as M. Cannone's *Harmonic Analysis Tools for Solving the Incompressible Navier–Stokes Equations* [14], H. Bahouri, J.Y. Chemin and R. Danchin's *Fourier Analysis and Nonlinear Partial Differential Equations* [2], or P.G. Lemarié-Rieusset's *Recent Developments in the Navier–Stokes Problem* [70].

But we shall try to show that the interaction of harmonic analysis with Navier–Stokes equations is broader than the scope of Littlewood–Paley decomposition and that many other ideas of E.M. Stein can be useful for future works. We shall pay a few words on the Clay problem but as well on some points of the Navier–Stokes

[1]The book is the one I published in 2002 [70].

theory in more general settings such as Kato's mild solutions in L^3 (existence and uniqueness), or Serrin criteria for weak–strong uniqueness or regularity of Leray weak solutions.

2 Fourier Transform

2.1 Fourier–Navier–Stokes Equations

Naturally, Fourier transform plays an important role in the study of our problem, as it is a differential problem with constant coefficients and defined on the whole space. If we note $\mathscr{F}_x$ the spatial Fourier transform

$$\mathscr{F}_x(f)(t,x) = \int_{\mathbb{R}^3} f(t,x)\, e^{-ix\cdot\xi}\, dx,$$

the Navier–Stokes equations are turned into

$$\partial_t \mathscr{F}_x \mathbf{u}(t,\xi) + \sum_{j=1}^{3} i\xi_j \mathscr{F}_x(u_j \mathbf{u})(t,\xi) = -\nu|\xi|^2 \mathscr{F}_x \mathbf{u}(t,\xi) - i\mathscr{F}_x p(t,\xi)\, \boldsymbol{\xi}$$

$$\boldsymbol{\xi} \cdot \mathscr{F}_x \mathbf{u}(t,\xi) = 0$$

$$\mathscr{F}_x \mathbf{u}(0,\xi) = \mathscr{F}_x \mathbf{u}^0(\xi) = \mathbf{U}^0(\xi).$$

Moreover, we have

$$\mathscr{F}_x(u_j \mathbf{u})(t,\xi) = \frac{1}{(2\pi)^3} \int_{\mathbb{R}^3} \mathscr{F}_x u_j(t,\eta) \mathscr{F}_x \mathbf{u}(t,\xi-\eta)\, d\eta$$

and

$$|\xi|^2 \mathscr{F}_x p(t,\xi) = -\sum_{j=1}^{3}\sum_{k=1}^{3} \xi_j \xi_k \mathscr{F}_x(u_j u_k)(t,\xi).$$

This gives the following simple system on the vector $\mathscr{F}_x \mathbf{u} = (\mathscr{F}_x u_1, \mathscr{F}_x u_2, \mathscr{F}_x u_3)$

$$\begin{aligned}\partial_t \mathscr{F}_x u_l(t,\xi) =& -|\xi|^2 \mathscr{F}_x u_l(t,\xi) \\ & - \sum_{j=1}^{3}\sum_{k=1}^{3} \int_{\mathbb{R}^3} \frac{i\xi_j \xi_k}{(2\pi)^3|\xi|^2} (\xi_l \mathscr{F}_x u_k(t,\eta) - \xi_k \mathscr{F}_x u_l(t,\eta)) \mathscr{F}_x u_j(t,\xi-\eta) d\eta.\end{aligned}$$

This is turned into an integral equation on $\mathbf{U} = \mathscr{F}_x \mathbf{u}$:

$$\mathbf{U}(t,\xi) = e^{-t|\xi|^2}\mathbf{U}^0(\xi) - B(\mathbf{U},\mathbf{U})(t,\xi)$$

with

$$B(U,V)_l(t,\xi) = \int_0^t e^{-(t-s)|\xi|^2} \sum_{j=1}^3 \sum_{k=1}^3 \int_{\mathbb{R}^3} \frac{i\xi_j\xi_k}{(2\pi)^3|\xi|^2}(\xi_l U_k(s,\eta) \\ -\xi_k U_l(s,\eta))V_j(s,\xi-\eta)\,d\eta\,ds.$$

This allows very simple computations for the search of solutions. Indeed, let us assume that $\mathbf{U}^0$ is controlled by a function W^0:

$$|\mathbf{U}^0(\xi)| \le W^0(\xi)$$

and that $W(t,\xi)$ is measurable, almost everywhere finite, and is a non-negative solution of the integral inequation for every $t \in [0,T]$ and every $\xi \in \mathbb{R}^3$

$$e^{-t|\xi|^2}W^0(\xi) + B_0(W,W)(t,\xi) \le W(t,\xi)$$

with

$$B_0(W,V)(t,\xi) = \frac{18}{(2\pi)^3}\int_0^t e^{-(t-s)|\xi|^2}|\xi| \int_{\mathbb{R}^3} W(s,\eta)V(s,\xi-\eta)\,d\eta\,ds.$$

Define $W^{[0]}(t,\xi){:=}e^{-t|\xi|^2}W^0(\xi)$, $W^{[n+1]}(t,\xi){:=}W^{[0]}(t,\xi){+}B_0(W^{[n]},W^{[n]})(t,\xi)$ and similarly $\mathbf{U}^{[0]} := e^{-t|\xi|^2}\mathbf{U}^0(\xi)$ and $\mathbf{U}^{[n+1]}(t,\xi) := \mathbf{U}^{[0]}(t,\xi) - B(\mathbf{U}^{[n]},\mathbf{U}^{[n]})(t,\xi)$. By induction on n, we find that we have the pointwise inequalities

- $0 \le W^{[n]}(t,\xi) \le W^{[n+1]}(t,\xi) \le W(t,\xi)$
- $|U^{[n]}(t,\xi)| \le W^{[n]}(t,\xi)$
- $|U^{[n+1]}(t,\xi) - U^{[n]}(t,\xi)| \le W^{[n+1]}(t,\xi) - W^{[n]}(t,\xi)$.

We find that $W^{[n]}$ is pointwise convergent to a function $W^{[\infty]} \le W$. By monotonous convergence, we have

$$W^{[\infty]} = W^{[0]} + B_0(W^{[\infty]}, W^{[\infty]}).$$

Then, by dominated convergence, we find that $\mathbf{U}^{[n]}$ converges to a limit $\mathbf{U}^{[\infty]}$ such that

$$\mathbf{U}^{[\infty]} = \mathbf{U}^{[0]} - B(\mathbf{U}^{[\infty]}, \mathbf{U}^{[\infty]}).$$

$\mathbf{U}^{[\infty]}$ is then the Fourier transform of a solution to the Navier–Stokes problem with initial value $\mathbf{u}_0$.

2.2 Gevrey Analyticity

This formalism allows one to get Gevrey-type analyticity estimates. If we assume more precisely that

$$|\mathbf{U}^0(\xi)| \le \frac{1}{2e} W^0(\xi)$$

, then we find that for $0 \le t \le T$,

$$|\mathbf{U}^{[\infty]}(t,\xi)| \le \frac{1}{2\sqrt{e}} e^{-\sqrt{t}|\xi|} W^{[\infty]}(\frac{1}{2}t,\xi). \tag{1}$$

Indeed, we write

$$\sup_{z\ge 0} e^{z-\frac{1}{2}z^2} = \sqrt{e}$$

and for $0 \le s \le t$

$$e^{\sqrt{t}|\xi|} e^{-\sqrt{s}|\xi-\eta|} e^{-\sqrt{s}|\eta|} \le e^{(\sqrt{t}-\sqrt{s})|\xi|} \le e^{\sqrt{t-s}|\xi|}.$$

We define

$$\mathbf{Z}^{[n]}(t,\xi) = e^{\sqrt{t}|\xi|}\mathbf{U}^{[n]}(t,\xi).$$

We have

$$Z^{[n+1]} = Z^{[0]} - B_*(\mathbf{Z}^{[n]}, \mathbf{Z}^{[n]})$$

with, for $\mathbf{Z} = e^{\sqrt{t}|\xi|}\mathbf{U}$,

$$\begin{aligned} B_*(\mathbf{Z},\mathbf{Z})_l(t,\xi) = e^{\sqrt{t\xi}|} \int_0^t e^{-(t-s)|\xi|^2} \sum_{j=1}^3 \sum_{k=1}^3 \int_{\mathbb{R}^3} \frac{i\xi_j\xi_k}{(2\pi)^3|\xi|^2} (\xi_l U_k(s,\eta) \\ - \xi_k U_l(s,\eta)) U_j(s,\xi-\eta)\, d\eta\, ds \\ = \int_0^t e^{-(t-s)|\xi|^2} \sum_{j=1}^3 \sum_{k=1}^3 \int_{\mathbb{R}^3} e^{\sqrt{t}|\xi|-\sqrt{s}|\xi-\eta|-\sqrt{s}|\eta|} \frac{i\xi_j\xi_k}{(2\pi)^3|\xi|^2} (\xi_l Z_k(s,\eta) \\ - \xi_k Z_l(s,\eta)) Z_j(s,\xi-\eta)\, d\eta\, ds \end{aligned}$$

If $|\mathbf{Z}(t,\xi)| \le A(t/2,\xi)$, we find

$$\begin{aligned}|B_*(\mathbf{Z},\mathbf{Z})(t,\xi)| &\le \sqrt{e}\frac{18}{(2\pi)^3}\int_0^t e^{-\frac{1}{2}(t-s)|\xi|^2}|\xi|\int_{\mathbb{R}^3} A(s/2,\eta)A(s/2,\xi-\eta)\,d\eta\,ds\\ &\le 2\sqrt{e}\frac{18}{(2\pi)^3}\int_0^{t/2} e^{-(\frac{t}{2}-\sigma)|\xi|^2}|\xi|\int_{\mathbb{R}^3} A(\sigma,\eta)A(\sigma,\xi-\eta)\,d\eta\,d\sigma\\ &= 2\sqrt{e}B_0(A,A)(\frac{t}{2},\xi).\end{aligned}$$

By induction on n, we then find that

$$|\mathbf{Z}^{[n]}(t,\xi)| \le \frac{1}{2\sqrt{e}}W^{[n]}(\frac{t}{2},\xi).$$

Thus, we have proved the Gevrey estimate (1).

2.3 *Cheap Navier–Stokes Equation*

Thus far, we have introduced, as a tool for studying the Navier–Stokes problem

$$\begin{cases} \partial_t \mathbf{u} + (\mathbf{u}\cdot\nabla)\mathbf{u} = \Delta\mathbf{u} - \nabla p \\ \nabla\cdot\mathbf{u} = 0 \\ \mathbf{u}(0,.) = \mathbf{u}^0, \end{cases} \tag{2}$$

the study of the equation

$$W = W^{[0]} + B_0(W,W) \tag{3}$$

or, taking the inverse Fourier transform $w = \mathscr{F}_x^{-1}W$ of W, the equation

$$\begin{cases} \partial_t w = \Delta w + 18\sqrt{-\Delta}(w^2) \\ w(0,.) = w^0. \end{cases} \tag{4}$$

Equation (4) is known as the *cheap Navier–Stokes equation*. It has been introduced in 2001 by S. Montgomery-Smith [82] as a toy model for the Navier–Stokes equations. He gave an example of an initial value w^0 in the Schwartz class ($w^0 \in \mathscr{S}(\mathbb{R}^3)$) with a non-negative Fourier transform W^0 such that the solution w blows up in finite time.

The study of equation (3) has provided simple classes of solutions to the Navier–Stokes equations. For instance, if $Z(\xi)$ is a non-negative measurable function that satisfies the following inequation

$$W^0(\xi) + \frac{18}{(2\pi)^3|\xi|}\int_{\mathbb{R}^3} Z(\xi-\eta)Z(\eta)\,d\eta \le Z(\xi),$$

we get, by induction on n, that $W^{[n]}(t,\xi) \le Z(\xi)$. This means that if W^0 belongs to a lattice Banach space of functions E such that the operator $(Z, V) \mapsto \frac{1}{|\xi|}(Z * V)$ is bounded on E and if $\|W^0\|_E$ is small enough, then the Navier–Stokes equations (2) with initial value $\mathbf{u}^0$ with $|\mathscr{F}_x\mathbf{u}^0| \le W^0$ has a global solution $\mathbf{u}$ with $\sup_{0<t<+\infty}|\mathscr{F}_x\mathbf{u}| \in E$. Two simple instances can be found in the literature:

- The case where $E = L^2(|\xi|\,d\xi)$: if $Z \in E$, it means that $Z = \frac{1}{|\xi|^{1/2}}Z_0$ with $Z_0 \in L^2$; thus Z belongs to the Lorentz space $L^{3/2,2}$ (as a product of a function in $L^{6,\infty}$ by a function in L^2); thus $Z*Z$ belongs to $L^{3,1} \subset L^{3,2}$ and $\frac{1}{|\xi|^{1/2}}Z*Z \in L^{2,2} = L^2$. Thus, we find that if the initial value $\mathbf{u}_0$ has a small norm in the homogeneous Sobolev space $\dot H^{1/2} = \mathscr{F}_x^{-1}(L^2(|\xi|\,d\xi))$, then the Navier–Stokes problem with initial value $\mathbf{u}_0$ has a global solution. This is the result of Fujita and Kato [43].
- The equality

$$\int \frac{1}{|\xi-\eta|^2}\frac{1}{|\eta|^2}\,d\eta = C_0\frac{1}{|\xi|}$$

  allowed Le Jan and Sznitman to consider the space E defined by

$$Z \in E \Leftrightarrow Z \in L^1_{\mathrm{loc}} \text{ and } |\xi|^2 Z \in L^\infty.$$

  Again, they found that if the initial value $\mathbf{u}_0$ has a small norm in the homogeneous Besov space $\dot B^{-2}_{PM,\infty} = \mathscr{F}_x^{-1}(\frac{1}{|\xi|^2}L^\infty(d\xi))$, then the Navier–Stokes problem with initial value $\mathbf{u}_0$ has a global solution [66].

If we look for local-in-time solutions, we must include the time variable in our estimations. For instance, since

$$e^{-(t-s)|\xi|^2} \le e^{\frac{3}{4}}\left(\frac{3}{4}\right)^{3/2}\frac{1}{(t-s)^{3/4}}\frac{1}{|\xi|^{3/2}},$$

then, if $Z(\xi)$ and $\alpha(t)$ are non-negative measurable functions that satisfy on $(0,T)$ the following inequation

$$W^{[0]}(\xi) + e^{\frac{3}{4}}\left(\frac{3}{4}\right)^{3/2}\frac{18}{(2\pi)^3}\int_0^t \frac{\alpha(s)^2}{(t-s)^{3/4}}\,ds\,\frac{1}{|\xi|^{1/2}}\int_{\mathbb{R}^3} Z(\xi-\eta)Z(\eta)d\eta \le \alpha(t)\,Z(\xi),$$

we get, by induction on n, that $W^{[n]}(t,\xi) \le \alpha(t)\,Z(\xi)$. Thus, if F is a lattice Banach space of functions such that the operator $(Z, V) \mapsto \frac{1}{|\xi|^{1/2}}(Z * V)$ is bounded on F and if $\sup_{0<t<T} t^{1/4}e^{-t|\xi|^2}W^0(\xi) \in F$ and $\|\sup_{0<t<T} t^{1/4}e^{-t|\xi|^2}W^0(\xi)\|_F$ is small

enough, then the Navier–Stokes equations (2) with initial value $\mathbf{u}^0$ with $|\mathscr{F}_x\mathbf{u}^0| \leq W^0$ has a global solution $\mathbf{u}$ with $\sup_{0<t<T} t^{1/4}|\mathscr{F}_x\mathbf{u}| \in F$. Let us look at our two simple instances:

- The case where $E = L^2(|\xi|\,d\xi)$ and $F = L^2(|\xi|^2\,d\xi)$: if $Z \in F$, it means that $Z = \frac{1}{|\xi|}Z_0$ with $Z_0 \in L^2$; thus Z belongs to the Lorentz space $L^{6/5,2}$ (as a product of a function in $L^{3,\infty}$ by a function in L^2) and $V = |\xi|^{1/2} \in L^{3/2,2}$, thus, writing

$$\frac{1}{|\xi|^{1/2}}|Z * Z| \leq \frac{2}{|\xi|}(|Z| * |V|),$$

we get that $|Z| * |V|$ belongs to $L^{6/5,2} * L^{3/2,2} \subset L^{2,1} \subset L^{2,2} = L^2$, so that we have $\frac{1}{|\xi|^{1/2}}(Z * Z) \in F$. Moreover, if $A > 0$ and $W_0 \in E$, we find that for $t > 0$,

$$|t^{1/4}e^{-t|\xi|^2}W_0(\xi)| \leq 1_{|\xi|\leq A}t^{\frac{1}{4}}W_0(\xi) + 1_{|\xi|>A}\frac{1}{|\xi|^{1/2}}W_0(\xi)$$

so that

$$\|\sup_{0<t<T} t^{1/4}e^{-t|\xi|^2}W^0(\xi)\|_F \leq T^{1/4}A^{1/2}\|W^0\|_E + \|1_{|\xi|>A}W^0\|_E$$

and

$$\lim_{T\to 0^+}\|\sup_{0<t<T} t^{1/4}e^{-t|\xi|^2}W^0(\xi)\|_F = 0.$$

Thus, we find that if the initial value $\mathbf{u}^0$ belongs to the homogeneous Sobolev space $\dot{H}^{1/2} = \mathscr{F}_x^{-1}(L^2(|\xi|\,d\xi))$, then the Navier–Stokes problem with initial value $\mathbf{u}^0$ has a local in time solution. This is the result of Fujita and Kato [43]. Gevrey regularity estimates for a data in the Sobolev space were first given by Foias and Temam [37].
- The case where $E = \frac{1}{|\xi|^2}L^\infty(d\xi)$ and $F = \frac{1}{|\xi|^{5/2}}L^\infty(d\xi)$: the equality

$$\int \frac{1}{|\xi-\eta|^{5/2}}\frac{1}{|\eta|^{5/2}}\,d\eta = C_0\frac{1}{|\xi|^2}$$

shows that $\frac{1}{|\xi|^{1/2}}(F * F) \subset F$. Moreover, if $A > 0$ and $W_0 \in E$, we write again that for $t > 0$,

$$|t^{1/4}e^{-t|\xi|^2}W_0(\xi)| \leq 1_{|\xi|\leq A}t^{\frac{1}{4}}W_0(\xi) + 1_{|\xi|>A}\frac{1}{|\xi|^{1/2}}W_0(\xi)$$

so that

$$\| \sup_{0<t<T} t^{1/4} e^{-t|\xi|^2} W^0(\xi)\|_F \le T^{1/4} A^{1/2} \|W^0\|_E + \|1_{|\xi|>A} W^0\|_E$$

and

$$\limsup_{T\to 0^+} \| \sup_{0<t<T} t^{1/4} e^{-t|\xi|^2} W^0(\xi)\|_F \le \limsup_{A\to+\infty} \sup_{|\xi|>A} |\xi|^2 |W_0(\xi)|.$$

Again, we find that if the initial value $\mathbf{u}_0$ belongs to the homogeneous Besov space $\dot{B}^{-2}_{PM,\infty}$ and if $\limsup_{A\to+\infty} \sup_{|\xi|>A} |\xi|^2 |\mathscr{F}_x \mathbf{u}^0(\xi)|$. is small enough, then the Navier–Stokes problem with initial value $\mathbf{u}_0$ has a local-in-time solution.

We have considered the basic examples of $E = L^2(|\xi|\, d\xi)$ or $E = \frac{1}{|\xi|^2} L^\infty(d\xi)$. But many other examples are known. In particular, the theory has been developed for W^0 in certain Herz spaces. Recall that the Herz space $\mathscr{B}^s_{p,q}$ [54] is defined by

$$W \in \mathscr{B}^s_{p,q} \Leftrightarrow (2^{js} \|1_{2^j \le |\xi| < 2^{j+1}} W\|_p)_{j\in\mathbb{Z}} \in l^q.$$

For instance, we have $L^2(|\xi|\, d\xi) = \mathscr{B}^{1/2}_{2,2}$ and $\frac{1}{|\xi|^2} L^\infty(d\xi) = \mathscr{B}^2_{\infty,\infty}$. In 2012, Cannone and Wu [15] have studied the Navier–Stokes problem with an initial value $\mathbf{u}_0$ such that $\mathscr{F}_x \mathbf{u}_0 \in \mathscr{B}^{-1}_{1,q}$ with $1 \le q \le 2$. The case $q = 1$ corresponds to the case $\mathscr{F}_x \mathbf{u}_0 \in L^1(\frac{d\xi}{|\xi|})$, a case studied by Lei and Lin in 2011 [67].

3 Singular Integrals

3.1 Helmholtz Decomposition

Modern history of harmonic analysis begins with the study of singular integrals, from the work of M. Riesz on the Hilbert transform in 1924 [86] to the fundamental paper of Calderón and Zygmund on singular integrals in 1952 [10] (and its extension to vector-valued integrals by Benedek, Calderón, and Panzone in 1962 [5]). Basic accounts of the theory are to be found in the first chapters of the books of Stein [89, 90] or Grafakos [50].

The paradigm of Calderón–Zygmund convolution operators on $\mathbb{R}^d$ is given by Marcinkiewicz multipliers: if K is the inverse Fourier transform of a function $m(\xi)$ such that, for every $\alpha \in \mathbb{N}_0^d$ with $|\alpha| \le d+2$,

$$\sup_{\xi\neq 0} \left| |\xi|^{|\alpha|} \frac{\partial^\alpha m}{\partial \xi^\alpha}(\xi) \right| < \infty,$$

then convolution with K is a Calderón–Zygmund operator. The most classical example is given by the Riesz transforms R_j, $j = 1, \ldots, d$:

$$R_j f = \frac{\partial_j}{\sqrt{-\Delta}} f = \mathscr{F}_x^{-1}\left(\frac{i\xi_j}{|\xi|}\mathscr{F}_x f\right).$$

Riesz transforms are naturally encountered when studying the Helmholtz decomposition of a vector field defined on the whole space $\mathbb{R}^3$. One considers a vector field $\mathbf{u}$, and we want to decompose it as a sum of a divergence-free vector field $\mathbf{v}$ and an irrotational vector field $\mathbf{w}$:

$$\mathbf{u} = \mathbf{v} + \mathbf{w} \text{ with } \boldsymbol{\nabla} \cdot \mathbf{v} \text{ and } \boldsymbol{\nabla} \wedge \mathbf{w} = 0.$$

Basic formulas of vector analysis link the divergence and the curl of a vector $\mathbf{u}$ to its Laplacian by

$$\boldsymbol{\nabla} \wedge (\boldsymbol{\nabla} \wedge \mathbf{u}) = -\Delta \mathbf{u} + \boldsymbol{\nabla}(\boldsymbol{\nabla} \cdot \mathbf{u}).$$

In particular, we have

$$-\Delta \mathbf{v} = \boldsymbol{\nabla} \wedge (\boldsymbol{\nabla} \wedge \mathbf{v}) = \boldsymbol{\nabla} \wedge (\boldsymbol{\nabla} \wedge \mathbf{u})$$

and

$$-\Delta w = -\boldsymbol{\nabla}(\boldsymbol{\nabla} \cdot w) = -\boldsymbol{\nabla}(\boldsymbol{\nabla} \cdot \mathbf{u}).$$

If $\mathbf{u}$ belongs to a function space E on which the Riesz transforms operate continuously, we find a particular solution $(\mathbf{v}, \mathbf{w})$ by the formulas

$$\mathbf{v} = \mathscr{R} \wedge (\mathscr{R} \wedge \mathbf{u}) \text{ and } \mathbf{w} = -\mathscr{R}(\mathscr{R} \cdot \mathbf{u})$$

where the vectorial Riesz transform is given as

$$\mathscr{R} = \frac{1}{\sqrt{-\Delta}} \boldsymbol{\nabla}.$$

If E contains no other harmonic function than the null function, then this decomposition $\mathbf{u} = \mathbf{v} + \mathbf{w}$ is unique.

The operator $\mathbf{u} \mapsto \mathscr{R} \wedge (\mathscr{R} \wedge \mathbf{u})$ is called the Leray projection operators (for $E = L^2$, it is the orthogonal projection of square integrable vector fields on divergence-free square integrable vector fields) and is usually written as $\mathbb{P}$. This allows to get rid of the pressure in the Navier–Stokes equations and to rewrite the system as

$$\mathbf{u} = \Delta \mathbf{u} - \mathbb{P}((\mathbf{u} \cdot \boldsymbol{\nabla})\mathbf{u})$$

with $\mathbf{u}(0,.) = \mathbf{u}_0$ where $\nabla \cdot \mathbf{u}_0 = 0$.

This way of eliminating the pressure p (or expressing it as a function of the velocity $\mathbf{u}$ by the formula $\nabla p = \mathscr{R}(\mathscr{R}\cdot(\mathbf{u}\cdot\nabla)\mathbf{u}))$ is quite general and is applied to the study of (weak) solution $\mathbf{u}$ in a large variety of function spaces. The justification for such computations has been given, for instance, by Furioli, Lemarié-Rieusset, and Terraneo in the case of uniformly square integrable solutions (vanishing at infinity) [45, 70] or recently by Fernandez-Dalgo and Lemarié-Rieusset in the case of locally square integrable solutions with low increase at infinity [36].

The nature of the Leray projection operator has a deep impact on the properties of solutions to the Navier–Stokes equations. Main features of the convolution kernel of the operator are that the kernel is not compactly supported, meaning that the operator is non-local and involves integration over the whole space and that it has a slow decay at infinity (as $\mathbb{P}$ has a kernel homogeneous of degree -3, the kernel decays only as $|x|^{-3}$ and its derivatives as $|x|^{-4}$). Writing, for a divergence-free vector field $\mathbf{u}$,

$$\mathscr{R}(\mathscr{R}\cdot(\mathbf{u}\cdot\nabla)\mathbf{u}) = \nabla((\mathscr{R}\otimes\mathscr{R})\cdot(\mathbf{u}\otimes\mathbf{u})),$$

Dobrokhotov and Shafarevich [28] proved that the spatial decay at infinity of “rapidly” decaying solutions was governed by the kernels $\partial_j\partial_k\partial_l G$ of the operators $\partial_j R_k R_l$ (where G is the Green function, fundamental solution of the Laplacian operator: $G(x) = \frac{1}{4\pi|x|}$, $(-\Delta)G = \delta$). More precisely, if $\lim_{x\to\infty}|x|^4|\mathbf{u}^0(x)| = 0$ (as it is the case, for instance, for the Millennium Prize problem) and if $(\mathbf{u}, p)$ is a classical solution of the Navier–Stokes problem on a strip $[0, T]\times\mathbb{R}^3$, then, for $0 < t < T$,

$$u(t,x) = -\sum_{j=1}^{3}\sum_{l=l}^{3} d_{j,l}(t)\nabla\partial_j\partial_l G(x) + o(|x|^{-4})$$

with $d_{j,l}(t) = \int_0^t \int u_j(s,x)u_l(s,x)\,dx\,ds$. This means that the good decay of $\mathbf{u}^0$ (as $o(|x|^{-4})$ is instantaneously lost whenever one of the integrals $\int u_j^0 u_l^0\,dx$ (with $j \neq l$) or $\int (u_j^0(x))^2 - (u_l^0(x))^2\,dx$ (with $j \neq l$) is not equal to 0; in that case, we have $\liminf_{x\to+\infty}|x|^4|\mathbf{u}(t,x)| > 0$ for t close enough to 0. This instantaneous spreading has been studied by Brandolese and Meyer in [8].

3.2 Lebesgue–Gevrey Estimates

A less direct application of singular integrals to the study of the Navier–Stokes equations can be found in the treatment of Gevrey regularity of solutions in the Lebesgue space $L^3(\mathbb{R}^3)$ that has been proposed by Lemarié-Rieusset [69, 71].

The idea starts from the result of Kato on existence of solutions for initial data in L^3 [55]. One transforms the differential problem

$$\partial_t \mathbf{u} = \Delta \mathbf{u} - \mathbb{P}\nabla \cdot (\mathbf{u} \otimes \mathbf{u}), \quad \mathbf{u}(0,.) = \mathbf{u}^0$$

into an integro-differential problem by solving

$$\mathbf{u} = e^{t\Delta}\mathbf{u}^0 - \int_0^t e^{(t-s)\Delta}\mathbb{P}\nabla \cdot (\mathbf{u} \otimes \mathbf{u})\, ds = e^{t\Delta}\mathbf{u}^0 - B(\mathbf{u}, \mathbf{u})$$

where the bilinear operator B is defined as

$$B(\mathbf{u}, \mathbf{v})(t,.) = \int_0^t e^{(t-s)\Delta}\mathbb{P}\nabla \cdot (\mathbf{u} \otimes \mathbf{v})\, ds. \tag{5}$$

By the contraction principle, if B is bounded on a Banach space $\mathbb{E}_T$ (with operator norm $C_{\mathbb{E}_T}$) of functions defined on $(0, T) \times \mathbb{R}^3$, then, for $\mathbf{u}^0$ small enough ($\|e^{t\Delta}\mathbf{u}^0\|_B < \frac{1}{4C_{\mathbb{E}_T}}$), one can find a solution $\mathbf{u} \in \mathbb{E}_T$. Now, if $\mathbf{u}^0 \in L^3$, we have

$$\sup_{t>0} t^{1/4}\|e^{t\Delta}\mathbf{u}^0\|_6 < +\infty \text{ and } \lim_{t\to 0} t^{1/4}\|e^{t\Delta}\mathbf{u}^0\|_6 = 0. \tag{6}$$

On the other hand, for every $t > 0$, the operator $e^{t\Delta}\mathbb{P}\nabla\cdot$ is given by convolutions with kernels $e^{t\Delta}\partial_j\partial_k\partial_l G$ that are in L^1 with

$$\|e^{t\Delta}\partial_j\partial_k\partial_l G\|_1 \le C\frac{1}{\sqrt{t}}.$$

We then use the regularizing properties of the heat kernel in Lebesgue spaces: for $1 \le p \le q$ and for $t > 0$

$$\|e^{t\Delta} f\|_q \le C_{p,q} t^{\frac{3}{2}(\frac{1}{q}-\frac{1}{p})}\|f\|_p.$$

We then have

$$\begin{aligned}\|B(\mathbf{u}, \mathbf{v})\|_6 \le& \int_0^t \|e^{\frac{t-s}{2}\Delta}\left(e^{\frac{t-s}{2}\Delta}\mathbb{P}\nabla \cdot (\mathbf{u} \otimes \mathbf{v})\right)\|_6\, ds \\ \le& C\int_0^t \frac{1}{(t-s)^{1/4}}\frac{1}{(t-s)^{1/2}}\frac{1}{s^{1/2}}\|s^{1/4}\mathbf{u}(s,.)\|_6\|s^{1/4}\mathbf{v}(s,.)\|_6\, ds.\end{aligned}$$

It means that B is bounded on

$$\mathbb{E}_T = \{\mathbf{u}(t,x) \,/\, \sup_{0<t<T} t^{1/4}\|e^{t\Delta}\mathbf{u}^0\|_6 < +\infty \text{ and } \lim_{t\to 0} t^{1/4}\|e^{t\Delta}\mathbf{u}^0\|_6 = 0\}$$

(with an operator norm that does not depend on T). Moreover, a solution in $\mathbb{E}_T$ will satisfy

$$\|B(\mathbf{u},\mathbf{u})\|_3 \le \int_0^t \|e^{\frac{t-s}{2}\Delta}\left(e^{\frac{t-s}{2}\Delta}\mathbb{P}\nabla\cdot(\mathbf{u}\otimes\mathbf{u})\right)\|_3\,ds$$
$$\le C\int_0^t \frac{1}{(t-s)^{1/2}}\frac{1}{s^{1/2}}(\|s^{1/4}\mathbf{u}\|_6)^2\,ds$$

so that $\mathbf{u} \in L^\infty((0,T),L^3)$. (As a matter of fact, one even finds that $\mathbf{u} \in \mathscr{C}([0,T),L^3)$).

Now, if we want to mimic the proof of the Gevrey regularity we saw for Fourier-Herz spaces, one must factor out in the Fourier–Navier–Stokes equations a term $e^{-\sqrt{t}\|\xi\|}$ and control the action of the factor $e^{\sqrt{t}\|\xi\|-\sqrt{s}\|\eta\|-\sqrt{t}\|\xi-\eta\|}$. It is not enough to control the size of the factor, as we are dealing now with Fourier transforms of functions in L^3 or in L^6 that are no longer functions but singular distributions. The control is then given by the theory of singular integrals and more precisely of Marcinkiewicz multipliers (as described in [89], for instance). More precisely, we factor out in the Fourier transform a term of the form $e^{\sqrt{t}\|\xi\|_1}$, where $\|\xi\|_1 = |\xi_1|+|\xi_2|+|\xi_3|$. We shall write $e^{-\sqrt{t}D_1}$ for the convolution operator with symbol $e^{-\sqrt{t}\|\xi\|_1}$ and $e^{\sqrt{t}D_1}$ for the convolution operator with symbol $e^{\sqrt{t}\|\xi\|_1}$. We then have to study the equation for $e^{\sqrt{t}D_1}\mathbf{u} = \mathbf{U}$ which is given by

$$\mathbf{U} = e^{\frac{t}{2}\Delta}\left(e^{\frac{t}{2}\Delta}e^{\sqrt{t}D_1}\right)\mathbf{u}_0$$
$$-\int_0^t e^{\frac{(t-s)}{2}\Delta}\mathbb{P}\nabla\cdot\left(e^{\frac{(t-s)}{2}\Delta}e^{\sqrt{t-s}D_1}e^{(\sqrt{t}-\sqrt{t-s}-\sqrt{s})D_1}\right.$$
$$\left.\left(e^{\sqrt{s}D_1}(e^{-\sqrt{s}D_1}\mathbf{U}\otimes e^{-\sqrt{s}D_1}\mathbf{U})\right)\right)\,ds.$$

The operator $e^{\frac{t}{2}\Delta}e^{\sqrt{t}D_1}$ is a tensor product of one-dimensional convolution operators associated to Marcinkiewicz multipliers $e^{-\frac{t}{2}\xi_j^2+\sqrt{t}|\xi_j|}$. Similarly, the operator $e^{(\sqrt{t}-\sqrt{t-s}-\sqrt{s})D_1}$ is a tensor product of one-dimensional convolution operators associated to Marcinkiewicz multipliers $e^{(\sqrt{t}-\sqrt{t-s}-\sqrt{s})|\xi_j|}$. The bilinear operator

$$T(f,g) = e^{\sqrt{s}D_1}(e^{-\sqrt{s}D_1}f\times e^{-\sqrt{s}D_1}g)$$

can similarly be written as a sum of tensor products of one-dimensional convolution operators associated to Marcinkiewicz multipliers: if S_j is associated to the multiplier $1_{\xi_j>0}$, T_j to the multiplier $1_{\xi_j} < 0$, and Z_j to the multiplier $e^{-\sqrt{s}|\xi_j|}$ and if W_j is the unbounded operator associated to the multiplier $e^{+\sqrt{s}|\xi_j|}$, then, for $f_j, g_j \in L^p(\mathbb{R})$,

$$W_j(Z_jf_j\times Z_jg_j) = S_jf_j\times S_jg_j + T_jf_j\times T_jg_j + S_j(S_jf\times Z_j^2T_jg)$$
$$+ S_j(Z_j^2T_jf\times S_jg) + T_j(Z_j^2S_jf\times T_jg) + T_j(T_jf\times Z_j^2S_jg).$$

Thus, using the contraction principle, we find that, if $\mathbf{u}_0 \in L^3$, we have a solution $\mathbf{u} = e^{-\sqrt{t}D_1}\mathbf{U}$ of the Navier–Stokes equations on a small enough time interval $(0, T)$ such that $\sup_{0<t<T} t^{1/4}\|\mathbf{U}\|_6 < +\infty$ and $\mathbf{U} \in L^\infty((0, T), L^3)$.

3.3 Maximal Regularity for the Heat Kernel

Another way of using singular integrals for the study of solutions to the Navier–Stokes equations is the proof proposed by Monniaux in [81] for the uniqueness of solutions in $\mathscr{C}([0, T), L^3)$. (Local) existence of solutions in L^3 (for an initial value $\mathbf{u}^0 \in L^3$) had been proved by Kato in 1984 [55], but uniqueness remained open until 1997, when Furioli, Lemarié-Rieusset, and Terraneo [45] proved uniqueness by using Besov spaces.

The proof by Monniaux is very simple. If $\mathbf{u}$ is a solution in $\mathscr{C}([0, T), L^3)$ and $\mathbf{u}_K$ is the solution provided by Kato in $\mathscr{C}([0, T), L^3)$ with the additional property that $\lim_{t\to 0} t^{1/4}\|\mathbf{u}_K\|_6 = 0$, then the function $\mathbf{w} = \mathbf{u} - \mathbf{u}_K$ satisfies the identity

$$\mathbf{w} = -B(\mathbf{u}_K, \mathbf{w}) - B(\mathbf{w}, \mathbf{u}_K) - B(\mathbf{w}, \mathbf{w}).$$

We shall write that $\mathbf{w}$ is an eigenvector of the linear transform

$$\mathbf{v} \mapsto L(\mathbf{v}) = -B(\mathbf{u}_K, \mathbf{v}) - B(\mathbf{v}, \mathbf{u}_K) - B(\mathbf{w}, \mathbf{v}).$$

We want to estimate $L(\mathbf{v})$ in $L^3((0, S), L^3)$, for $S < T$. We have

$$\|B(\mathbf{u}_K, \mathbf{v})(t, .)\|_3 \le C \int_0^t \frac{1}{(t-s)^{1/4}} \frac{1}{\sqrt{t-s}} \frac{1}{s^{1/4}} \|\mathbf{v}(s, .)\|_3 \|s^{1/4}\mathbf{u}_K(s, .)\|_6\, ds$$

so that, since multiplication is bounded from $L^{4,\infty} \times L^3$ to $L^{12/7,3}$ and convolution is bounded from $L^{12/7,3} \times L^{4/3,\infty}$ to $L^{3,3} = L^3$,

$$\|B(\mathbf{u}_K, \mathbf{v})\|_{L^3((0,S),L^3)} \le C\|\mathbf{v}\|_{L^3((0,S),L^3)} \sup_{0<s<S} s^{1/4}\|\mathbf{u}_K(s, .)\|_6.$$

Similarly, we have

$$\|B(\mathbf{v}, \mathbf{u}_K)\|_{L^3((0,S),L^3)} \le C\|\mathbf{v}\|_{L^3((0,S),L^3)} \sup_{0<s<S} s^{1/4}\|\mathbf{u}_K(s, .)\|_6.$$

For estimating $B(\mathbf{w}, \mathbf{v})$, we write

$$\partial_j R_k R_l (w_k v_l) = -\Delta R_j R_k R_l \frac{1}{\sqrt{-\Delta}} (w_k v_l$$

and use the inequality on Riesz potential

$$\|\frac{1}{\sqrt{-\Delta}}(f)\|_3 \leq \|f\|_{3/2}.$$

Thus, we find that

$$\frac{1}{\Delta}(\mathbb{P}\nabla \cdot (\mathbf{w} \otimes \mathbf{v})) \in L^3L^3.$$

Maximal regularity in L^3L^3 for the heat kernel states that

$$\|\int_0^t e^{(t-s)\Delta}\Delta f\, ds\|_{L^3((0,S),L^3)} \leq C\|f\|_{L^3((0,S),L^3)}$$

where the constant C does not depend on S. Thus, we have

$$\|B(\mathbf{w}, \mathbf{v})\|_{L^3((0,S),L^3)} \leq C\|\mathbf{v}\|_{L^3((0,S),L^3)}\|\mathbf{w}\|_{L^\infty((0,S),L^3)}$$

By continuity of $\mathbf{w}$ in L^3, we find that $\lim_{S\to+\infty}\|\mathbf{w}\|_{L^\infty((0,S),L^3)} = 0$. Thus, for S small enough, L is contractive on $L^3((0, L^3), L^3)$. Hence, the fixed point $\mathbf{w}$ is equal to 0, and $\mathbf{u} = \mathbf{u}_K$ on $(0, S)$. The end of the proof follows by a bootstrap argument.

The maximal regularity property is linked to singular integrals but no longer on $\mathbb{R}^3$ but on the space $\mathbb{R} \times \mathbb{R}^3$ endowed with the parabolic distance $\delta((t, x), (s, y)) = \sqrt{|t-s| + |x-y|^2}$. Together, with the Lebesgue measure on $X = \mathbb{R} \times \mathbb{R}^3$, δ provides X with a structure of homogeneous space (as studied by Coifman and Weiss) [27]. We have, for f supported in $[0, +\infty) \times \mathbb{R}^3$,

$$1_{t>0}\int_0^t e^{(t-s)\Delta}\Delta f\, ds = \iint_X K(t-s, x-y) f(s, y)\, ds\, dy$$

where K is a convolution operator associated to the Fourier multiplier

$$m(\tau, \xi) = -\frac{\xi^2}{\xi^2 + i\tau}$$

(where we consider the Fourier transform $\mathscr{F}_{t,x} f(\tau, \xi) = \iint_X f(s, y) e^{-i(t\tau + x\cdot\xi)} dt dx$). We have

$$\sup_{\alpha\in\mathbb{N}_0^3, \beta\in\mathbb{N}_0} \sup_{(\tau,\xi)\neq(0,0)} (|\tau|^{1/2} + |\xi|)^{|\alpha|+2\beta} |\frac{\partial^\alpha}{\partial\xi^\alpha}\frac{\partial^\beta}{\partial\tau^\beta} m(\tau, \xi)| < +\infty.$$

Thus, m can be seen as a Marcinkiewicz multiplier on the parabolic space $\mathbb{R} \times \mathbb{R}^3$.

3.4 Marcinkiewicz Multipliers for Bilinear Operators

In 1978, Coifman and Meyer extended the theory of multipliers to the setting of bilinear operators [24, 26]. They consider a smooth function σ on $\mathbb{R}^d \times \mathbb{R}^d$ such that, for all $\alpha, \beta \in \mathbb{N}_0^d$,

$$\sup_{(\xi,\eta)\neq(0,0)} (|\xi| + |\eta|)^{|\alpha|+|\beta|} \left| \frac{\partial^\alpha}{\partial\xi^\alpha} \frac{\partial^\beta}{\partial\eta^\beta} \sigma(\tau, \xi) \right| < +\infty$$

and they define

$$T_\sigma(f, g) = \frac{1}{(2\pi)^{2d}} \iint e^{i(\xi+\eta)\cdot x} \mathcal{F}_x f(\xi) \mathcal{F}_x g(\eta)\, d\xi\, d\eta.$$

T is bounded from $L^\infty \times L^p$ to L^p for every $1 < p < +\infty$. One key property is that for fixed $f \in L^\infty$, $g \mapsto T(f, g)$ is not a convolution operator but is a generalized Calderón–Zygmund operator (in the sense of [25]).

This theory has been applied by Kato and Ponce to derive a useful commutator estimate that they applied to the study of the regularity of solutions to the Navier–Stokes equations or to the Euler equations [57].

4 The Hardy–Littlewood Maximal Function

4.1 Kato's Mild Solutions and Maximal Functions

Our treatment of the Navier–Stokes equations through the cheap Navier–Stokes equation was very elementary, using absolute values and convolution inequalities in the frequency variables. C. Calderón [11] noticed that we can deal with the equations in the space variable in an equivalently elementary way through the use of the maximal function, another basic tool in harmonic analysis introduced by Hardy and Littlewood in 1930 [52]. We shall write $\mathcal{M}_f$ for the maximal function of f:

$$\mathcal{M}_f(x) = \sup_{r>0} \frac{1}{|B(x, r)|} \int_{B(x,r)} |f(y| \, dy.$$

A basic result on maximal functions [50] is the control of convolution with radial kernels. More precisely, if a function g admits a majorant k ($|g(x)| \leq k(x)$) such that k is integrable, radial, and radially non-increasing, then

$$|g * f(x)| \leq \|k\|_1 \mathcal{M}_f(x).$$

In order to deal with the integro-differential problem

$$\mathbf{u} = e^{t\Delta}\mathbf{u}^0 - \int_0^t e^{(t-s)\Delta}\mathbb{P}\nabla\cdot(\mathbf{u}\otimes\mathbf{u})\,ds = e^{t\Delta}\mathbf{u}^0 - B(\mathbf{u},\mathbf{u}),$$

Calderón writes

$$e^{t\Delta}\mathbf{u}^0 = e^{t\Delta}(-\Delta)^{1/4}(-\Delta)^{-1/4}\mathbf{u}^0$$

and

$$B(\mathbf{u},\mathbf{v}) = \int_0^t e^{\frac{(t-s)}{2}\Delta}\mathbb{P}\nabla\cdot e^{\frac{(t-s)}{2}\Delta}(-\Delta)^{1/4}(-\Delta)^{-1/4}(\mathbf{u}\otimes\mathbf{v})\,ds.$$

and then uses the inequalities

- $|e^{t\Delta}f(x)| \le C\int \frac{\sqrt{t}}{(\sqrt{t}+|x-y|)^4}|f(y)|\,dy$
- $|e^{t\Delta}(-\Delta)^{1/4}f(x)| \le C\int \frac{1}{(\sqrt{t}+|x-y|)^{7/2}}|f(y)|\,dy$
- $|e^{t\Delta}(-\Delta)^{1/4}R_jR_k\partial_l f(x)| \le C\int \frac{1}{(\sqrt{t}+|x-y|)^4}|f(y)|\,dy$
- $|(-\Delta)^{-1/4}(fg)(x)| \le C\int \frac{1}{|x-y|^{5/2}}|f(y)g(y)|\,dy$

The last inequality is just a consequence of the correspondence of fractional integration $(-\Delta)^{-\alpha/2}$ $(0<\alpha<3)$ with the Riesz potentials I_α:

$$(-\Delta)^{-\alpha/2}f(x) = I_\alpha(f)(x) = c_\alpha\int_{\mathbb{R}^3}\frac{1}{|x-y|^{3-\alpha}}f(y)\,dy.$$

With the first two inequalities, we get that

$$\sup_{t>0}|e^{t\Delta}\mathbf{u}^0(x)| \le C\mathcal{M}_{|\mathbf{u}_0|}(x)$$

and

$$\sup_{t>0} t^{1/4}|e^{t\Delta}\mathbf{u}^0(x)| \le C\mathcal{M}_{I_{1/2}(|\mathbf{u}_0|)}(x).$$

In particular, if $\mathbf{u}^0$ belongs to L^3, then (as $I_{1/2}$ maps L^3 to L^6) we have $e^{t\Delta}\mathbf{u}^0 \in \mathbb{E}$ where

$$f\in\mathbb{E} \Leftrightarrow \sup_{t>0} f(t,x)\in L^\infty(\mathbb{R}^3) \text{ and } \sup_{t>0} t^{1/4}f(t,x)\in L^6(\mathbb{R}^3).$$

Moreover, as $L^6\cap L^3$ is dense in L^3, we have

$$\lim_{T\to 0}\|\sup_{0<t<T} t^{1/4}e^{t\Delta}\mathbf{u}^0\|_6 = 0.$$

Now if $\mathbf{u} \in \mathbb{E}_T$ and $\mathbf{v} \in \mathbb{E}_T$ where

$$f \in \mathbb{E}_T \Leftrightarrow A_T(f) = \sup_{0<t<T} t^{1/4}|f(t,x)| \in L^6,$$

we find (using the inequalities on $e^{t\Delta}(-\Delta)^{1/4}$ and on $e^{t\Delta}R_j R_k \partial_l$), the inequalities (for $0 < t < T$)

$$\begin{aligned}|B(\mathbf{u},\mathbf{v})(t,x)| &\leq C \int_0^t \frac{1}{(t-s)^{3/4}} \frac{1}{\sqrt{s}} \mathcal{M}_{I_{1/2}(A_T(\mathbf{u})A_T(\mathbf{v}))}(x)\, ds \\ &\leq C' t^{-\frac{1}{4}} \mathcal{M}_{I_{1/2}(A_T(\mathbf{u})A_T(\mathbf{v}))}(x)\end{aligned}$$

and

$$|B(\mathbf{u},\mathbf{v})(t,x)| \leq C \int_0^t \frac{1}{(t-s)^{1/2}} \frac{1}{\sqrt{s}} \mathcal{M}_{A_T(\mathbf{u})A_T(\mathbf{v})}(x)\, ds \leq C' \mathcal{M}_{A_T(\mathbf{u})A_T(\mathbf{v})}(x).$$

The first inequality gives that $B(\mathbf{u},\mathbf{v})$ still belongs to $\mathbb{E}_T$, and thus, if T is small enough (to grant that $e^{t\Delta}\mathbf{u}^0$ is small in $\mathbb{E}_T$), we find a solution $\mathbf{u}$ to the Navier–Stokes problem; the second inequality gives us a control in L^3 norm for this solution $\mathbf{u}$. Thus, we recover a Kato-type solution $\mathbf{u}$ such that

$$\sup_{0<t<T} |\mathbf{u}(t,.)| \in L^3 \text{ and } \sup_{0<t<T} t^{1/4}|\mathbf{u}(t,.)| \in L^6.$$

The main difference with Kato's formalism is that, now, we first take the supremum on t before integrating in x.

When $\mathbf{u}^0$ is small in L^3, we have an even simpler proof of existence of a mild solution $\mathbf{u}$ such that $\sup_{t>0} |\mathbf{u}(t,.)|$ belongs to L^3. This result of Calderón is based on the fact that the bilinear operator B, which is not bounded on $L^\infty_t L^3_x$ [84], is actually bounded on $L^3_x L^\infty_t$. If $\mathbf{u} \in L^3_x L^\infty_t$ and $\mathbf{v} \in L^3_x L^\infty_t$, then

$$\begin{aligned}|B(\mathbf{u},\mathbf{v})(t,x)| \leq& C \int_0^t \int_{\mathbb{R}^3} \frac{1}{(\sqrt{t}+|x-y|)^4} |\mathbf{u}(s,y)|\,|\mathbf{v}(s,y)|\, dy \\ \leq& C \int_{\mathbb{R}^3} \sup_{s>0} |\mathbf{u}(s,y)| \sup_{s>0} |\mathbf{v}(s,y)| \left[\int_0^t \frac{1}{(\sqrt{t}+|x-y|)^4}\, ds\right] dy \\ =& C' \int_{\mathbb{R}^3} \frac{1}{|x-y|^2} \sup_{s>0} |\mathbf{u}(s,y)| \sup_{s>0} |\mathbf{v}(s,y)|\, dy \\ =& C'' I_{1/2}(\sup_{s>0} |\mathbf{u}(s,y)| \sup_{s>0} |\mathbf{v}(s,y)|)(x).\end{aligned}$$

Thus, if $\mathcal{M}_{|\mathbf{u}^0|} \leq U^0$ and if U is a solution of the cheap equation

$$U = U^0 + C'' I_{1/2}(U^2)$$

(solution which exists if U^0 is small enough in L^3), then we have a solution $\mathbf{u}$ with $|\mathbf{u}(t,x)| \leq U(x)$.

4.2 Hardy Spaces and Molecules

We can rewrite the Hardy–Littlewood maximal function as

$$\mathcal{M}_f(x) = \sup_{t>0} K_t * |f|(x)$$

where

$$K(x) = \frac{1}{|B(0,1)|}\mathbf{1}_{B(0,1)}(x) \text{ and } K_t(x) = \frac{1}{t^3}K(\frac{x}{t}).$$

One can see clearly the role of scaling in this definition of the operator. Basic features for this operator are the boundedness on L^p for $1 < p \leq +\infty$

$$\|\mathcal{M}_f\|_p \approx \|f\|_p$$

and the lack of control in L^1 norm:

$$f \neq 0 \implies \|\mathcal{M}_f\|_1 = +\infty.$$

The theory of Hardy spaces developed by Fefferman and Stein [35, 90] involves a modified maximal function: taking $\Phi \in \mathcal{S}$ a radially non-increasing smooth function, and defining $\Phi_t(x) = \frac{1}{t^3}\Phi(\frac{x}{t})$, one defines

$$\mathcal{M}_f^{[\Phi]}(x) = \sup_{t>0} |\Phi_t * f(x)|.$$

From the properties that $\mathcal{M}_f^{[\Phi]}(x) \leq \mathcal{M}_f(x)$ and that $\lim_{t\to 0}\Phi_t * f = f$ in $\mathcal{S}'$ for every distribution $f \in \mathcal{S}'(\mathbb{R}^3)$, one finds that we have again, for $1 < p \leq +\infty$,

$$\|\mathcal{M}_f^{[\Phi]}\|_p \approx \|f\|_p.$$

But, now, it turns out that there are many distributions f such that $\mathcal{M}_f^{[\Phi]}$ is integrable or belongs to L^p for some $p \in (0,1)$. The Hardy space $\mathcal{H}^p$ is defined for $0 < p < +\infty$ by the property:

$$f \in \mathcal{H}^p \Leftrightarrow \mathcal{M}_f^{[\Phi]} \in L^p.$$

An important feature of Hardy spaces is their duality property with BMO or with homogeneous Hölder spaces: the dual of $\mathcal{H}^1$ can be identified with BMO and the dual of $\mathcal{H}^p(\mathbb{R}^3)$ for $\frac{3}{4} < p < 1$ can be identified with the homogeneous Hölder space $\dot{B}^{\alpha}_{\infty,\infty}$ with $\alpha = \frac{3}{p} - 1$. This has been used by Kozono and Taniuchi [62] to prove weak–strong uniqueness solutions when the Navier–Stokes problem with initial value $\mathbf{u}^0 \in L^2$ generates a weak Leray solution in $L^\infty_t L^2_x \cap L^2_t \dot{H}^1_x$ (with Leray energy inequality) and a solution in $L^\infty_t L^2_x \cap L^2_t \dot{H}^1_x \cap L^2_t \mathrm{BMO}_x$: the proof relies on the proof by Coifman, Lions, Meyer, and Semmes [22] that for a vector field $\mathbf{u} \in L^2$ that is divergence-free (***nabla*** $\cdot\, \mathbf{u}$) and a vector field $\mathbf{v} \in L^2$ that is curl-free ($\nabla \wedge \mathbf{v} = 0$), we have $\mathbf{u} \cdot \mathbf{v} \in \mathcal{H}^1$. Thus, the usual estimate for weak–strong uniqueness

$$\|\mathbf{v} - \mathbf{u}\|_2^2 + 2\int_0^t \|\nabla \otimes (\mathbf{u} - \mathbf{v})\|_2^2\, ds \leq 2\int_0^t \int_{\mathbb{R}^3} \mathbf{u} \cdot ((\mathbf{u} - \mathbf{v}) \cdot \nabla(\mathbf{u} - \mathbf{v}))\, dx\, ds$$

is turned to

$$\|\mathbf{v} - \mathbf{u}\|_2^2 + 2\int_0^t \|\nabla \otimes (\mathbf{v} - \mathbf{u})\|_2^2\, ds \leq C\int_0^t \|\mathbf{v} - \mathbf{u}\|_2 \|\nabla(\mathbf{v} - \mathbf{u})\|_2 \|\mathbf{u}\|_{\mathrm{BMO}}\, ds$$

which leads to a Gronwall estimate.

Another important feature of Hardy spaces is their atomic decomposition, as described, for instance, in [27]. For $3/4 < p \leq 1$, we have that a distribution f belongs to $\mathcal{H}^p(\mathbb{R}^3)$ if and only if it can be written as a sum $f = \sum_{j\in\mathbb{N}} \lambda_j a_j$ where $\sum_{j\in\mathbb{N}} |\lambda_j|^p < +\infty$ and a_j is a $\mathcal{H}^p$ atom: there exists some $r_j > 0$ and some $x_j \in \mathbb{R}^3$ such that a_j is supported in the ball $B(x_j, r_j)$, $\|a_j\|_2 \leq |B(x_j, r_j)|^{-\frac{2}{2-p}}$, and $\int a_j\, dx = 0$.

Atoms are not stable under the action of Calderón–Zygmund convolution operators with a non-local singular kernel, because compactness of supports is destroyed by the convolution. But if we relax the conditions on a_j into, for some $r_j > 0$ and $x_j \in \mathbb{R}^3$, $\|a_j\|_2 \leq |B(x_j, r_j)|^{-\frac{2}{2-p}}$, $\|\, |x - x_j|\, a_j\|_2 \leq r_j |B(x_j, r_j)|^{-\frac{2}{2-p}}$, and $\int a_j\, dx = 0$ [a_j are no longer an atom for $\mathcal{H}^p$, but it is called a molecule; the situation is much better. If T is a convolution operator with a Calderón–Zygmund kernel, then there exists a constant $C > 0$ such that the image $\frac{1}{C} T(a_j)$ of a molecule is still a molecule (associated to the same center x_j and the same radius r_j).

There are very few examples of the use of Hardy molecular decompositions in fluid mechanics. We may quote a paper of Chamorro on advection–diffusion in the setting of a non-local diffusion and a rough drift [16]. Futioli and Terraneo [46] studied the Cauchy problem for the Navier–Stokes equations when the Laplacian of the initial value $\mathbf{u}_0$ is a $\mathcal{H}^1$ molecule. Their results were extended by Brandolese in [7].

4.3 *Wavelets*

Atomic or molecular decompositions lead quite naturally to wavelets. However, the basic atoms that generate wavelet decompositions are usually more regular than simply Lebesgue measurable and are assumed to have some Höder regularity. In that case, one works more in the setting of Besov spaces than of Hardy spaces. A systematic approach of Besov space through atomic decompositions has been proposed by many authors, including the seminal paper of Frazier and Jawerth [40].

However, the first approach of the Navier–Stokes equations with a decomposition on wavelet bases was performed by Federbush [32] in yet another space, the Morrey space $\dot{M}^{2,3}$ (see the subsection on Morrey spaces in section 5). The study by Federbush was based on the use of divergence-free vector wavelet bases [4, 68]. Divergence-free wavelet bases were also used by Urban [94] for the numerical approximation of the equations of fluid mechanics.

There have been many claims that wavelet analysis of turbulent signals may provide valuable insights in the actual structure of turbulent fluids [31, 41], especially in the frame of self–similar universality laws such as studied by Frisch [42]. But Meyer proved that the claim that wavelets were asymptotically decorrelated in the nonlinearity of the Navier–Stokes equations was unfounded [80].

5 Function Spaces

Many function spaces of measurable or differentiable functions have close relationships with harmonic analysis, and their theory was developed quite extensively in the books of Stein: Lorentz spaces in *Introduction to Fourier Analysis on Euclidean Spaces* [91], Besov spaces in *Introduction to Fourier Analysis on Euclidean Spaces* [89], BMO, tent spaces, or Muckenhoupt weights in *Harmonic Analysis* [90]. As a matter of fact, all those spaces are met in the modern study of Navier–Stokes equations developed in the 1990s. More recently, use of Morrey spaces has been developed as well by many authors (see [74] for references).

5.1 *Lorentz Spaces*

Sobolev spaces $W^{k,p}$ of functions in L^p such that their derivatives (in the sense of distributions) up to order k are still in L^p can be extended for $1 < p < +\infty$ to the scale of spaces H^s_p defined, for $s \in \mathbb{R}$, by

$$f \in H^s_p \Leftrightarrow f \in \mathcal{S}' \text{ and } \mathcal{F}_x^{-1}\left((1+|\xi|^2)^{s/2}\mathcal{F}_x f\right) \in L^p.$$

For $1 < p < +\infty$ and $k \in \mathbb{N}_0$, we have $W^{k,p} = H_p^k$ [89]. The Sobolev embeddings then state that for $0 \leq s < \frac{3}{p}$ (and $1 < p < +\infty$), we have

$$H_p^s \subset L^r \text{ with } \frac{1}{r} = \frac{1}{p} - \frac{s}{3}.$$

The *sharp Sobolev embedding* states more precisely that

$$H_p^s \subset L^{r,p} \subset L^r \text{ with } \frac{1}{r} = \frac{1}{p} - \frac{s}{3}$$

where $L^{r,p}$ is a Lorentz space. This can be done through various methods that have been developed by Stein; for instance:

- Let $\mathcal{J}^s$ be the convolution with the Bessel kernel associated with the Fourier multiplier $(1+|\xi|^2)^{-s/2}$; convolution with $\mathcal{J}^s$ maps L^p onto H_p^s for $1 < p < +\infty$; r-the Sobolev embeddings state that it maps L^p to L^r with $r = \frac{3p}{3-sp}$ when $0 \leq s < \frac{3}{p}$; then, picking p_0 and p_1 with $1 < p_0 < p < p_1 < \frac{3}{s}$, the Marcinkiewicz interpolation theorem (as extended by Stein and Weiss [77, 91]) gives the boundedness of $\mathcal{J}^s$ from L^p to $L^{\frac{3p}{3-sp},p}$ as an interpolation of the boundedness of $\mathcal{J}^s$ from L^{p_0} to $L^{\frac{3p_0}{3-sp_0}}$ and from L^{p_1} to $L^{\frac{3p_1}{3-sp_1}}$.
- For $0 < s < 3$, the kernel K_s of the convolution operator $\mathcal{J}^s$ satisfies

$$|K_s(x)| \leq C \frac{1}{|x|^{3-s}}$$

  , and thus K_s belongs to the Lorentz space $L^{\frac{3}{3-s},\infty}$. Convolution in Lorentz spaces has been studied by O'Neil [83] following ideas of Stein. In particular, we have $L^{p,q} * L^{r,s} \subset L^{t,u}$ with $\frac{1}{t} = \frac{1}{p} + \frac{1}{r} - 1$ and $\frac{1}{u} = \min(\frac{1}{q} + \frac{1}{s}, 1)$ (whenever $1 < t < +\infty$). Applying this to $L^p = L^{p,p}$ and $L^{\frac{3}{3-s},\infty}$ gives the desired embedding.

Due to their good properties of interpolation and to their simple convolution and product laws, Lorentz spaces have turned out to be very efficient tools for providing sharp estimates in Lebesgue norms. For instance, the Hardy inequality

$$\int_{\mathbb{R}^3} \frac{|f|^p}{|x|^{sp}}\, dx \leq C_{s,p} \int_{\mathbb{R}^3} |(-\Delta)^{s/2} f|^p\, dx$$

for $f \in H_p^s$, $1 < p < +\infty$, and $0 < s < 3/p$ is a direct consequence of the facts that the kernel of $(-\Delta)^{-s/2}$ (i.e., the Riesz potential $\mathcal{I}^s$) belongs to $L^{\frac{3}{3-s},\infty}$, the multiplier $\frac{1}{|x|^s}$ belongs to $L^{\frac{3}{s},\infty}$, the convolution maps $L^p \times L^{\frac{3}{3-s},\infty}$ to $L^{\frac{3p}{3-sp},p}$, and the pointwise product maps $L^{\frac{3p}{3-sp},p} \times L^{\frac{3}{s},\infty}$ to $L^{p,p} = L^p$.

Besides being a useful tool for refining inequalities, Lorentz spaces occur as a natural setting in various problems in the study of the Navier–Stokes equations, especially in problems with critical scaling. For instance, the bilinear operator B defined by equation (5) is bounded on $L^\infty((0, T), L^p)$ for $p > 3$, with a norm

$$\|B\|_{\mathscr{B}(L^\infty L^p \times L^\infty L^p \mapsto L^\infty L^p)} = C_p T^{\frac{1}{2}(1-\frac{3}{p})}.$$

But it is no longer bounded on $L^\infty L^3$ [84]. It turns out that, however, it is bounded on $L^\infty L^{3,\infty}$, as proved by Meyer [80].

Kozono and Nakao [59] studied time-periodic solutions for the Navier–Stokes equations with a time-periodic forcing. They found solutions in $L^\infty L^{3,\infty}$. More precisely, one studies the equations

$$\partial_t \mathbf{u} = \Delta \mathbf{u} - \mathbb{P}\nabla \cdot (\mathbf{u} \otimes \mathbf{u}) + \mathbb{P}\nabla \cdot \mathbb{F}$$

where the forcing tensor $\mathbb{F}$ is time-periodic, and one seeks for a solution $\mathbf{u}$ which is still time-periodic. If we assume that $\mathbb{F}$ belongs to $L^\infty L^{3/2,\infty}$, then we define $\mathbf{U}^0$ as

$$\mathbf{U}^0 = \int_{-\infty}^t e^{(t-s)\Delta} \mathbb{P}\nabla \cdot \mathbb{F}\, ds$$

, and we find that $\mathbf{U}^0$ belongs to $L^\infty L^{3,\infty}$. Thus, looking for time-periodic solutions of the Navier–Stokes solutions with time-periodic tensor $\mathbb{F}$ is turned into the solving of the integro-differential problem

$$\mathbf{u} = \mathbf{U}^0 - \int_{-\infty}^t e^{(t-s)\Delta} \mathbb{P}\nabla \cdot (\mathbf{u} \otimes \mathbf{u})\, ds = \mathbf{U}^0 - B_\infty(\mathbf{u}, \mathbf{u})$$

where the bilinear operator B_∞ is defined as

$$B_\infty(\mathbf{u}, \mathbf{v})(t, .) = \int_{-\infty}^t e^{(t-s)\Delta} \mathbb{P}\nabla \cdot (\mathbf{u} \otimes \mathbf{v})\, ds.$$

As B_∞ is bounded on $L^\infty L^{3,\infty}$, the Banach contraction principle will give us a solution as soon as $\mathbb{F}$ is small enough.

Meyer [80] applied the boundedness of B on $L^\infty L^{3,\infty}$ to another problem, namely, uniqueness of solutions of the Navier–Stokes equations in $\mathscr{C}([0, T), L^3)$. We already discussed this problem. Let $\mathbf{u}^0 \in L^3$. If $\mathbf{u}$ is a solution in $\mathscr{C}[0, T), L^3)$ and $\mathbf{u}_K$ is the solution provided by Kato in $\mathscr{C}[0, T), L^3)$ with the additional property that $\lim_{t\to 0} t^{1/4} \|\mathbf{u}_K\|_6 = 0$, then the function $\mathbf{w} = \mathbf{u} - \mathbf{u}_K$ satisfies the identity

$$\mathbf{w} = -B(\mathbf{u}_K, \mathbf{w}) - B(\mathbf{w}, \mathbf{u}_K) - B(\mathbf{w}, \mathbf{w}).$$

The main idea of the proof of uniqueness initially given by Furioli, Lemarié-Rieusset, and Terraneo [45] was to establish a contractive estimate (locally in time) on $\mathbf{w}$ to prove that $\mathbf{w}$ is equal to 0. The difficult term is $B(\mathbf{w}, \mathbf{w})$, as B is not bounded on $L^\infty L^3$. But, as B is bounded on $L^\infty L^{3,\infty}$, it is easy to prove that for $0 < S < T$,

$$\sup_{0<t<S} \|\mathbf{w}(t, .)\|_{L^{3,\infty}} \leq C \sup_{0<t<S} \|\mathbf{w}(t, .)\|_{L^{3,\infty}} \left(\sup_{0<s<S} s^{1/4} \|\mathbf{u}_K(s, .)\|_6 + \sup_{0<s<S} \|\mathbf{w}(t, .)\|_{L^{3,\infty}} \right).$$

By continuity of both $\mathbf{u}$ and $\mathbf{u}_K$ in L^3 norm, and by the embedding $L^3 \subset L^{3,\infty}$, we have $\lim_{t\to 0} \|\mathbf{w}(t, .)\|_{L^{3,\infty}} = 0$, and we find that we have a contractive estimate for $\mathbf{w}$ if S is small enough. Hence, the fixed point $\mathbf{w}$ is equal to 0, and $\mathbf{u} = \mathbf{u}_K$ on $(0, S)$. The end of the proof follows by a bootstrap argument.

Another example where one naturally deals with Lorentz spaces is the study of self-similar solutions. If $\mathbf{u}$ and p are solutions of

$$\partial_t \mathbf{u} + (\mathbf{u} \cdot \nabla)\mathbf{u} = \Delta \mathbf{u} - \nabla p,$$

$$\nabla \cdot \mathbf{u} = 0,$$

$$\mathbf{u}(0, .) = \mathbf{u}^0,$$

then, for $\lambda > 0$, defining $\mathbf{u}_\lambda(t, x) = \lambda \mathbf{u}(\lambda^2 t, \lambda x)$, $p_\lambda(t, x) = \lambda^2 p(\lambda^2 t, \lambda x)$, and $\mathbf{u}^0_\lambda(x) = \lambda \mathbf{u}^0(\lambda x)$, we find that $\mathbf{u}_\lambda$ and p_λ are solutions of

$$\partial_t \mathbf{u}_\lambda + (\mathbf{u}_\lambda \cdot \nabla)\mathbf{u}_\lambda = \Delta \mathbf{u}_\lambda - \nabla p_\lambda,$$

$$\nabla \cdot \mathbf{u}_\lambda = 0,$$

$$\mathbf{u}_\lambda(0, .) = \mathbf{u}^0_\lambda.$$

Thus, provided that $\mathbf{u}_0$ is homogeneous (so that $\mathbf{u}^0_\lambda = \mathbf{u}^0$), one may look for self-similar solutions (such that $\mathbf{u}_\lambda = \mathbf{u}$ and $p_\lambda = p$). However, if $\mathbf{u}^0$ is homogeneous and is not equal to 0, it cannot belong to the usual spaces (Lebesgue spaces L^p or Sobolev spaces H^s), by lack of integrability either at $x = 0$ or at $x = \infty$. But $L^{3,\infty}$ contains non-trivial homogenous functions, so that the problem of looking for self-similar solutions is meaningful in the setting of this Lorentz space (this has been done by Barraza in 1996 [3]).

5.2 Besov Spaces

Besov spaces are usually viewed as the main tool of real variable harmonic analysis methods for the Navier–Stokes equations [2, 14, 70]. However, the role played by Besov spaces has various aspects.

The most obvious occurrence of Besov spaces, more related to the classical theory of parabolic equations than to real methods in harmonic analysis, is linked to the analysis of the heat kernel and the thermic characterization of Besov spaces. If $s < 0$, then a distribution $f \in \mathscr{S}'$ will belong to $B^s_{p,q}$ ($1 \leq p, q \leq +\infty$) if and only if $t^{s/2}\|e^{t\Delta} f\|_q$ belongs to $L^p((0, T), \frac{dt}{t})$ (where $0 < T < +\infty$). If $T = +\infty$, then f belongs to the (realization of) homogeneous Besov space $\dot{B}^s_{p,q}$. Thus, the result of Fabes, Jones, and Rivière [30] that B is bounded on $L^p((0, T), L^q)$ when T is finite, $3 < q < +\infty$, and $\frac{2}{p} + \frac{3}{q} \leq 1$, or when $T = +\infty$, $3 < q < +\infty$, and $\frac{2}{p} + \frac{3}{q} = 1$, implies that one may find a (local in time) solution $\mathbf{u} \in L^p L^q$ to the Cauchy problem for the Navier–Stokes equations with initial value $\mathbf{u}_0$ if and only if $\mathbf{u}_0 \in B^{-\frac{2}{p}}_{q,p}$; this solution will be global if $\frac{2}{p} + \frac{3}{q} = 1$ and $\mathbf{u}_0$ is small enough in $\dot{B}^{-\frac{2}{p}}_{q,p}$.

Similarly, the inequality (6) we used to construct Kato solutions to the problem with $\mathbf{u}_0 \in L^3$ can be seen as the embedding $L^3 \subset \dot{B}^{-\frac{1}{2}}_{6,\infty}$.

Homogeneous Besov spaces occur naturally in the study of the Navier–Stokes equations, due to the scaling invariance of the equations. If we want to study the initial value problem in a Banach space $\mathbb{E}$ of distributions that respects symmetries of the problem, we shall ask the norm of $\mathbb{E}$ to be invariant under translations in $\mathbb{R}^3$ ($\|f(x - x_0)\|_{\mathbb{E}} = \|f\|_{\mathbb{E}}$ for $x_0 \in \mathbb{R}^3$) and under dilations ($\lambda\|f(\lambda x)\|_{\mathbb{E}} = \|f\|_{\mathbb{E}}$ for $\lambda > 0$). In that case, Meyer [80] remarked that we have the embedding $\mathbb{E} \subset \dot{B}^{-1}_{\infty,\infty}$, a Besov space that plays a prominent role in the study of the Navier–Stokes equations.

The use of Besov spaces in fluid mechanics relies essentially on the dyadic Littlewood–Paley decomposition (sometimes called the Littlewood–Paley–Stein decomposition, as Stein is one of the first analysts to use it). This decomposition makes easy dealing with the nonlinearity $\mathbf{u} \cdot \boldsymbol{\nabla}\mathbf{u}$ of the equations, by using the paraproduct operators of Bony [6]. Seminal works on Besov spaces and fluid mechanics appeared in the 1990s, as the paper of Chemin in 1992 [18] or the book of Cannone in 1995 [13]; applications of the Littlewood-Paley decomposition to the borderline cases of regularity for solutions of Euler equations were given by Vishik in 1998–1999 [95, 96]. Chemin developed a theory of time-space Besov spaces where the nonlinear evolution partial differential equations are treated more efficiently after localization by means of Littlewood–Paley decomposition [2, 19] (especially in the borderline cases of regularity).

An interesting example of the use of Besov spaces for Navier–Stokes equations is the proof of uniqueness of solutions in $\mathscr{C}([0, T], L^3)$ to the Cauchy problem for initial value $\mathbf{u}^0 \in L^3$. The first proof of such uniqueness has been given by Furioli, Lemarié-Rieusset, and Terraneo [44, 45]. If $\mathbf{u}$ is a solution in $\mathscr{C}[0, T), L^3)$ and $\mathbf{u}_K$

is the solution provided by Kato in $\mathscr{C}[0,T), L^3)$ with the additional property that $\lim_{t\to 0} t^{1/4}\|\mathbf{u}_K\|_6 = 0$, then the function $\mathbf{w} = \mathbf{u} - \mathbf{u}_K$ satisfies the identity

$$\mathbf{w} = -B(\mathbf{u}_K, \mathbf{w}) - B(\mathbf{w}, \mathbf{u}_K) - B(\mathbf{w}, \mathbf{w}).$$

The main step of the proof of uniqueness by Furioli, Lemarié-Rieusset, and Terraneo was to establish a contractive estimate (locally in time) on $\mathbf{w}$ to prove that $\mathbf{w}$ is equal to 0, in spite of the fact that B is not bounded on $L^\infty L^3$. They remarked that $\mathbf{w}$ is more regular than $\mathbf{u}$ and $\mathbf{u}_K$: $\mathbf{u} - e^{t\Delta}\mathbf{u}^0$ and $\mathbf{u}_K - e^{t\Delta}\mathbf{u}^0$ belong to $L^\infty \dot{B}^{1/2}_{2,\infty}$; the contractive estimate they found is then the following one: for $0 < S < T$,

$$\sup_{0<t<S} \|\mathbf{w}(t,.)\|_{\dot{B}^{1/2}_{2,\infty}} \le C \sup_{0<t<S} \|\mathbf{w}(t,.)\|_{\dot{B}^{1/2}_{2,\infty}} \left(\sup_{0<s<S} s^{1/8}\|\mathbf{u}_K(s,.)\|_4 + \sup_{0<s<S} \|\mathbf{w}(t,.)\|_3 \right).$$

By continuity of both $\mathbf{u}$ and $\mathbf{u}_K$ in L^3 norm, we have $\lim_{t\to 0} \|\mathbf{w}(t,.)\|_{L^{3,\infty}} = 0$, and we find that we have a contractive estimate for $\mathbf{w}$ if S is small enough. Hence, the fixed point $\mathbf{w}$ is equal to 0, and $\mathbf{u} = \mathbf{u}_K$ on $(0, S)$. The end of the proof follows by a bootstrap argument.[2]

In some points of the study of the Navier–Stokes equations, Besov spaces appear to be optimal. Let us quote three examples concerning the Leray solutions. We consider a solution $\mathbf{u} \in L^\infty((0,T), L^2) \cap L^2((0,T), \dot{H}^1)$ of the Navier–Stokes equations with initial value $\mathbf{u}^0 \in L^2$, satisfying Leray's energy inequality:

$$\|\mathbf{u}(t,.)\|_2^2 + 2\int_0^t \|\nabla \otimes \mathbf{u}\|_2^2\, ds \le \|\mathbf{u}^0\|_2^2.$$

- Regularity: a well-known result of Serrin [88] states that if $\mathbf{u}^0$ belongs more precisely to H^1, then $\mathbf{u}$ will remain in H^1 as long as $\int_0^T \|\mathbf{u}\|_q^p\, dt < +\infty$ with $3 < q \le \infty$ and $\frac{2}{p} + \frac{3}{q} = 1$. The space L^q has been replaced by many larger spaces with the same scaling properties. The largest one is $\dot{B}^{-\frac{3}{q}}_{\infty,\infty}$. Serrin's criterion has been proved to hold for $\mathbf{u} \in L^p \dot{B}^\sigma_{\infty,\infty}$ for $\frac{2}{p} = 1 + \sigma$ and $1 \le p < +\infty$ (Kozono and Shimada [61] for $p > 2$, Chen and Zhang [20] for $1 < p \le 2$, Kozono, Ogawa, and Taniuchi [60] for $p = 1$).
- Weak–strong uniqueness: a well-known result of Prodi [85] and Serrin [87] states that if the Cauchy problem for $\mathbf{u}^0$ has another solution $\mathbf{v}$ in $L^\infty L^2 \cap L^2 \dot{H}^1$ and if moreover $\mathbf{v} \in L^p_t L^q_x$ with $\frac{2}{p} + \frac{3}{q} = 1$ and $3 < q \le +\infty$, then $\mathbf{u} = \mathbf{v}$. Again, this

[2] This was after this result that Brezis asked me to write a book on Besov estimates for Navier–Stokes equations [70].

has been extended by replacing L^q by many larger spaces with the same scaling properties. Serrin's criterion has been proved to hold for $\mathbf{u} \in L^p X_\sigma$ for $\frac{2}{p} = 1+\sigma$ and $1 < p < +\infty$, where $X_\sigma = \dot{B}^\sigma_{\infty,\infty}$ if $\sigma > 0$ (i.e., $p < 2$) (Chen, Miao, and Zhang [21]), $X_0 = \text{BMO}$ (Kozono and Taniuchi [62]), and $X_\sigma = \dot{M}^{2,q}$ if $\sigma < 0$ and $\sigma = -\frac{3}{q}$ (see the subsection on Morrey spaces).

- Energy (in)equality: a classical result of Lions [76] states that if the Leray solution $\mathbf{u}$ satisfies $\mathbf{u} \in L^4L^4$, then Leray's energy inequality for $\mathbf{u}$ is indeed an equality. The assumption $\mathbf{u} \in L^4L^4$ has been weakened by Duchon and Robert [29] to $\mathbf{u} \in L^3 b^{1/3}_{3,\infty}$ where $b^{1/3}_{3,\infty}$ is the closure of test functions in $\dot{B}^{1/3}_{3,\infty}$. (Remark that $L^2\dot{H}^1 \cap L^4L^4 \subset L^3 b^{1/3}_{3,\infty}$.)

5.3 Morrey Spaces and Morrey–Campanato Spaces

When dealing with scaled estimates in spaces of measurable functions, one is naturally driven to use Morrey spaces. The Morrey space $\dot{M}^{p,q}$, $1 < p < +\infty$, $p \le q \le \infty$ is defined by

$$f \in \dot{M}^{p,q} \Leftrightarrow f \in L^p_{\text{loc}} \text{ and } \sup_{x_0 \in \mathbb{R}^3, r>0} \frac{1}{|B(x_0,r)|^{\frac{1}{p}-\frac{1}{q}}} \|\mathbf{1}_{B(x_0,r)} f\|_p < +\infty.$$

Again, we define $\sigma = -\frac{3}{q}$ and we find equivalently

$$f \in \dot{M}^{p,q} \Leftrightarrow f \in L^p_{\text{loc}} \text{ and } \sup_{x_0 \in \mathbb{R}^3, r>0} \frac{1}{r^{\frac{3}{p}+\sigma}} \|\mathbf{1}_{B(x,r)} f\|_p < +\infty.$$

The restriction $p \le q \le +\infty$ implies that we have $-\frac{3}{p} \le \sigma \le 0$. Remark that if $\sigma < -\frac{3}{p}$ or $\sigma > 0$, then $f = 0$. Moreover, $\dot{M}^{p,\infty} = L^\infty$.

Morrey–Campanato spaces are quite similar, except that we correct f with its mean value $m_{B(x_0,r)} f = \frac{1}{|B(x_0,r)|} \int_{B(x_0,r)} f\, dx$:

$$f \in \mathcal{M}^{p,\sigma} \Leftrightarrow f \in L^p_{\text{loc}} \text{ and } \sup_{x_0 \in \mathbb{R}^3, r>0} \frac{1}{r^{\frac{3}{p}+\sigma}} \|\mathbf{1}_{B(x_0,r)} f - m_{B(x_0,r)} f\|_p < +\infty.$$

This time, σ will be in the range $-\frac{3}{p} \le \sigma < 1$. Moreover, $\mathcal{M}^{p,0} = \text{BMO}$ and, for $0 < \sigma < 1$, $\mathcal{M}^{p,\sigma} =$ (the realization of) $\dot{B}^\sigma_{\infty,\infty}$ [12]. Thus, $\mathcal{M}^{p,\sigma}$ is the dual of the Hardy space $\mathcal{H}^r$ for $\frac{3}{4} < r \le 1$ and $\sigma = \frac{3}{r} - 1$ [39].

If $\sigma < 0$ and if $(\psi_{\epsilon,j,k})_{1\le\epsilon\le7, j\in\mathbb{Z}, k\in Z^3}$ is a compactly supported wavelet bases with regularity $\mathcal{C}^3$, then we find that

$$|\langle f|\psi_{\epsilon,j,k}\rangle| \le C 2^{3j(\frac{1}{p}-\frac{1}{2})} \|\psi_\epsilon\|_{\frac{p}{p-1}} \|f\|_{\mathcal{M}^{p,\sigma}} 2^{-j(\frac{3}{p}+\sigma)}$$

and we find that $\sum_{\epsilon,j,k}\langle f|\psi_{\epsilon,j,k}\rangle\psi_{\epsilon,j,k}$ is (*-weakly) convergent in (the realization of) $\dot{B}^{\sigma}_{\infty,\infty}$. Moreover, the series $\sum_{\epsilon,j,k}\langle f|\psi_{\epsilon,j,k}\rangle\nabla\psi_{\epsilon,j,k}$ converges in $\mathscr{D}'$ to ∇f. Thus, we have a decomposition $\mathscr{M}^{p,\sigma} = \dot{M}^{p,-\frac{3}{\sigma}} \oplus \mathbb{R}\mathbf{1}$ and an identification $\dot{M}^{p,-\frac{3}{\sigma}} = \mathscr{M}^{p,\sigma} \cap \dot{B}^{\sigma}_{\infty,\infty}$.

The first occurrence of Morrey spaces in the study of Navier–Stokes equations was in a paper by Giga and Miyakawa [48] on self-similar solutions. Then, in the early 1990s, there has been results on mild solutions in Morrey spaces given by Kato [56], Taylor [93], and Federbush [32]. In 1994, Kozono and Yamazaki [63] introduced Besov–Morrey spaces in order to give examples of singular initial values (or of initial values with large L^3 norms) leading to global mild solutions. Cannone's book [13] or Lemarié-Rieusset's one [70] gave a systematic treatment of those spaces.

The flourishing of various classes of mild solutions for the Navier–Stokes equations that occurred in the 1990s opened the question of the largest space that would lead, through Picard iterations, to solutions. This space is included in $B^{-1}_{\infty,\infty}$ but is smaller, as the regularization by the heat kernel is not sufficient to give a meaning to the nonlinear term. This space was identified by Koch and Tataru [58] and named bmo^{-1}: this is the space of distributions that are a sum of a bounded function $f_0 \in L^\infty$ and of derivatives $\partial_j f_j$ of functions f_j in the bmo space of Goldberg (a local version of BMO) [49]. The homogeneous version of this space is $\mathrm{BMO}^{-1} = \sqrt{-\Delta}(\mathrm{BMO})$. Recently, Auscher and Frey [1] gave a new proof of the theorem of Koch and Tataru, based on the duality between the Hardy space $\mathscr{H}^1$ and BMO.

Variations on the Koch and Tataru theorem led May [78] and Xiao [97] to consider initial values in $(\sqrt{-\Delta})^{1-\sigma}\mathscr{M}^{2,-\sigma} = (\sqrt{-\Delta})^{1-\sigma}\dot{M}^{2,\frac{3}{\sigma}}$. Xiao linked his results to his theory of Q-spaces and to Carleson measures and the tent spaces of Coifman, Meyer, and Stein [23].

Morrey spaces appear in many papers on the Navier–Stokes equations, extending results involving Lebesgue spaces where scaling properties prevail over global integrability. For instance, uniqueness of mild solutions in $\mathscr{C}([0,T],L^3)$ proven by Furioli, Lemarié-Rieusset, and Terraneo holds as well in $\mathscr{C}([0,T],\dot{m}^{p,3})$ for $2 < p \le 3$, where $\dot{m}^{p,3}$ is the closure of test functions in the space $\dot{M}^{p,3}$ [45, 72]. The case of $\mathscr{C}([0,T],\dot{m}^{2,3})$ remains open. The space $\dot{X}^1 = \mathscr{M}(\dot{H}^1 \mapsto L^2)$ of pointwise multipliers from the Sobolev space $\dot{H}^1(\mathbb{R}^3)$ to $L^2(\mathbb{R}^3)$ satisfies the embeddings, for $2 < p \le 3$, $\dot{M}^{p,3} \subset \dot{X}^1 \subset \dot{M}^{2,3}$ (Fefferman [33]). May [75] proved uniqueness of solutions in $\mathscr{C}([0,T],\dot{x}^1)$, where $\dot{x}^1$ is the closure of test functions in the space $\dot{X}^1$.

Another interesting occurrence of Morrey spaces in the study of the Navier–Stokes equations is the extension of the criterion of weak–strong uniqueness of Prodi [85] and Serrin [87]. The key point in the proof of the criterion is an inequality of the type

$$|\int \mathbf{u}\cdot(\mathbf{v}\cdot\nabla)\mathbf{v}\,dx| \le C\|\mathbf{u}\|_{X_\sigma}\|\mathbf{v}\|_2^{1+\sigma}\|\nabla\otimes\mathbf{v}\|_2^{1-\sigma}$$

for two divergence-free vector fields $\mathbf{u}$ and $\mathbf{v}$. For $-1 \leq \sigma \leq 0$, a simple approach is to use an inequality of the type $\|\mathbf{u}\otimes\mathbf{v}\|_2 \leq C\|\mathbf{u}\|_{X_\sigma}\|\mathbf{v}\|_{H^{-\sigma}}$ together with $\|\mathbf{v}\|_{H^{-\sigma}} \leq \|\mathbf{v}\|_2^{1+\sigma}\|\nabla\otimes\mathbf{v}\|_2^{-\sigma}$. Let us write $\dot X^r = \mathcal{M}(\dot H^r \mapsto L^2)$ for the set of pointwise multipliers from the Sobolev space $\dot H^r(\mathbb{R}^3)$ to $L^2(\mathbb{R}^3)$ (for a characterization of $\dot X^1$, see Maz'ya [79]); we get

$$|\int \mathbf{u}\cdot(\mathbf{v}\cdot\nabla)\mathbf{v}\,dx| \leq C\|\mathbf{u}\|_{\dot X^r}\|\mathbf{v}\|_2^{1-r}\|\nabla\otimes\mathbf{v}\|_2^{1+r}$$

and find weak–strong uniqueness for Leray solutions of the Cauchy problem with initial value $\mathbf{u}_0$ if one of those solutions belongs moreover to $L^p\dot X^r$ ($0 \leq r < 1$ and $\frac{2}{p} = 1 - r$) or to $\mathcal{C}([0,T),\dot x^1)$ (for $r = 1$) [70].

For $0 < r < 1$, a better approach is to use an inequality of the type $\|\mathbf{u}\otimes\mathbf{v}\|_2 \leq C\|\mathbf{u}\|_{X_\sigma}\|\mathbf{v}\|_{\dot B^r_{2,1}}$ together with $\|\mathbf{v}\|_{\dot B^r_{2,1}} \leq C\|\mathbf{v}\|_2^{1+r}\|\nabla\otimes\mathbf{v}\|_2^r$. Thus, we are interested in the space $\mathcal{M}(\dot B^r_{2,1} \mapsto L^2)$ of pointwise multipliers from the Besov space $\dot B^r_{2,1}(\mathbb{R}^3)$ to $L^2(\mathbb{R}^3)$; this space turns out to be the Morrey space $\dot M^{2,\frac{3}{r}}$ [72, 74], which is larger than $\dot X^r$. Hence, we get

$$|\int \mathbf{u}\cdot(\mathbf{v}\cdot\nabla)\mathbf{v}\,dx| \leq C\|\mathbf{u}\|_{\dot M^{2,\frac{3}{r}}}\|\mathbf{v}\|_2^{1-r}\|\nabla\otimes\mathbf{v}\|_2^{1+r}$$

and find weak–strong uniqueness for Leray solutions of the Cauchy problem with initial value $\mathbf{u}_0$ if one of those solutions belongs moreover to $L^p\dot M^{2,\frac{3}{r}}$ ($0 < r < 1$ and $\frac{2}{p} = 1 - r$).

For $\sigma \leq 0$, one uses the fact that $\mathbf{v}$ is divergence free. Recall that for $\sigma = 0$, Kozono and Taniuchi [62] wrote

$$|\int \mathbf{u}\cdot(\mathbf{v}\cdot\nabla)\mathbf{v}\,dx| \leq C\|\mathbf{u}\|_{\mathrm{BMO}}\|\mathbf{v}\cdot\nabla\mathbf{v}\|_{\mathcal{H}^1} \leq C'\|\mathbf{u}\|_{\mathrm{BMO}}\|\mathbf{v}\|_2\|\nabla\otimes\mathbf{v}\|_2$$

and got weak–strong uniqueness for Leray solutions of the Cauchy problem with initial value $\mathbf{u}_0$ if one of those solutions belongs moreover to $L^2\,\mathrm{BMO}$.

For $0 < \sigma < 1$, we use product laws in Sobolev spaces to estimate the (positive) regularity of $\mathbf{v}\otimes\mathbf{v}$:

$$\|\mathbf{v}\otimes\mathbf{v}\|_{\dot B^{1-\sigma}_{1,1}} \leq C\|\mathbf{v}\|^2_{\dot H^{\frac{1-\sigma}{2}}} \leq C\|\mathbf{v}\|_2^{1+\sigma}\|\nabla\otimes\mathbf{v}\|_2^{1-\sigma}$$

so that

$$|\int \mathbf{u}\cdot(\mathbf{v}\cdot\nabla)\mathbf{v}\,dx| \leq C\|\nabla\otimes\mathbf{u}\|_{\dot B^{\sigma-1}_{\infty,\infty}}\|\mathbf{v}\otimes\mathbf{v}\|_{\dot B^{1-\sigma}_{1,1}} \leq C\|\mathbf{u}\|_{\dot B^{\sigma}_{\infty,\infty}}\|\mathbf{v}\|_2^{1+\sigma}\|\nabla\otimes\mathbf{v}\|_2^{1-\sigma}.$$

Thus, find weak–strong uniqueness for Leray solutions of the Cauchy problem with initial value $\mathbf{u}_0$ if one of those solutions belongs moreover to $L^p\dot B^{\sigma}_{\infty,\infty}$ ($0 < \sigma < 1$

and $\frac{2}{p} = 1 + \sigma$). The limit case $\sigma = 1$ gives weak–strong uniqueness when one of the solutions belongs moreover to L^1 Lip.

For $-1 < \sigma < 1$, those results may be unified in the following way: weak–strong uniqueness for Leray solutions of the Cauchy problem with initial value $\mathbf{u}_0$ if one of those solutions belongs moreover to $L^p \mathcal{M}^{2,\sigma}$ ($-1 < \sigma < 1$ and $\frac{2}{p} = 1 + \sigma$).

Another point where scaling plays an important role is the theory of partial regularity for suitable weak solutions of the Navier–Stokes solutions developed by Caffarelli, Kohn, and Nirenberg [9]. In order to simplify the proof given by Caffarelli, Kohn, and Nirenberg in 1982, Ladyzhenskaya and Seregin [65] used Morrey spaces as a basic tool for elliptic or parabolic equations. A systematic and inspiring proof wholly given ion terms of Morrey spaces has been given by Kukavica in 2011 [64].

6 Conclusion

We have given many examples of the interaction of harmonic analysis with the study of Navier–Stokes equations, beyond the simple use of Littlewood–Paley decomposition. The usefulness of such tools can be nicely illustrated by the case of the refined Gagliardo–Nirenberg inequalities of Gérard, Meyer, and Oru [47]. This inequality states that if $1 < p \leq +\infty$, $\alpha > 0$, and $\beta > 0$, then the control of $(\sqrt{-\Delta})^\alpha f$ in L^p norm and of f in $\dot{B}^{-\beta}_{\infty,\infty}$ gives a control of f in L^q, with $\frac{1}{q} = \frac{\beta}{\alpha+\beta} \frac{1}{p}$:

$$\|f\|_q \leq \|(\sqrt{-\Delta})^\alpha f\|_p^{\frac{\beta}{\alpha+\beta}} \|f\|_{\dot{B}^{-\beta}_{\infty,\infty}}^{\frac{\alpha}{\alpha+\beta}} \tag{7}$$

The original proof is given in terms of the Littlewood–Paley decomposition of f and of the characterization of L^q as a Triebel–Lizorkin space $\dot{F}^0_{p,2}$. But there is a very shorter and simpler proof based on Hedberg's inequality [53]. (More precisely a variant of Hedberg's inequality, where one replaces the role played by the Hardy–Littlewood maximal function by Stein's maximal function, in order to be able to deal with distributions in $\dot{B}^{-\beta}_{\infty,\infty}$.) More precisely, if $N > \alpha/2$, one writes (for $f \in \dot{B}^{-\beta}_{\infty,\infty}$)

$$f = \frac{(-1)^N}{\Gamma(N)} \int_0^{+\infty} (t\Delta)^N e^{t\Delta} f \frac{dt}{t}$$

and uses the inequalities

$$|(t\Delta)^N e^{t\Delta} f(x)| \leq C t^{\frac{\alpha}{2}} \mathcal{M}_{(\sqrt{-\Delta})^\alpha f}(x)$$

and

$$|(t\Delta)^N e^{t\Delta} f(x)| \leq C t^{-\frac{\beta}{2}} \|f\|_{\dot{B}^{-\beta}_{\infty,\infty}}$$

to find Hedberg's inequality

$$|f(x)| \leq C \left(\mathcal{M}_{(\sqrt{-\Delta})^\alpha f}(x)\right)^{\frac{\beta}{\alpha+\beta}} \left(\|f\|_{\dot{B}^{-\beta}_{\infty,\infty}}\right)^{\frac{\alpha}{\alpha+\beta}}.$$

Inequality (7) is then obvious.

Hedberg's inequality, combined with basic theory of singular integrals and maximal functions, should be a powerful tool to deal with some nonlinear PDEs, avoiding the rigidity of the Littlewood–Paley decomposition or of wavelet decompositions and in a way replacing it by a molecular approach (molecules in Hardy spaces [where only size of the molecules is controlled) or in Besov spaces (where size and regularity of the molecules are controlled]). This was the claim in [73] and the basis for the book [74]. Indeed, a Littlewood-Paley decomposition is stable neither through a transport equation nor under the action of a singular integral convolution operator. On the other hand, a molecular decomposition will be stable, since a molecule is preserved under a transport equation with Lipschitzian drift (moving the center along the characteristic curve and deforming the profile of the molecule but without altering too much its scale) or through the action of a singular integral convolution operator (with roughly speaking the same center and the same scale but with a deformation of the profile). Similarly, a wavelet decomposition is not preserved but transformed into a vaguelette decomposition [70]. An interesting example of what can be done with molecules is the paper by Chamorro and Menozzi establishing regularization properties for an advection–diffusion problem with non-local diffusion and rough drift [17]; the title of their paper is quite programmatic for the use of real methods in harmonic analysis when studying nonlinear PDEs: *Nonlinear Singular Drifts and Fractional Operators: When Besov Meets Morrey and Campanato*.

References

1. P. Auscher, D. Frey, On well-posedness of parabolic equations of Navier-Stokes type with $BMO^{-1}(\mathbb{R}^n)$ data. J. Inst. Math. Jussieu **16**, 947–985 (2017)
2. H. Bahouri, J.Y. Chemin, R. Danchin, *Fourier Analysis and Nonlinear Partial Differential Equations* (Springer, Berlin/Heidelberg, 2011)
3. O. Barraza, Self-similar solutions in weak L^p-spaces of the Navier–Stokes equations. Rev. Mat. Iberoam. **12**, 411–439 (1996)
4. G. Battle, P. Federbush, Divergence–free vector wavellets. Mich. Math. J. **40**, 181–195 (1995)
5. A. Benedek, A.P. Calderón, R. Panzone, Convolution operators on Banach space valued functions. Proc. Nat. Acad. Sci. USA **48**, 356–365 (1962)
6. J.M. Bony. Calcul symbolique et propagation des singularités pour les équations aux dérivées partielles non linéaires. Ann. Sci. Ec. Norm. Sup. **14**, 209–246 (1981)

7. L. Brandolese. *Localisation, oscillations et comportement asymptotique pour les équations de Navier–Stokes* (Thèse, ENS Cachan, 2001)
8. L. Brandolese, Y. Meyer, On the instantaneous spreading for the Navier–Stokes system in the whole space. ESAIM Contr. Optim. Calc. Var. **8**, 273–285 (2002)
9. L. Caffarelli, R. Kohn, L. Nirenberg, Partial regularity of suitable weak solutions of the Navier–Stokes equations. Commun. Pure Appl. Math. **35**, 771–831 (1982)
10. A.P. Calderón, A. Zygmund, On the existence of certain singular integrals. Acta Math. **88**, 85–139 (1952)
11. C. Calderón, Initial values of Navier–Stokes equations. Proc. Am. Math. Soc. **117**, 761–766 (1993)
12. S. Campanato, Proprietà di hölderianità di alcune classi di funzioni. Ann. Scuola Norm. Sup. Pisa **17**, 175–188 (1963)
13. M. Cannone, *Ondelettes, Paraproduits et Navier–Stokes* (Diderot Editeur, Paris, 1995)
14. M. Cannone, Harmonic analysis tools for solving the incompressible Navier–Stokes equations, in *Handbook of Mathematical Fluid Mechanics*, vol. III, ed. by S.J. Friedlander, D. Serre (Elsevier, Amsterdam, 2004)
15. M. Cannone, G. Wu, Global well–posedness for Navier–Stokes equations in critical Fourier–Herz spaces. Nonlinear Anal. **75**, 3754–3760 (2012)
16. D. Chamorro, A molecular method applied to a non-local PDE in stratified Lie groups. J. Math. Anal. Appl. **413**, 583–608 (2014)
17. D. Chamorro, S. Menozzi, Non linear singular drifts and fractional operators: when Besov meets Morrey and Campanato. Potential Anal. **49**, 1–35 (2018)
18. J.M. Chemin, Remarques sur l'existence globale pour le système de Navier–Stokes incompressible. SIAM J. Math. Anal. **23**, 20–28 (1992)
19. J.Y. Chemin, N. Lerner, Flot de champs de vecteurs non-lipschitziens et équations de Navier–Stokes. J. Diff. Equ. **12**, 314–326 (1995)
20. Q. Chen, Z. Zhang, Space-time estimates in the Besov spaces and the 3D Navier–Stokes equations. Methods Appl. Anal. **13**, 107–122 (2006)
21. Q. Chen, C. Miao, Z. Zhang, On the uniqueness of weak solutions for the 3D Navier–Stokes equations. Ann. Inst. H. Poincaré Anal. Non Linéaire **26**, 2165–2180 (2009)
22. R. Coifman, P.L. Lions, Y. Meyer, S. Semmes, Compensated compactness and Hardy spaces. J. Math. Pures et Appl. **72**, 247–286 (1992)
23. R. Coifman, Y. Meyer, E.M. Stein, Some new function spaces and their applications to harmonic analysis. J. Funct. Anal. **62**, 304–335 (1985)
24. R. Coifman, Y. Meyer, Commutateurs d'intégrales singulières et opérateurs multilinéaires. Ann. Inst. Fourier **28**, 177–202 (1978)
25. R. Coifman, Y. Meyer, Au delà des opérateurs pseudo-différentiels. Astérisque **57**, Société Mathématique de France (1978)
26. R. Coifman, Y. Meyer, *Ondelettes et Opérateurs*, vol. III (Hermann, Paris, 1991)
27. R. Coifman, G. Weiss, *Analyse Harmonique Non-commutative sur Certains Espaces Homogènes*, Lecture Notes in Mathematics (Springer, Berlin/Heidelberg, 1971)
28. S. Dobrokhotov, A. Shafarevich, Some integral identities and remarks on the decay at infinity of the solutions to the Navier–Stokes equations in the entire space. Russ. J. Math. Phys. **2**, 133–135 (1994)
29. J. Duchon, R. Robert, Dissipation d'énergie pour des solutions faibles des équations d'Euler et Navier–Stokes incompressibles. C. R. Acad. Sci. Paris (Série I) **329**, 243–248 (1999)
30. E. Fabes, B.F. Jones, N. Rivière, The initial value problem for the Navier–Stokes equations with data in L^p. Arch. Ration. Mech. Anal. **45**, 222–240 (1972)
31. M. Farge, N. Kevlahan, V. Perrier, K. Schneider, Turbulence analysis, modelling and computing using wavelets, in *Wavelets and Physics*, ed. by van der Berg (Cambridge University Press, Cambridge, 1999)
32. P. Federbush, Navier and Stokes meet the wavelet. Commun. Math. Phys. **155**, 219–248 (1993)
33. C. Fefferman, The uncertainty principle. Bull. Am. Math. Soc. **9**, 129–206 (1983)

34. C. Fefferman, Existence and smoothness of the Navier–Stokes equation, in *The Millennium Prize Problems*, ed, by J.A. Carlson, A. Jaffe, A. Wiles (American Mathematical Society, Cambridge, 2006), pp. 57–67
35. C. Fefferman, E.M. Stein, H^p spaces of several variables. Acta Math. **129**, 137–193 (1972)
36. P.G. Fernandez-Dálgo, P.G. Lemarié–Rieusset, Weak solutions for Navier–Stokes equations with initial data in weighted L^2 spaces. Preprint, Univ. Évry (2019)
37. C. Foias, R. Temam, Gevrey class regularity for the solutions of the Navier–Stokes equations. J. Funct. Anal. **87**, 359–369 (1989)
38. G.B. Folland, Some topics in the history of harmonic analysis in the twentieth century. Indian J. Pure Appl. Math. **48**, 1–58 (2017)
39. G.B. Folland, E.M. Stein, *Hardy Spaces on Homogeneous Groups* (Princeton University Press, Princeton, 1982)
40. M. Frazier, B. Jawerth, A discrete transform and decomposition of distribution spaces. J. Funct. Anal. **93**, 34–170 (1990)
41. P. Frick, V. Zimin, Hierarchical models of turbulence, in *Wavelets, Fractals and Fourier Transforms*, ed. by M. Farge et al. (Oxford University Press, Oxford, 1993)
42. U. Frisch. *Turbulence. The Legacy of A.N. Kolmogorov* (Cambridge University Press, Cambridge, 1995)
43. H. Fujita, T. Kato, On the non-stationary Navier-Stokes system. Rend. Sem. Math. Univ. Padova **32**, 243–260 (1962)
44. G. Furioli, P.G. Lemarié-Rieusset, E. Terraneo, Sur l'unicité dans $L^3(\mathbb{R}^3)$ des solutions "mild" de l'équation de Navier–Stokes. C. R. Acad. Sci. Paris, Série I **325**, 1253–1256 (1997)
45. G. Furioli, P.G. Lemarié-Rieusset, E. Terraneo, Unicité dans $\mathrm{L}^3(\mathbb{R}^3)$ et d'autres espaces limites pour Navier–Stokes. Rev. Mat. Iberoam. **16**, 605–667 (2000)
46. G. Furioli, E. Terraneo, Molecules of the Hardy space and the Navier–Stokes equations. Funkcial. Ekvac. **45**, 141–160 (2002)
47. P. Gérard, Y. Meyer, F. Oru, *Inégalités de Sobolev précisées*, in *Séminaire X-EDP* (1996)
48. Y. Giga, T. Miyakawa, Navier–Stokes flow in $\mathbb{R}^3$ with measures as initial vorticity and Morrey spaces. Commun. Partial Diff. Equ. **14**, 577–618 (1989)
49. D. Goldberg, A local version of real Hardy spaces. Duke Math. J. **46**, 27–42 (1979)
50. L. Grafakos, *Classical Harmonic Analysis*, 2nd edn. (Springer, New York, 2008)
51. L. Grafakos, *Modern Harmonic Analysis*, 2nd edn. (Springer, London, 2009)
52. G.H. Hardy, J.E. Littlewood, A maximal theorem with function-theoretic applications. Acta Math. **54**, 81–116 (1930)
53. L. Hedberg, On certain convolution inequalities. Proc. Am. Math. Soc. **10**, 505–510 (1972)
54. C. Herz, Lipschitz spaces and Bernstein's theorem on absolutely convergent Fourier transforms. J. Math. Mech. **18**, 283–323 (1968/1969)
55. T. Kato, Strong L^p solutions of the Navier–Stokes equations in $\mathbb{R}^m$ with applications to weak solutions. Math. Z. **187**, 471–480 (1984)
56. T. Kato, Strong solutions of the Navier–Stokes equations in Morrey spaces. Boletim da Sociedade Brasileira de Matemática **22**, 127–155 (1992)
57. T. Kato, G. Ponce, Commutator estimates and the Euler and Navier–Stokes equations. Commun. Pure Appl. Math. **41**, 891–907 (1988)
58. H. Koch, D. Tataru, Well-posedness for the Navier–Stokes equations. Adv. Math. **157**, 22–35 (2001)
59. H. Kozono, M. Nakao, Periodic solutions of the Navier–Stokes equations in unbounded domains. Tohoku Math. J. **48**, 33–50 (1996)
60. H. Kozono, T. Ogawa, Y. Taniuchi, The critical Sobolev inequalities in Besov spaces and regularity criterion to some semi-linear evolution equations. Math. Z. **242**, 251–278 (2002)
61. H. Kozono, Y. Shimada, Bilinear estimates in homogeneous Triebel–Lizorkin spaces and the Navier–Stokes equations. Math. Nachr. **276**, 63–74 (2004)
62. H. Kozono, Y. Taniuchi, Bilinear estimates in BMO and Navier–Stokes equations. Math. Z. **157**, 173–194 (2000)

63. H. Kozono, M. Yamazaki, Semilinear heat equations and the Navier–Stokes equations with distributions in new function spaces as initial data. Commun. Partial Differ. Equ. **19**, 959–1014 (1994)
64. I. Kukavica, Partial regularity for the Navier–Stokes equations with a force in a Morrey space. J. Math. Anal. Appl. **374**, 573–584 (2011)
65. O.A. Ladyzhenskaya, G.A. Seregin, On partial regularity of suitable weak solutions to the three-dimensional Navier–Stokes equations. J. Math. Fluid Mech. **1**, 356–387 (1999)
66. Y. Le Jan, A.S. Sznitman, Cascades aléatoires et équations de Navier–Stokes. C. R. Acad. Sci. Paris **324 Série I**, 823–826 (1997)
67. Z. Lei, F. Lin, Global mild solutions of Navier–Sokes equations. Commun. Pure Appl. Math. **64**, 297–1304 (2011)
68. P.G. Lemarié-Rieusset. Analyses multi-résolutions non orthogonales, commutation entre projecteurs et dérivations et ondelettes vecteurs à divergence nulle, Revista Mat. Iberoamer. **8**, 221–237 (1992)
69. P.G. Lemarié-Rieusset, Une remarque sur l'analyticité des solutions milds des équations de Navier–Stokes dans $\mathbb{R}^3$. C. R. Acad. Sci. Paris Serie I **330**, 183–186 (2000)
70. P.G. Lemarié-Rieusset, *Recent Developments in the Navier–Stokes Problem* (CRC Press, Boca Raton, 2002)
71. P.G. Lemarié-Rieusset, Nouvelles remarques sur l'analyticité des solutions milds des équations de Navier–Stokes dans $\mathbb{R}^3$. C. R. Acad. Sci. Paris Serie I **338**, 443–446 (2004)
72. P.G. Lemarié–Rieusset, The Navier–Stokes equations in the critical Morrey-Campanato space. Revista Matematica Iberoamericana **23**, 897–930 (2007)
73. P.G. Lemarié-Rieusset, Euler equations and real harmonic analysis. Arch. Rat. Mech. Anal. **204**, 355–386 (2012)
74. P.G. Lemarié–Rieusset, *The Navier–Stokes Problem in the 21st Century* (Chapman & Hall/CRC, Boca Raton, 2016)
75. P.G. Lemarié–Rieusset, R. May, Uniqueness for the Navier–Stokes equations and multipliers between Sobolev spaces. Nonlinear Anal. **66**, 813–838 (2007)
76. J.L. Lions, Sur la régularité et l'unicité des solutions turbulentes des équations de Navier–Stokes. Rendiconti del Seminario Matematico della Università di Padova **30**, 16–23 (1960)
77. J. Marcinkiewicz, Sur l'interpolation d'opérateurs. C. R. Acad. Sci. Paris **208**, 1272–1273 (1939)
78. R. May, *Régularité et unicité des solutions milds des équations de Navier–Stokes*, Ph.D. Thesis, Université d'Évry (2002)
79. V. Maz'ya, On the theory of the n-dimensional Schrödinger operator [in Russian]. Izvestya Akademii Nauk SSSR (ser. Mat.,) **28**, 1145–1172 (1964)
80. Y. Meyer, *Wavelets, Paraproducts and Navier–Stokes Equations*, Current developments in Mathematics 1996 (International Press, Cambridge, 1999), pp. 02238–2872
81. S. Monniaux, Uniqueness of mild solutions of the Navier–Stokes equation and maximal L^p-regularity. C. R. Acad. Sci. Paris, Série I **328**, 663–668 (1999)
82. S. Montgomery–Smith, Finite time blow up for a Navier–Sokes like equation. Proc. Am. Math. Soc. **129**, 3017–3023 (2007)
83. R. O'Neil, Convolution operators and $L(p, q)$ spaces. Duke Math. J. **30**, 129–142 (1963)
84. F. Oru, *Rôle des oscillations dans quelques problèmes d'analyse non linéaire* (Thèse, École Normale Supérieure de Cachan, 1998)
85. G. Prodi, Un teorema di unicitá per le equazioni di Navier–Stokes. Ann. Mat. Pura Appl. **48**, 173–182 (1959)
86. M. Riesz, Sur les fonctions conjuguées. Math. Z. **27**, 218–244 (1927)
87. J. Serrin, The initial value problem for the Navier–Stokes equations, in *Nonlinear Problems (Proceedings of Symposium, Madison, 1962)* (University of Wisconsin Press, Madison, 1963), pp. 69–98
88. J. Serrin, On the interior regularity of weak solutions of the Navier–Stokes equations. Arch. Ration. Mech. Anal. **9**, 187–195 (1962)

89. E.M. Stein, *Singular Integrals and Differentiability Properties of Functions* (Princeton University Press, Princeton, 1971)
90. E.M. Stein, *Harmonic Analysis* (Princeton University Press, Princeton, 1993)
91. E.M. Stein, G. Weiss, *Introduction to Fourier Analysis on Euclidean Spaces* (Princeton University Press, Princeton, 1971)
92. L. Tartar, *An Introduction to Navier–Stokes Equation and Oceanography* (Springer, Berlin/New York, 2006)
93. M.E. Taylor, *Analysis on Morrey Spaces and Applications to Navier–Stokes Equations and Other Evolution Equations* Commun. Partial Differ. Equ. **17**, 1407–1456 (1992)
94. K. Urban, *Multiskalenverfahren für das Stokes-Problem und angepasste Wavelet-Basen* (Verlag der Augustinus-Buchhandlung, Aachen, 1995)
95. M. Vishik, Hydrodynamics in Besov spaces. Arch. Ration. Mech. Anal. **145**, 197–214 (1998)
96. M. Vishik, Incompressible flows of an ideal fluid with vorticity in borderline spaces of Besov type. Annales Scientifiques de l'École Normale Supérieure **32**, 769–812 (1999)
97. J. Xiao, Homothetic variant of fractional Sobolev space with application to Navier–Stokes system. Dyn. Partial Differ. Equ. **4**, 227–245 (2007)

Explore Intrinsic Geometry of Sleep Dynamics and Predict Sleep Stage by Unsupervised Learning Techniques

Gi-Ren Liu, Yu-Lun Lo, Yuan-Chung Sheu, and Hau-Tieng Wu

Abstract We propose a novel unsupervised approach for sleep dynamics exploration and automatic annotation by combining modern harmonic analysis tools. Specifically, we apply diffusion-based algorithms, diffusion map (DM), and alternating diffusion (AD) algorithms, to reconstruct the intrinsic geometry of sleep dynamics by reorganizing the spectral information of an electroencephalogram (EEG) extracted from a nonlinear-type time frequency analysis tool, the synchrosqueezing transform (SST). The visualization is achieved by the nonlinear dimension reduction properties of DM and AD. Moreover, the reconstructed nonlinear geometric structure of the sleep dynamics allows us to achieve the automatic annotation purpose. The hidden Markov model is trained to predict the sleep stage. The prediction performance is validated on a publicly available benchmark database, Physionet Sleep-EDF [extended] SC* and ST*, with the leave-one-subject-out cross-validation. The overall accuracy and macro F1 achieve 82.57% and 76% in Sleep-EDF SC* and 77.01% and 71.53% in Sleep-EDF ST*, which is compatible with the state-of-the-art results by supervised learning-based algorithms. The results suggest the potential of the proposed algorithm for clinical applications.

G.-R. Liu
Department of Mathematics, National Chen-Kung University, Tainan, Taiwan

Y.-L. Lo
Department of Thoracic Medicine, Healthcare Center, Chang Gung Memorial Hospital, Chang Gung University, School of Medicine, New Taipei, Taiwan

Y.-C. Sheu
Department of Applied Mathematics, National Chiao Tung University, Hsinchu, Taiwan
e-mail: sheu@math.nctu.edu.tw

H.-T. Wu (✉)
Department of Mathematics and Department of Statistical Science, Duke University, Durham, NC, USA
e-mail: hauwu@math.duke.edu

M. Th. Rassias (ed.), *Harmonic Analysis and Applications*, Springer Optimization and Its Applications 168, https://doi.org/10.1007/978-3-030-61887-2_11

1 Introduction

Sleep is a recurring physiological dynamical activity in mammals. Since 1968, the Rechtschaffen and Kales (R&K) criteria [1] is the gold standard when researchers study human sleep dynamics, and this criteria was further generalized by American Academy of Sleep Medicine (AASM) [2]. According to the AASM criteria, the sleep dynamics is quantified by finite discrete stages, and those stages can be divided into two broad categories, the rapid eye movement (REM) and the non-rapid eye movement (NREM), and the NREM stage is further divided into shallow sleep (stages N1 and N2) and deep sleep (stage N3). Based on this quantification, up to now, we have accumulated plenty of knowledge about sleep dynamics [3], and sleep dynamics is nowadays an active research field due to more unknowns [4]. Despite those unknowns, it has been well-known that a distortion of sleep dynamics could lead to catastrophic outcomes. For example, REM disturbance slows down the perceptual skill improvement [5], insufficient N2 sleep is associated with weaning failure [6], deprivation of slow wave sleep is associated with Alzheimer's disease [7], etc. Moreover, several public disasters are caused by low sleep quality [8].

The polysomnography (PSG) is the gold standard of evaluating the sleep dynamics. The PSG usually records multiple channels from a subject, ranging from electroencephologram (EEG), electrooculogram (EOG), electrocardiogram (ECG), electromyogram (EMG), photoplethysmogram (PPG), several respiratory signals, etc. While each channel contains its own information about sleep, to apply the AASM criteria [2] to study sleep dynamics, sleep experts mainly count on EEG, EOG, and EMG. To simplify the discussion, and better explore how much sleep dynamics is captured by EEG signals, we focus on exploring sleep dynamics via analyzing EEG signals.

An EEG is a complicated time series. It is non-stationary in nature. A common model to study a given EEG assumes that it is composed of diverse spectra [9], each of which depicts a portion of the brain dynamics. Furthermore, we assume that the brain dynamics, as a dynamical system, is supported on a low-dimensional manifold. Based on this model, a natural question to ask is if we are able to recover the low-dimensional manifold for the sleep dynamics study. The first step toward answering this question is to quantify the spectral content in an EEG signal by the time-frequency analysis. Synchrosqueezing transform (SST) [10, 11] is a nonlinear-type time-frequency analysis technique, which allows us to combine the phase information into the spectrogram (the squared magnitude of the short-time Fourier transform (STFT)). Note that the phase information is ignored in the spectrogram. A direct consequence is "sharpening" the spectra of the EEG signal and prevents the "energy leakage" that is commonly encountered in the spectrogram due to the uncertainty principle [12]. Specifically, the spectrogram is sharpened by taking the phase information of STFT into account to nonlinearly deform the spectrogram, and the blurring effect/energy leakage of the spectrogram caused by the uncertainty principle is alleviated. We call the outcome the *synchrosqueezed EEG spectral features*, which contains not only the spectrogram but also the EEG phase information. The synchrosqueezed EEG spectral feature is in general different from the intrinsic sleep dynamics. This difference comes from various resources including, for example, the distortion caused by the transform and inevitable noise and artifact. Thus, an extra step is needed to refine/reorganize the synchrosqueezed EEG spectral feature and define the final intrinsic features for the sleep dynamics. We suggest to conquer the distortion by the local Mahalanobis distance (MD) framework [13, 14]. Via local MD, we reorganize the synchrosqueezed EEG spectral features by the diffusion-based machine learning algorithms, including diffusion maps (DM) [15], alternating diffusion (AD) [16], and co-clustering [17], depending on how many EEG channels we have, for this purpose. We call the resulting features, the *intrinsic sleep dynamical features*. Based on the established theory, the intrinsic sleep dynamical features recover the intrinsic geometry of sleep dynamics and hence provides a visualization tool to explore sleep dynamics.

A direct application of discovering the geometry of sleep dynamics is an automatic annotation system. Scoring the overnight sleep stage from the PSG outputs by sleep experts is time-consuming [18] and error-prone [19] due to the huge signal loading. Due to its importance for the whole healthcare system, an accurate automatic sleep dynamics scoring system is critical in the current clinical environment. For this automatic annotation purpose, we consider the standard hidden Markov model (HMM) [20] to learn the sleep experts' knowledge. Based on the physiological knowledge, we first take the available phenotype information of a new-arriving subject to determine "similar subjects" from the existing database. Then, based on the intrinsic sleep dynamical features and the annotations of these similar subjects, a prediction model for the new subject is established for the new-arriving subject. To evaluate the performance of the proposed algorithm for the prediction purpose, we consider the publicly available benchmark database, the Physionet Sleep-EDF database [21]. This database contains two subsets, the SC* and ST*. The SC* subset consists of 20 healthy subjects without any sleep-related medication, and the ST* subset consists of 22 subjects who had mild difficulty falling asleep. The overall accuracy and macro F1 achieve 82.57% and 76% in Sleep-EDF SC* and 77.01% and 71.53% in Sleep-EDF ST*, which is compatible with the state-of-the-art results by supervised learning-based algorithms. The results suggest the potential of the proposed algorithm for clinical applications.

The rest of this paper is organized as follows. In Section 2, we summarize the theoretical background of SST and demonstrate how SST works in an EEG signal. In Section 3, the local MD and the empirical intrinsic geometry framework are summarized. In Section 4, the diffusion-based algorithm, DM, is discussed, and its theoretical support is summarized. In Section 5, diffusion-based sensor fusion algorithms, including AD, co-clustering, and multiview DM, as well as known theoretical background, are provided. In Section 6, the implementation details of the proposed algorithm for recovering the geometry of sleep dynamics, including feature extraction algorithm for the synchrosqueezed EEG spectrum and diffusion based feature organization and sensor fusions, are provided. The HMM for the automatic sleep stage annotation is also summarized. In Section 7, we describe the publicly available benchmark database, the Sleep-EDF Database [Expanded] from PhysioNet, and the statistics for the performance evaluation purpose. In Section 8, the results of applying the proposed algorithm to the Sleep-EDF Database [Expanded] are shown. The paper is closed with the discussion and conclusion in Section 9, with a comparison with existing relevant literature in automatic sleep stage annotation.

2 Synchrosqueezing Transform

In this subsection, we introduce the synchrosqueezed EEG spectrogram based on the STFT-based SST [11] algorithm. The basic idea underlying SST is utilizing the phase information in the STFT of the EEG signal to sharpen the spectrogram. There are two benefits. First, the phase information that is commonly ignored in the spectrogram is preserved. Second, the uncertainty principle intrinsic to the spectrogram is alleviated and the spectrogram is sharpened, which can prevent the "energy leakage" caused by the blurring effect inherited in the uncertainty principle associated with the STFT [12]. These two benefits allow us a better quantification of the EEG dynamics. The SST algorithm is composed of three steps – extract the local dynamics of the EEG signal by the STFT, manipulate the phase information, and sharpen the spectrogram according to the extracted phase. Below we summarize the SST based on STFT.

Take a continuously recorded signal f, for example, an EEG in this work. In practice, f can be as general as a tempered distribution function. Take a Schwartz function h to be the chosen window. The STFT of f is then defined by

$$V_f^{(h)}(t,\omega) = \int_{-\infty}^{\infty} f(s)h(s-t)e^{-i2\pi\omega(s-t)}ds, \tag{1}$$

where $t \in \mathbb{R}$ is the time, and $\omega \in \mathbb{R}$ is the frequency. We call $|V_f^{(h)}|^2 : \mathbb{R}\times\mathbb{R} \to \mathbb{R}^+ \cup \{0\}$ the *spectrogram*. As a complex-valued function, the local phase information of f could be approximated by $\partial_t V_f^{(h)}(t,\omega)$. Intuitively, the phase function records the number of oscillations, and this intuition leads us to calculate the following *reassignment rule*:

$$\Omega_f^{(h)}(t,\omega) := \begin{cases} \Im\dfrac{\partial_t V_f^{(h)}(t,\omega)}{2\pi V_f^{(h)}(t,\omega)} & \text{when} \quad |V_f^{(h)}(t,\omega)| \neq 0; \\ -\infty & \text{when} \quad |V_f^{(h)}(t,\omega)| = 0\,, \end{cases} \tag{2}$$

where $\Im$ means taking the imaginary part. The spectrogram of f is finally sharpened by re-allocating the coefficients of the spectrogram according to the reassignment rule [22]:

$$S_f^{(h)}(t,\xi) := \int |V_f^{(h)}(t,\omega)|^2 \frac{1}{\alpha} g\Big(\frac{|\omega - \Omega_f^{(h)}(t,\omega) - \xi|}{\alpha}\Big) d\omega\,, \tag{3}$$

where $\xi > 0$ is the frequency and g is a Schwartz function so that $g(\cdot/\alpha)/\alpha$ converges weakly to the Dirac measure supported at 0 when $\alpha \to 0$. Theoretically, α controls the resolution of the frequency axis in the SST. This seemingly complicated transform has an intuitive interpretation. At each time t, we identify all spectrogram entities that contain oscillatory components at frequency ξ by reading $\frac{1}{\alpha} g\big(\frac{|\omega - \Omega_f^{(h)}(t,\omega) - \xi|}{\alpha}\big)$ and put all identified entities in the (t, ξ) slot. We call $S_f^{(h)}(t,\xi)$ the *synchrosqueezed spectrogram*.

Note that the SST defined in (3) is slightly different from that introduced in [11] – in [11], it is the STFT that is re-allocated, but here, it is spectrogram that is re-allocated. We choose to re-allocate the spectrogram since we do not need to reconstruct the components in this work, and it is the sharpened energy distribution that encodes the phase information that we are interested in. In addition to providing a sharp and concentrated spectrogram, it has been proved in [23] that SST is robust to different kinds of noise. This property is desirable since the EEG signal is commonly noisy. For theoretical developments and more discussions, we refer readers to [10, 23].

Example 1 To have a better insight of the SST algorithm, look at the following *toy* example that motivates the design of SST. Take a harmonic function $f(t) = e^{i2\pi\omega_0 t}$, where $\omega_0 > 0$ is constant. By a direct calculation, we have $V_f^{(h)}(\omega, t) = e^{i2\pi\omega_0 t}\exp(-2\pi^2(\omega - \omega_0)^2 H^2)$. From (2), we know that the instantaneous frequency of $f(t)$ can be recovered from the phase information of $V_f^{(h)}(\omega, t)$; that is, $\omega_0 = \frac{1}{2\pi}\Im\big(\frac{\partial}{\partial t}\ln V_f^{(h)}(t,\omega)\big)$. A direct calculation leads to $\omega_0 = \omega - \mathrm{Im}\big(\frac{1}{2\pi H}\frac{\mathbf{S}_{g'}(\omega,t)}{\mathbf{S}_g(\omega,t)}\big)$, which means that the spectra energy near ω is spread from $\omega - \mathrm{Im}\big(\frac{1}{2\pi H}\frac{\mathbf{S}_{g'}(\omega,t)}{\mathbf{S}_g(\omega,t)}\big)$ for any $t > 0$. Thus, to obtain a sharp spectrogram, we only need to reallocate the spectrogram to the right frequency ω_0.

To better appreciate the effect of the reassignment step, see Figures 1 and 2 for a comparison of the synchrosqueezed spectrogram and the spectrogram of an EEG signal during different sleep stages. We call the spectrogram and synchrosqueezed spectrogram of an EEG signal *EEG spectrogram* and *synchrosqueezed EEG spectrogram*, respectively. In these figures, the EEG signal is superimposed as red curves for a visual comparison. Compared with the EEG spectrogram, the synchrosqueezed EEG spectrogram is sharper since the phase information is taken into account for the reassignment; for example, the alpha wave from 13-th second to 21-th second of REM that oscillates at about 10 Hz (blue arrows in the middle right plot in Figure 1) and the spindles around 8-th, 13-th, and 17-th seconds of N2 that oscillates at about 14 Hz (blue arrows in the top right plot in Figure 2) can be clearly visualized in the synchrosqueezed EEG spectrogram (magenta arrows in the middle left plot in Figure 1 and magenta arrows in the top left plot in Figure 2). These oscillation

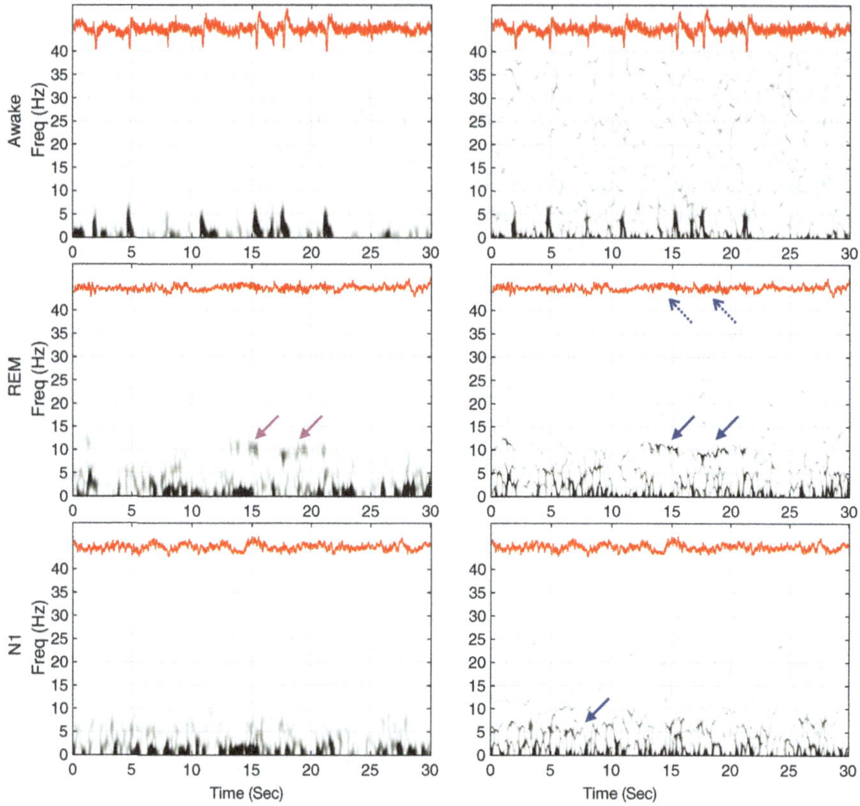

Fig. 1 An illustration of EEG spectrogram (left) and synchrosqueezed EEG spectrogram (right) of sleep stages Awake, REM, and N1 from Sleep-EDF SC* database. The EEG signals are superimposed as red curves for visual comparison

behavior can also be found in the EEG signal (blue dashed arrows in the middle right plot in Figure 1 and blue dashed arrows in the top right plot in Figure 2) but harder to quantify. In the EEG spectrogram, these curves are blurred and not easy to directly identify the variation of the time-varying frequency. Quantifying the time-vary frequency is critical for further understanding the physiology of sleep dynamics, and its study will be reported in the future work. Although the theta wave (blue arrows in the bottom right plot in Figure 1) and delta wave (blue arrows in the bottom right plot in Figure 2) in N1 and N3 stages have less regular oscillatory pattern, the synchrosqueezed EEG spectrogram is more concentrated.

To sum up, the SST-based approach takes the phase information into account and helps avoid energy leakage due to uncertainty principle. As a result, it prevents the possibility of failing to sum up all the signal energy corresponding to one mode centered in a given subband, in the case that the STFT energy leaks into another subband.

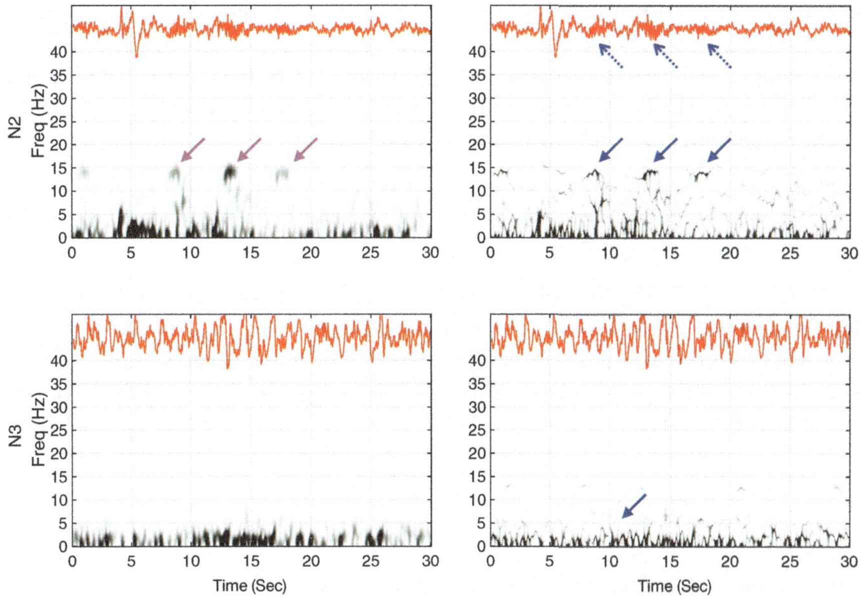

Fig. 2 An illustration of EEG spectrogram (left) and synchrosqueezed EEG spectrogram (right) of sleep stages N2 and N3 from Sleep-EDF SC* database. The EEG signals are superimposed as red curves for visual comparison

3 Local Mahalanobis Distance and Empirical Intrinsic Geometry

To recover the intrinsic sleep dynamics, we need a sophisticated metric to help organize features obtained from SST. In this paper, we consider the *local Mahalanobis distance* as the metric. To motive how the local Mahalanobis distance is designed and how it is incorporated into the inter-individual prediction framework, below we review the dynamics system model and the empirical intrinsic geometry (EIG) model [13, 14]. Then we show how to generalize the EIG model and detail the desired local Mahalanobis distance.

We start from recalling the EIG model as the motivation. We assume that the point cloud, or features, denoted as $\{\mathbf{u}^{(j)}\}_{j=1}^{n} \subset \mathbb{R}^q$, comes from a diffeomorphic deformation $\Phi : \mathbb{R}^p \to \mathbb{R}^q$ which maps the latent intrinsic state space that hosts the dynamical system $\boldsymbol{\theta}^{(j)}$ that describes the dynamics we have interest; that is, $\mathbf{u}^{(j)} = \Phi(\boldsymbol{\theta}^{(j)})$, where $p > 0$ is the dimension of the inaccessible dynamical space. We call Φ the *observation transform*. Furthermore, assume that the inaccessible intrinsic state $\boldsymbol{\theta}^{(j)}$ is the value of the process $\boldsymbol{\theta}$ at the j-th sampling time stamp and $\boldsymbol{\theta}$ satisfies the stochastic differential equation

$$d\boldsymbol{\theta}(t) = a(\boldsymbol{\theta}(t))dt + d\boldsymbol{\omega}(t) \tag{4}$$

where a is an unknown drifting function and ω is the standard d-dim Brownian motion. This latent space model is called EIG in [13, 14], and it has been widely considered to designed other algorithms, like Kalman filter. Based on (4), it is shown in [13, Section 3] that when the intrinsic distance of $\mathbf{u}^{(i)}$ and $\mathbf{u}^{(j)}$ is sufficiently small, we could recover the intrinsic distance between $\boldsymbol{\theta}^{(i)}$ and $\boldsymbol{\theta}^{(j)}$ by

$$\|\boldsymbol{\theta}^{(i)} - \boldsymbol{\theta}^{(j)}\|_{\mathbb{R}^d}^2 = \frac{1}{2}\big(\mathbf{u}^{(i)} - \mathbf{u}^{(j)}\big)^\top \big[(\mathbf{C}^{(i)})^{-1} + (\mathbf{C}^{(j)})^{-1}\big]\big(\mathbf{u}^{(i)} - \mathbf{u}^{(j)}\big), \tag{5}$$

where $\mathbf{C}^{(i)} = \nabla\Phi(\boldsymbol{\theta}^{(i)})[\nabla\Phi(\boldsymbol{\theta}^{(i)})]^\top$ is the covariance matrix associated with the deformed Brownian motion, up to the error term $O(\|\mathbf{u}^{(i)} - \mathbf{u}^{(j)}\|^4)$. Furthermore, it is shown in [13, 14] that $\mathbf{C}^{(i)}$ can be estimated by the covariance matrix of $\{\mathbf{u}^{(k)}\}_{k=i-\delta}^{i+\delta}$, where δ is a predetermined integer. This face comes from the Ito's formula. Note that by the Ito's formula, we immediately have

$$\mathrm{d}\mathbf{u}_t = \left(\frac{1}{2}\Delta\Phi|_{\boldsymbol{\theta}_t} + \nabla\Phi|_{\boldsymbol{\theta}_t} a(\boldsymbol{\theta}_t)\right)\mathrm{d}t + \nabla\Phi|_{\theta_t}\mathrm{d}\omega_t. \tag{6}$$

Since $(\frac{1}{2}\Delta\Phi|_{\boldsymbol{\theta}_t} + \nabla\Phi|_{\boldsymbol{\theta}_t} a(\boldsymbol{\theta}_t))\mathrm{d}t$ is the drifting term, we know

$$\mathrm{Cov}\,(\mathrm{d}\mathbf{u}_t) = \nabla\Phi|_{\boldsymbol{\theta}_t}\nabla\Phi|_{\boldsymbol{\theta}_t}^\top. \tag{7}$$

We call $\|\boldsymbol{\theta}^{(i)} - \boldsymbol{\theta}^{(j)}\|_{\mathbb{R}^d}$ the *local Mahalanobis distance* between $\mathbf{u}^{(i)}$ and $\mathbf{u}^{(j)}$, which is a generalization of Mahalanobis distance from the statistical viewpoint.

While the EIG model works well for a single subject [24], it may not be suitable for the inter-subject sleep assessment mission in which we have interest. Indeed, since different subjects have different sleep dynamics, and the observation transforms are different, physiologically it is not reasonable to quantify the intrinsic state dynamics of different subjects by a single equation like (4). Indeed, it does not make sense from the physiological viewpoint to integrate the temporal sleep dynamics of different subjects into one; that is, there is no temporal relationship between epochs from different subjects. In this work, motivated by the success of (4), we consider the following generalization of the EIG model to study the synchrosqueezed EEG spectral features. Based on the physiological assumption that the EEG signals of different subjects share similar spectral behavior, we assume that the synchrosqueezed EEG spectral features from different subjects come from the same map $\Psi : \mathbb{R}^p \to \mathbb{R}^{10}$, which maps the inaccessible intrinsic state $\boldsymbol{\theta}^{(j)}$, where $0 < d \le p$ is the dimension of the inaccessible space hosting the state space of the dynamics, which is assumed to be the same among subjects, to the space hosting the synchrosqueezed EEG spectral features. We further assume that $p = 10$ and Ψ is an identity from the state space to its range perturbed by a stationary random perturbation that has mean 0 and the covariance $I_{p\times p}$. To simplify the terminology, we still call Ψ the *observation transform*. Note that this is the simplified EIG model with noise in the observation considered in [25].

Under this simplified EIG model, we can get an estimate for $\|\boldsymbol{\theta}^{(i)} - \boldsymbol{\theta}^{(j)}\|^2_{\mathbb{R}^p}$ similar to (5) by modifying the proof of [13]. Denote the K-neighborhood of $\mathbf{u}^{(j)}$ by $\mathcal{N}_j$ for each $j \in \{1, 2, \ldots, n\}$. Based on the assumption of Ψ, when the data is supported on a d-dimensional smooth manifold embedded in $\mathbb{R}^p$, we have

$$\|\boldsymbol{\theta}^{(i)} - \boldsymbol{\theta}^{(j)}\|^2_{\mathbb{R}^p} \approx (\mathbf{u}^{(i)} - \mathbf{u}^{(j)})^\top \mathscr{T}_d(\nabla\Psi|_{\boldsymbol{\theta}^{(j)}} \nabla\Psi|^\top_{\boldsymbol{\theta}^{(j)}})(\mathbf{u}^{(i)} - \mathbf{u}^{(j)}), \tag{8}$$

where $\mathscr{T}_d(\nabla\Psi|_{\boldsymbol{\theta}^{(j)}} \nabla\Psi|^\top_{\boldsymbol{\theta}^{(j)}})$ means the truncated pseudo-inverse of $\nabla\Psi|_{\boldsymbol{\theta}^{(j)}} \nabla\Psi|^\top_{\boldsymbol{\theta}^{(j)}}$ defined by $U\Lambda^\dagger_d U^\top$, $\nabla\Psi|_{\boldsymbol{\theta}^{(j)}} \nabla\Psi|^\top_{\boldsymbol{\theta}^{(j)}} = U\Lambda U^\top$ is the eigendecomposition, $\Lambda^\dagger_d = \mathrm{diag}[\ell_1^{-1}, \ldots, \ell_d^{-1}, 0, \ldots, 0]$, and $\Lambda = \mathrm{diag}[\ell_1, \ldots, \ell_p]$ with $\ell_1 \geq \ell_2 \geq \ldots \geq \ell_p \geq 0$. Similarly, the covariance of the stationary random perturbation associated with Ψ becomes

$$\begin{aligned}
\Gamma_j &:= \frac{1}{K}\sum_{i\in\mathcal{N}_j} (\mathbf{u}^{(i)} - \mathbf{u}^{(j)})(\mathbf{u}^{(i)} - \mathbf{u}^{(j)})^\top \\
&= \frac{1}{K}\sum_{i\in\mathcal{N}_j} \nabla\Psi|_{\boldsymbol{\theta}^{(j)}}(\boldsymbol{\theta}^{(i)} - \boldsymbol{\theta}^{(j)})(\boldsymbol{\theta}^{(i)} - \boldsymbol{\theta}^{(j)})^\top[\nabla\Psi|_{\boldsymbol{\theta}^{(j)}}]^\top \\
&\approx \nabla\Psi|_{\boldsymbol{\theta}^{(j)}}[\nabla\Psi|_{\boldsymbol{\theta}^{(j)}}]^\top .
\end{aligned} \tag{9}$$

Combining (8) and (9) yields

$$\|\boldsymbol{\theta}^{(i)} - \boldsymbol{\theta}^{(j)}\|^2_{\mathbb{R}^d} \approx (\mathbf{u}^{(i)} - \mathbf{u}^{(j)})^\top \mathscr{T}_d[\Gamma_j](\mathbf{u}^{(i)} - \mathbf{u}^{(j)}).$$

Therefore, following the observation in (5), we thus consider the following metric:

$$d_{\mathrm{LMD}}(\mathbf{u}^{(i)}, \mathbf{u}^{(j)})^2 = \frac{1}{2}(\mathbf{u}^{(i)} - \mathbf{u}^{(j)})^\top(\mathscr{T}_d[\Gamma_i] + \mathscr{T}_d[\Gamma_j])(\mathbf{u}^{(i)} - \mathbf{u}^{(j)}), \tag{10}$$

which we call the *local Mahalanobis distance* (MD). We mention that this approach leads to a more accurate geodesic distance estimation by a direct hard threshold of the noisy covariance matrix to remove the influence of noise. A more general discussion can be found in [26].

4 Diffusion Map

Graph Laplacian (GL)-based algorithms have attracted a lot of attention in the machine learning society. DM [15] is one of those successful algorithms. To better understand the theoretical foundation of DM, in the past decade, several works have been done based on the differential geometry framework. The behavior of spectral embedding under the spectral geometry is discussed in [27]. Later, based on the manifold model, how the GL converges to the Laplace-Beltrami operator is studied in [28–30]. The spectral convergence of GL is studied in [31]. The spectral convergent rate is reported in [32, 33]. In [34] the central limit theory of GL is provided. The problem of embedding by finite eigenfunctions of Laplace-Beltrami operator is studied in [35–37]. Below we summarize those theoretical results.

We start from recalling the DM algorithm. Given a dataset $\mathscr{X} := \{x_i\}$. Construct a $n \times n$ affinity matrix W so that

$$W_{ij} = e^{-d(x_i, x_j)^2/\epsilon}, \quad \text{for } i, j = 1, \ldots, n, \tag{11}$$

where $d(\cdot, \cdot)$ is the chosen metric and the bandwidth $\epsilon > 0$ is chosen by the user.[1] The affinity is clearly the composition of the radial basis function kernel $K(t) = e^{-t^2/\epsilon}$ and the distance $d(x_i, x_j)$. In general, we can choose more general kernels with a sufficient decay rate. To simplify the discussion, we focus on the radial basis function kernel. Next, define a diagonal matrix D of size $n \times n$ as

$$D(i,i) = \sum_{j=1}^{n} W(i, j), \quad \text{for } i = 1, \ldots, n. \tag{12}$$

In general, D is called the degree matrix. With matrices W and D, define a random walk on the point cloud $\mathscr{X}$ with the transition matrix given by the formula

$$A := D^{-1} W\,. \tag{13}$$

Clearly, A is diagonalizable since A is similar to the symmetric matrix $D^{-1/2} W D^{-1/2}$. Therefore, it has a complete set of right eigenvectors $\phi_1, \phi_2, \cdots, \phi_n \in \mathbb{R}^n$ with corresponding eigenvalues $1 = \lambda_1 > \lambda_2 \geq \cdots \geq \lambda_n \geq 0$, where $\phi_1 = [1, 1, \ldots, 1]^\top \in \mathbb{R}^n$. Indeed, from the eigendecomposition $D^{-1/2} W D^{-1/2} = O \Lambda O^\top$, where $O \in O(n)$ and $\Lambda = \text{diag}[\lambda_1, \ldots, \lambda_n]$ is a $n \times n$ diagonal matrix, we have $A = U \Lambda V^\top$, where $U = D^{-1/2} O$ and $V = D^{1/2} O$. Note that $\lambda_1 > \lambda_2$ since we assume the graph is complete, and hence connected, and $\lambda_n \geq 0$ comes from the chosen kernel and the Bochner theorem [39]. With the decomposition $A = U \Lambda V^\top$, the DM is defined as

$$\Phi_t : x_j \mapsto \left(\lambda_2^t \phi_2(j), \lambda_3^t \phi_3(j), \ldots, \lambda_{\hat{d}+1}^t \phi_{\hat{d}+1}(j)\right) \in \mathbb{R}^{\hat{d}}\,, \tag{14}$$

where $j = 1, \ldots, n$, $t > 0$ is the diffusion time chosen by the user, and $\hat{d}$ is the embedding dimension chosen by the user. Note that λ_1 and ϕ_1 are ignored in the embedding since they are not informative. In practice, in addition to determining $\hat{d}$ by a direct assignment, $\hat{d}$ can be determined by a more adaptive way according to the

[1] According to the noise analysis in [38], when the signal-to-noise ratio is small, it is beneficial to set the diagonal terms of the affinity matrix to 0; that is, set $W_{ii} = 0$.

decay of the eigenvalue decay; for example, $\hat{d}$ can be chosen to be the largest j so that $\lambda_j^t > \delta > 0$, where δ is chosen by the user. Both can be obtained by optimizing some quantities of interest based on the problem at hand. Clearly, $\Phi_t(x_j)$ consists of the second to $(\hat{d}+1)$-th coordinates of $e_j^\top U \Lambda^t$, where e_j is the unit n-dim vector with the j-th entry 1. With the DM, we can define the *diffusion distance* (DD) with the diffusion time $t > 0$ between x_i and x_j as

$$D_t(x_i, x_j) = \|\Phi_t(x_i) - \Phi_t(x_j)\|_{\mathbb{R}^{\hat{d}}} . \tag{15}$$

We now summarize the theory behind DM under the manifold model. Supposing a dataset $\mathscr{X} = \{x_i\}_{i=1}^n \subset \mathbb{R}^p$ is independently and identically sampled from a random vector $X : (\Omega, \mathscr{F}, \mathbb{P}) \to \mathbb{R}^p$, where the range of X is supported on a low-dimensional manifold M embedded in $\mathbb{R}^p$. We assume that the manifold is compact and smooth and the metric g is induced from $\mathbb{R}^p$. We assume that the induced measure on the Borel sigma algebra on M, denoted as $X_*\mathbb{P}$, is absolutely continuous with respect to the Riemannian measure dV_g. Furthermore, we assume that the function $p = \frac{dX_*\mathbb{P}}{dV_g}$ by Radon-Nikodym theorem is bounded away from zero and is sufficiently smooth. When $n \to \infty$, the transition matrix A defined in (13) converges to a continuous diffusion operator defined on the manifold when the metric in (11) is chosen to be $d(x_i, x_j) = \|x_i - x_j\|_{\mathbb{R}^p}$ [28–30]. Under the smooth manifold setup, the geodesic distance between two close points can be well approximated by the Euclidean distance; see, for example, [26, Lemma 2,Theorem 2]. If the sampling scheme is uniform, we can further approximate the Laplace-Beltrami operator of the manifold, denoted as Δ_g, when $\epsilon \to 0$; if the sampling scheme is non-uniform but smooth enough, by estimating the density function, we could correct the diffusion process by the *α-normalization* scheme proposed in [15] and again approximate the Laplace-Beltrami operator of the manifold when $\epsilon \to 0$. The convergence happens both in the pointwise sense and spectral sense [31–33]. In summary, we can view the eigenvectors and eigenvalues of $\frac{A-I}{\epsilon}$ as approximation of the eigenfunctions and eigenvalues of the Laplace-Beltrami operator associated with the manifold.

With the Laplace-Beltrami operator, we could apply the spectral embedding theory [27, 40] to embed the manifold (and hence the data) using the eigenfunctions of the diffusion operator. Suppose the manifold is connected, and the l-th eigenvalue of Δ_g is $-\mu_l$ with the eigenfunction f_l; that is, $f_l = -\mu_l f_l$, where $\mu_1 = 0 < \mu_2 \leq \mu_3, \ldots$. Note that since the manifold is connected, $\mu_2 > 0$, and when μ_l has a non-trivial multiplicity, f_l might not be unique. With a given set of eigenvalues and eigenvectors $\{(\mu_l, f_l)\}_{l=1}^\infty$, the *spectral embedding* is defined as [27]

$$\phi_t : x \mapsto c(e^{-t\mu_l} f_l(x))_{l=2}^\infty \in \ell_2 , \tag{16}$$

where $t > 0$ is the *diffusion time*, and c is the normalization constant depending on t. Note that the embedding defined in (14) is a discretization of ϕ_t. Indeed, as is shown in [31, Theorem 5.4], for $t > 0$, $A^{t/\epsilon}$ will converge to the heat kernel of $e^{-t\Delta_g}$ in the spectral sense when $n \to \infty$ and $\epsilon = \epsilon(n) \to 0$ satisfying some mild conditions. Therefore, λ_l in (14) converges to μ_l, and ϕ_l in (14) converges to f_l.[2] We also define $d_t(x, y) = \|\phi_t(x) - \phi_t(y)\|_{\ell_2}$ to be the diffusion distance. Again, (15) is a discretization of $d_t(x, y)$. This embedding allows us to reveal the geometric and topological structure of the manifold and hence the structure of the dataset. In particular, it is shown in [27] that ϕ_t is an almost isometric embedding when t is small enough; in [41], it is shown that the local geodesic distance of the manifold could be well approximated by the diffusion distance, when the diffusion time is long enough compared with the geodesic distance. Lastly, when combined with the finite dimensional embedding result of the spectral embedding theory of the Laplace-Beltrami operator shown in [36, 37], we could guarantee that with finite sampling points, when n is large enough, we can reconstruct the manifold with a given accuracy. This is the *embedding property* that we expect to reorganize the dataset.

The ability to reconstruct the underlying intrinsic structure is not the only significant strength of DM. It has been shown in [38, 42] that DM is also robust to noise in the following sense. Suppose the data point $y_i \in \mathscr{Y}$ comes from contaminating a clean sample x_i by *some* noise ξ_i. Suppose ξ_i satisfies some mild conditions, like finite variance and reasonably controlled noise level. Note that we do not require ξ_i to be identical from point to point. Denote the transition matrix built up from $\{y_i\}_{i=1}^n$ as $W^{(\texttt{noisy})}$. Under this condition, the deviation of $W^{(\texttt{noisy})}$ from W is well controlled by the noise level in the norm sense. Thus, we conclude that the eigenvectors of W with sufficiently large eigenvalues could be well reconstructed from $W^{(\texttt{noisy})}$. This is the *robustness property* we expect from DM to analyze the noisy data.

With the embedding property and the robustness property, we can well approximate the underlying geometric structure from the noisy dataset. To show this, recall that the clean points $\{x_i\}_{i=1}^n$ are sampled from a manifold as discussed above. By Weyl's law [40], and the spectral convergence of DM, the eigenvalues of W_0 decay exponentially fast. Therefore, by the robustness property, the first few eigenvectors and eigenvalues could be well reconstructed. However, the eigenvectors with small eigenvalues will be highly contaminated. Since eigenvalues of the clean data decay fast, those eigenvectors with small eigenvalues are not that informative from the spectral embedding viewpoint – although DM is a nonlinear map, when $\hat{d}$ is chosen large enough, the finite dimensional embedding result guarantees that we can reconstruct the manifold, and hence DM is an almost isometric embedding. Therefore, we conclude that DM with the truncation scheme allows us to a well reconstruction of the clean data up to a tolerable error. We call this property the *recovery property*.

[2]Note that ϕ_l is a n-dim vector while f_l is a smooth function defined on M. To properly state the convergence, we need to convert ϕ_l into a continuous function defined on M. Also, when μ_l has a non-trivial multiplicity, the convergence should be stated using the eigenprojection operator. We refer these technical details to [31].

5 Sensor Fusion by Alternating Diffusion, Co-clustering, and Multiview DM

When we have two sensors collecting data simultaneously from the system of interest, a common question to ask is if we can integrate information from both sensors and achieve a better representation/feature for the system. This problem is understood as the *sensor fusion* problem. In our motivating sleep dynamics problem, while different EEG channels capture information from the same brain, the information recorded might vary, and they might be contaminated by brain-activity irrelevant artifacts from different sources, including noise and other sensor-specific nuisance. These artifacts not only deteriorate the quality of the extracted features but might also mislead the analysis result. The main purpose of sensor fusion is distilling the brain information and removing those unwanted artifacts.

To simplify the discussion, we assume that we have two simultaneously recorded datasets $\mathscr{X} = \{x_i\}_{i=1}^n$ and $\mathscr{Y} = \{y_i\}_{i=1}^n$ (e.g., two EEG channels); that is, x_i and y_i are recorded at the same time. While in general x_i and y_i can be of complicated data format, to simplify the discussion, assume $\mathscr{X} \subset \mathbb{R}^{d_x}$ and $\mathscr{Y} \subset \mathbb{R}^{d_y}$, respectively. In other words, we may view x_i and y_i as the i-th features captured by two sensors. A naive way to combine information from $\mathscr{X}$ and $\mathscr{Y}$ is simply concatenating x_i and y_i and form a new feature of size $d_x + d_y$. However, due to the inevitable sensor-specific artifacts or errors, such a concatenating scheme might not be the optimal route to fuse sensors [16, 43]. We consider the recently developed diffusion-based approaches to fuse sensor information, including AD [16, 43] and co-clustering [17] (a special case of multiview DM [44]). In short, the information from different sensors are "diffused" to integrate the nonlinear common information shared between different sensors and simultaneously eliminate artifacts or noise specific to each sensor. This is a nonlinear generalization of the well-known canonical correlation analysis (CCA) [45]. While we focus on the diffusion-based approaches in this work, there is a huge literature about sensor fusion, and we refer readers to [46] for a systematic review.

5.1 Alternating Diffusion

For AD, we first form two transition matrices $A_x \in \mathbb{R}^{n\times n}$ and $A_y \in \mathbb{R}^{n\times n}$ from $\mathscr{X}$ and $\mathscr{Y}$, respectively, following the definition (13). Then, form a new transition matrix

$$A_{xy} = A_x A_y \in \mathbb{R}^{n\times n}\,. \tag{17}$$

Note that $A_{xy}\mathbf{1} = A_x A_y \mathbf{1} = A_x\mathbf{1}$. Thus, A_{xy} can be viewed as a transition matrix associated with the random walk on the "joint dataset" $\{z_i\}_{i=1}^n$, where $z_i = (x_i, y_i)$, while the associated affinity graph is directed. In fact, if we write $A_x = D_x^{-1} W_x$ and

$A_y = D_y^{-1} W_y$, we have

$$A_{xy} = D_x^{-1}(W_x D_y^{-1} W_y).$$

Note that $W_x D_y^{-1} W_y$ is a non-negative matrix and is in general non-symmetric. If we view $W_x D_y^{-1} W_y$ as an affinity matrix, the associated degree matrix is D_x. Thus, the affinity graph associated with $W_x D_y^{-1} W_y$ is directed. On the other hand, A_{xy} can be viewed as starting a random walk on $\mathscr{Y}$, jumping to $\mathscr{X}$, and continuing another random walk on $\mathscr{X}$. This is the motivation of the terminology AD. Clearly, AD depends on the order of multiplication. In [16], it is proposed that we determine the intrinsic distance between the i-th sample pair, (x_i, y_i), and the j-th sample pair, (x_j, y_j), by the ℓ^2 norm of the i-th row of A_{xy} and the j-th row of A_{xy}. With this intrinsic distance, it is suggested in [16] to apply another DM.

To proceed, we need the "spectral decomposition" of A_{xy}. When A_{xy} is diagonalizable, we apply the spectral decomposition to get $A_{xy} = \Phi_{xy} \Lambda_{xy} \Psi_{xy}^{-1}$, where $\Phi_{xy}, \Psi_{xy} \in Gl(n)$. Then we may proceed with the *alternating diffusion map* (ADM) [43] by embedding the j-th sample to an Euclidean space

$$\Phi_{xy,t}:(x_j, y_j) \mapsto \big(\lambda_{xy,2}^t \phi_{xy,2}(j), \lambda_{xy,3}^t \phi_{xy,3}(j), \ldots, \lambda_{xy,d_{xy}+1}^t \phi_{xy,d_{xy}+1}(j)\big) \in \mathbb{R}^{d_{xy}},$$

where $\lambda_{xy,l}$ and $\phi_{xy,l}$ is the l-th right eigenpair, and d_{xy} is the embedding dimension chosen by the user. The discussion of the embedding is the same as that of DM. However, in general A_{xy} is not diagonalizable, even if A_x and A_y are both diagonalizable. In this case, we may consider the singular value decomposition (SVD) of A_{xy} and proceed with the embedding by taking the singular value and singular vector into account [43, 47]. We mention that AD is closely related to a nonlinear generalization of CCA, called *nonlinear CCA* (NCCA) [48]. In NCCA, when the kernel is chosen to be Gaussian, the main operator of interest is the SVD of

$$C_{xy} := A_x A_y^\top.$$

Unlike AD, C_{xy} is in general not a transition matrix. However, it is related to the mutual information between $\mathscr{X}$ and $\mathscr{Y}$. We refer readers with interest to [48] for more details. Note that both A_{xy} and C_{xy} are in general not diagonalizable and depend on the order of multiplication. To remedy this drawback, in [49], a symmetrized AD is proposed, that is,

$$A = A_x A_y^\top + A_y A_x^\top \in \mathbb{R}^{n \times n}.$$

Again, A is in general not a transition matrix. We refer readers with interest to [49] for more details. For more discussion about AD, ADM, and NCCA, we refer the reader with interest to [16, 43, 48]. When there are more than two channels, a possible generalization of the following discussion can be found in [24] or [16, (28)].

We now summarize some theoretical analyses of AD. Consider three latent random variables X, Y, and Z and assume that the associated joint probability density satisfies

$$f_{X,Y,Z}(x,y,z) = f_X(x) f_{Y|X}(y|x) f_{Z|X}(z|x),$$

that is, Y and Z are conditionally independent given X. The data collected from the first sensor is modeled as

$$S^{(1)} = g_1(X, Y),$$

and that of the second sensor is modeled as

$$S^{(2)} = g_2(X, Z),$$

where we assume g_1 and g_2 satisfy some regularity; for example, g_1 and g_2 are both bilipschitz. In this model, X is the system we have interest to study, while Y and Z are *noises* (or artifacts, nuisances) associated with the first and second sensor, respectively. We call X the *common information.* When the first (resp. second) sensor collects data from the system X, the sensor-dependent noise Y (resp. Z) comes into play via g_1 (resp. g_2). Specifically, the sample datasets $\{s_j^{(i)}\}_{j=1}^n$, where $i = 1, 2$, come from mapping (x_i, y_i, z_i) i.i.d. sampled from the latent space (X, Y, Z) via g_i; that is, $s_j^{(1)} = g_1(x_j, y_j)$ and $s_j^{(2)} = g_2(x_j, z_j)$, $j = 1, \ldots, n$. See Figure 3 for an illustration of the model, where $(\Omega, \mathscr{F}, \mathbb{P})$ means the event space Ω, the sigma algebra $\mathscr{F}$, and the probability measure $\mathbb{P}$. Note that in general, the common information might be deformed via the observation process, g_1 or g_2. The same model can be easily generalized to the case when there are multiple sensors

Following the analysis in [16], the ℓ^2 distance between the i-th row and the j-th row of A_{xy} serves as a good estimate of the distance on the common information associated with the i-th and the j-th samples. We call this distance the *common metric*. In other words, although two sensors are contaminated by different noises, AD allows us a stable recovery of the common information. In [43], it is shown that when the common information can be modeled as a Riemannian manifold, AD recovers the Laplace-Beltrami operator of the manifold when g_1 and g_2 do not deform the common information. If either g_1 or g_2 deform the common information, under mild assumption about the deformation, AD leads to a "deformed" Laplace-Beltrami operator. In other words, it says that although A_{xy} is in general not symmetric, its asymptotic is a self-adjoint operator. We refer readers to [43] for technical details. The behavior of ADM can be analyzed by combining the recovery property based on spectral geometry summarized in Section 4. More results about symmetrized AD can be found in [49].

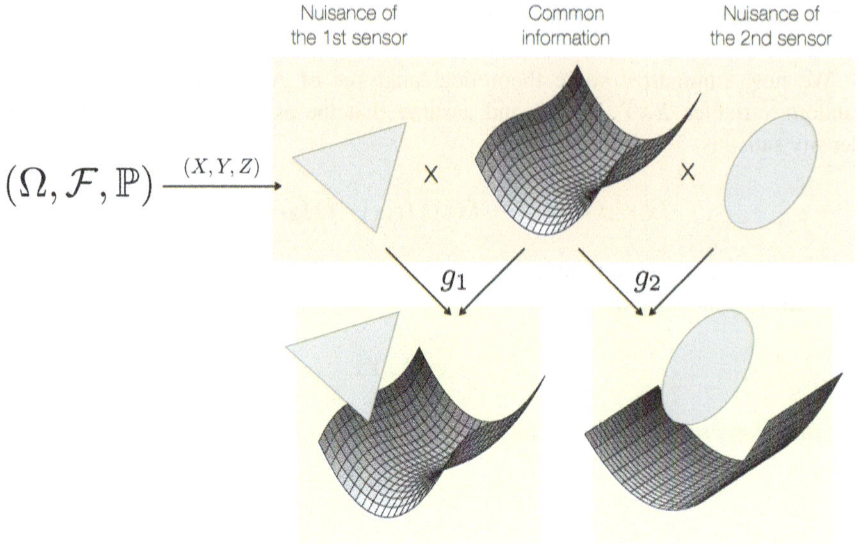

Fig. 3 Illustration of the mathematical model for the sensor fusion algorithm, alternating diffusion

5.2 Co-clustering and Multiview DM

We now discuss the well-known co-clustering algorithm [17] based on the bipartite graph model. Denote $V_1 = \mathscr{X} = \{x_i\}_{i=1}^n$ (resp. $V_2 = \mathscr{Y} = \{y_i\}_{i=1}^n$) to be the set of n vertices representing the n feature vectors extracted from the first (resp. second) sensor. Form a bipartite graph $G = (\mathbb{V}, \mathbb{E})$, where $\mathbb{V}$ consists of $2n$ vertices from V_1 and V_2 and $\mathbb{E}$ is the set of edges between vertices in V_1 and vertices in V_2; that is, edges only exist between V_1 and V_2, and there is no edge inside V_1, and there is no edge inside V_2. Then, by assigning affinities on all edges, (x_i, y_j) for all $i, j = 1, \ldots, n$, we obtain a bipartite affinity graph. The affinity is determined by the user. Denote

$$\mathsf{M} \in \mathbb{R}^{2n\times 2n}$$

to be the affinity matrix associated with the constructed bipartite affinity graph, where the first n columns and rows of M are associated with the first sensor, and the last n columns and rows are associated with the second sensor.

In [17], based on M, it is argued how we can determine the corresponding clusters in V_2 based on the clusters in V_1, and vice versa. More precisely, given disjoint clusters $V_{1,1}, \ldots, V_{1,K}$ for vertices in V_1, where K is a predetermined integer, the clusters $V_{2,1}, \ldots, V_{2,K}$ for vertices in V_2 can be formed by

$$V_{2,m} = \left\{ w_j \in V_2 : \sum_{x_i \in V_{1,m}} \mathsf{M}_{i,j} \geq \sum_{x_i \in V_{1,\ell}} \mathsf{M}_{i,j} \text{ for any } \ell = 1, \ldots, K \right\}, \tag{18}$$

where $m = 1, \ldots, K$. The motivation beyond this assignment is intuitive – a given recording x_i from the second channel has a higher chance to belong to the j-th cluster if it more likely belongs to the j-th cluster of the first channel than other clusters of the first channel. Similarly, given K disjoint clusters $V_{2,1}, \ldots, V_{2,K}$ for vertices in V_2, the induced clusters $V_{1,1}, \ldots, V_{1,K}$ for vertices in V_1 are determined by

$$V_{1,m} = \left\{ x_i \in V_1 : \sum_{y_j \in V_{2,m}} \mathsf{M}_{i,j} \geq \sum_{y_j \in V_{2,\ell}} \mathsf{M}_{i,j} \text{ for any } \ell = 1, \ldots, K \right\}. \tag{19}$$

The information extracted from the first and second sensors iteratively interacts through (18) and (19). This approach is called the "co-clustering" algorithm. We now summarize how the co-clustering of the vertices of a bipartite graph $G = (V, E)$ is related to the traditional spectral clustering algorithm and the well-known Cheeger's inequality [50].[3] Take the bi-clustering for the illustration purpose. Take two disjoint sets $\mathcal{U}_1$ and $\mathcal{U}_2$ with $V = \mathcal{U}_1 \cup \mathcal{U}_2$ as a set of clusters for vertices in V. First of all, define a loss function by

$$\mathcal{N}(\mathcal{U}_1, \mathcal{U}_2) := \frac{\text{cut}(\mathcal{U}_1, \mathcal{U}_2)}{\text{weight}(\mathcal{U}_1)} + \frac{\text{cut}(\mathcal{U}_1, \mathcal{U}_2)}{\text{weight}(\mathcal{U}_2)}, \tag{20}$$

where

$$\text{cut}(\mathcal{U}_1, \mathcal{U}_2) := \sum_{i \in \mathcal{U}_1, j \in \mathcal{U}_2} \mathsf{M}_{i,j}, \quad \text{weight}(\mathcal{U}) := \sum_{i \in \mathcal{U}} \mathsf{D}_{i,i}, \tag{21}$$

and D is a diagonal matrix so that $\mathsf{D}_{i,i} = \sum_{j=1}^{2n} \mathsf{M}_{i,j}$. $\mathcal{N}(\mathcal{U}_1, \mathcal{U}_2)$ is usually called a *normalized cut*. Note that $\mathsf{D}_{i,i}$ represents the degree of the vertex $x_i \in V$ (respectively $y_{i-n} \in V$) when $i \leq n$ (respectively $i > n$). The generalized partition vector $\mathbf{q} = [q_j]_{j=1,\ldots,2n} \in \mathbb{R}^{2n \times 1}$ is defined by

$$q_j = \begin{cases} +\sqrt{\frac{\text{weight}(\mathcal{U}_2)}{\text{weight}(\mathcal{U}_1)}}, & j \in \mathcal{U}_1, \\ -\sqrt{\frac{\text{weight}(\mathcal{U}_1)}{\text{weight}(\mathcal{U}_2)}}, & j \in \mathcal{U}_2. \end{cases} \tag{22}$$

It is shown in [17, Theorem 3] that

[3] Its multiway clustering is supported by the recently developed theory for the multiway spectral clustering algorithm [51].

$$\frac{\mathbf{q}^\top(\mathsf{D}-\mathsf{M})\mathbf{q}}{\mathbf{q}^\top\mathsf{D}\mathbf{q}}=\frac{\mathrm{cut}(\mathscr{U}_1,\mathscr{U}_2)}{\mathrm{weight}(\mathscr{U}_1)}+\frac{\mathrm{cut}(\mathscr{U}_1,\mathscr{U}_2)}{\mathrm{weight}(\mathscr{U}_2)}, \tag{23}$$

which implies that the problem of finding a balanced partition with small cut value can be relaxed and cast as an eigenvalue problem as follows

$$\min_{\mathbf{q}\neq\mathbf{0}}\frac{\mathbf{q}^\top(\mathsf{D}-\mathsf{M})\mathbf{q}}{\mathbf{q}^\top\mathsf{D}\mathbf{q}},\ \text{subject to } \mathbf{q}^\top\mathsf{D}[1\ 1\ \cdots 1]^\top=0. \tag{24}$$

The minimizer of (24) is the eigenvector q_2 of $\mathsf{D}^{-1}\mathsf{M}$ corresponding to the second largest eigenvalue, and the bipartition is achieved by running k-means on $\{q_2(i)\}_{i=1}^{2n}\subset\mathbb{R}$ with $k=2$. Note that the first n entries of q_2 (as well as other eigenvectors) are associated with the clustering of the first sensor and the last n entries are associated with the clustering of the second sensor. The result is the "co-cluster" of the two sensors.

The co-clustering algorithm is intimately related to the recently proposed sensor fusion algorithm, multiview DM [44], particularly when there are two sensors. This relationship is clear after we summarize the multiview DM when there are two sensors. Form two affinity matrices $W_{xy}:=W_xW_y\in\mathbb{R}^{n\times n}$ and $W_{yx}:=W_yW_x\in\mathbb{R}^{n\times n}$, where W_x and W_y are affinity matrices for $\mathscr{X}$ and $\mathscr{Y}$, respectively, that are defined as that in (11). Then, define the affinity matrix M by taking the product of affinities of two sensors by

$$\mathsf{M}=\begin{bmatrix}0_{n\times n} & W_{xy}\\ W_{yx} & 0_{n\times n}\end{bmatrix}\in\mathbb{R}^{2n\times 2n}. \tag{25}$$

Note that the (i,j)-th entry of W_xW_y describes how similar the information of the i-th sample captured by the x sensor is to the information of the j-th sample captured by the y sensor. Denote q_l to be the l-th eigenvector of $\mathsf{D}^{-1}\mathsf{M}$. By the above discussion, we know that $q_2(i),\cdots q_{\hat{d}+1}(i)$ provide the co-clustering information of the i-th sample captured by the first sensor and $q_2(n+i),\cdots,q_{\hat{d}+1}(n+i)$ provide the co-clustering information of the i-th sample captured by the second sensor. Since both channels provide information, for each $i\in\{1,\ldots,n\}$, we consider a concatenation of both, $[q_2(i)\ \cdots\ q_{\hat{d}+1}(i)\ q_2(n+i)\ \cdots\ q_{\hat{d}+1}(n+i)]^\top$, to be the new feature of the i-th sample. This approach is the multiview DM algorithm proposed in [44] when there are two sensors. Note that the multiview DM is more general than simply co-clustering since it can fuse information from multiple sensors. Its theoretical property and its relationship with other algorithms will be explored in the future work.

6 Proposed Algorithm to Explore Intrinsic Geometry of Sleep Dynamics and Predict Sleep Stage

The proposed algorithm is based on the abovementioned unsupervised feature extraction from the EEG signal. The feature extraction consists of two steps. First, we extract spectral information from the EEG signal (indicated by Part 1 in Figure 4). Second, we apply DM or ADM (indicated by Part 2-1 and Part 2-2 in Figure 4) with the local Mahalanobis distance to determine the final features. We can explore the sleep dynamics by visualizing the final features. For the sleep stage prediction purpose, we take the well-established HMM to build up the prediction model. Below, we detail the algorithm implementation step by step.

6.1 Step 1: Extract Synchrosqueezed EEG Spectral Feature

Take $x \in \mathbb{R}^n$ to be the digitalized EEG signal sampled uniformly every τ second from a subject during his/her sleep, where τ is the reciprocal of the sampling rate. For $j = 1, 2, \ldots$, the STFT at $(j\tau)$ seconds is directly implemented by the weighted Fourier transform:

$$\mathbf{V}_x^{(h)}(j,k) := \sum_{m=j-\lfloor 5/\tau \rfloor}^{j+\lceil 5/\tau \rceil} x_m \frac{1}{H} h\left(\frac{m}{H}\right) e^{-i2\pi \frac{k}{K} m}, \tag{26}$$

where $\lceil x \rceil$ means the smallest integer greater than $x > 0$, $\lfloor x \rfloor$ is the largest integer smaller than $x > 0$, $k \in \{0, 1, \ldots, K-1\}$, $K \in \mathbb{N}$ is a parameter used to adjust the frequency resolution, $h(z)$ is the standard Gaussian function, and $H > 0$ is the bandwidth. The SST is then implemented as

$$\mathbf{S}_x(j, \hat{k}) = \sum_{k \in \Lambda(\hat{k})} |\mathbf{V}_x^{(h)}(j,k)|^2 \tag{27}$$

where

$$\Lambda(\hat{k}) = \left\{ k \in \{0, 1, \ldots, K-1\} \,\middle|\, k - \Im\left(\frac{K}{2\pi H} \frac{\mathbf{V}_x^{(\mathscr{D}h)}(j,k)}{\mathbf{V}_x^{(h)}(j,k)} \right) \in \left[\hat{k} - \frac{1}{2}, \hat{k} + \frac{1}{2}\right) \right\}$$

and $\mathscr{D}h$ is the derivative of h. Note that $\mathbf{S}_x(j, \cdot) \in \mathbb{R}^K$ is the synchrosqueezed EEG spectrogram at $(j\tau)$ seconds, which can be viewed as a dynamical spectral feature for the sleep dynamics. Note that the numerical algorithm is a direct discretization of the continuous setup for SST in Section 2 and hence the synchrosqueezed EEG spectrogram.

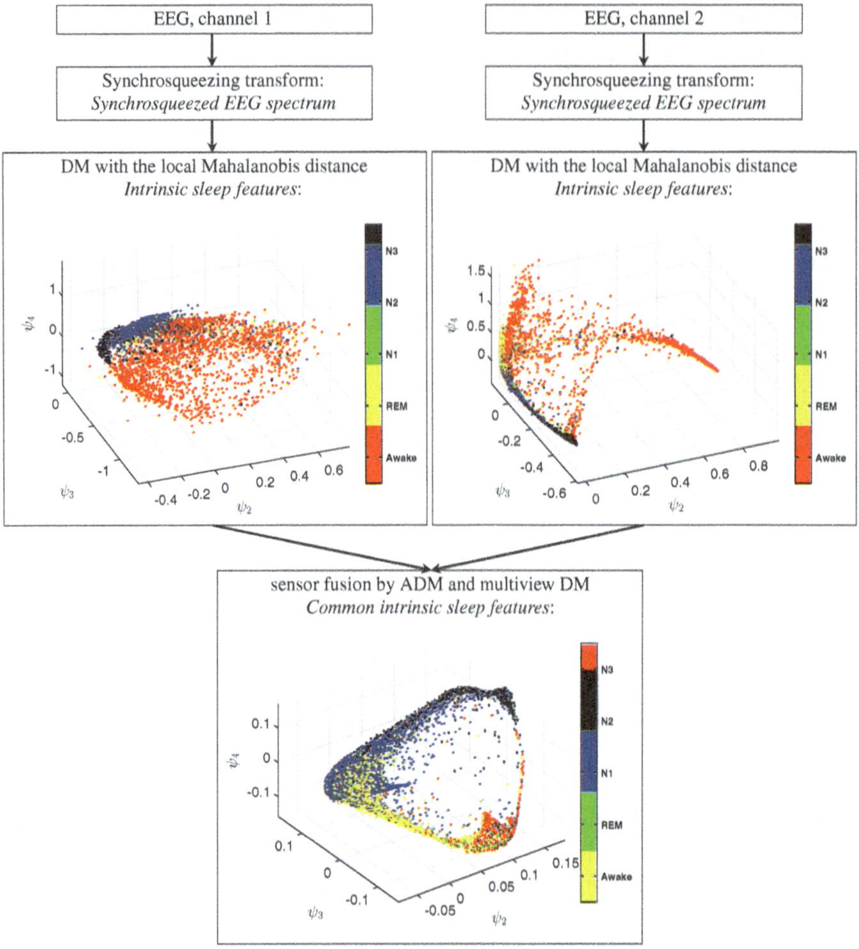

Fig. 4 The flow chart of the proposed feature extraction steps and a visualization of the extracted features by the diffusion map and sensor fusion are shown. The signal is from subject SC414 in Sleep-EDF SC* [21], where Channel 1 is Fpz-Cz and channel 2 is Pz-Oz. In the bottom figure, only the result of AD is shown. The ratios of the stages Awake, REM, N1, N2, and N3 are 14.8%, 21.9%, 5.7%, 47.5%, and 10.1%

Finally, we follow the common dimension reduction approach to convert the synchrosqueezed EEG spectrogram into the features that interest us. In this work, we follow the sleep stage scoring standard, the AASM criteria [2], and take an epoch for the sleep stage evaluation to be 30 seconds long. Therefore, we have $J := n\tau/30$ epochs. Consider nine frequency regions $R_1, \ldots, R_9 \subset \mathbb{R}$ in the spectral domain, defined as $R_1 = [0.5, 4]$ (R_1 is the delta band), $R_2 = [4, 7]$ (R_2 is the theta band), $R_3 = [7, 12]$ (R_3 is the alpha band), $R_4 = [12, 16]$ (R_4 catches the spindle), $R_5 = [16, 20]$, $R_6 = [20, 24]$, $R_7 = [24, 28]$, $R_8 = [28, 31]$ (R_5 to R_8 form the beta band), and $R_9 = [31, 49]$ (R_9 is the gamma band). These bands are chosen due to their well-known physiological meanings [9]. In practice we observed that suitably dividing the beta wave frequency range and the low-gamma wave frequency range ($25 \sim 49$ Hz) into finer bands leads to higher classification accuracy, so we consider the frequency bands $\{R_\ell\}_{\ell=1,\ldots,9}$ in this work.

For the j-th epoch, we get a ten-dimension vector $\mathbf{u}^{(j)} = [u_0^{(j)}\ u_1^{(j)}\ \cdots\ u_9^{(j)}]$ that include the total energy

$$u_0^{(j)} = \frac{\tau}{30} \sum_{\hat{k}:\hat{k}/(\tau K)\in[0.5,49]} \sum_{\hat{j}=30(j-1)/\tau+1}^{30j/\tau} \mathbf{S}_x(\hat{j}, \hat{k}) \tag{28}$$

and the band power ratios on $R_1, \ldots, R_9$:

$$u_\ell^{(j)} = \frac{\tau}{30} \frac{1}{u_0^{(j)}} \sum_{\hat{k}:\hat{k}/(\tau K)\in R_\ell} \sum_{\hat{j}=30(j-1)/\tau+1}^{30j/\tau} \mathbf{S}_x(\hat{j}, \hat{k}),\ \ell = 1, \ldots, 9. \tag{29}$$

We call $\mathbf{u}^{(j)}$, $j = 1, \ldots, J$, the *synchrosqueezed EEG spectral feature* of the j-th epoch.

6.2 Step 2: Convert Synchrosqueezed EEG Spectral Feature into Intrinsic Sleep Feature

In the optimal situation, the spectral content of the sleep dynamics can be well captured by the synchrosqueezed EEG spectral features. However, the synchrosqueezed EEG spectral features might be erroneous due to the inevitable noise, other sensor-specific artifacts, and the information distortion caused by the observation procedure. We then stabilize these features to better quantify the intrinsic sleep dynamics.

Take the synchrosqueezed EEG spectral features $\mathcal{U}^x := \{\mathbf{u}^{(j)}\}_{j=1}^J$, which is a point cloud in an Euclidean space. First, from the point cloud $\mathcal{U}^x$, we build a graph with $\mathcal{U}^x$ being vertices. The affinity between the features $\mathbf{u}^{(i)}$ and $\mathbf{u}^{(j)}$ is defined

$$W_x(i, j) = \exp\left\{-\frac{d_{\text{LMD}}^2(\mathbf{u}^{(i)}, \mathbf{u}^{(j)})}{\varepsilon}\right\},\ \text{ for } i, j = 1, \ldots, J,\ i \neq j, \tag{30}$$

where $\epsilon > 0$ is chosen by the user and $d_{\text{LMD}}(\cdot, \cdot)$ is the local MD. The local MD is chosen due to its scale-invariant property and stability property [14]. To calculate the local MD, denote the K-neighborhood of $\mathbf{u}^{(j)}$ by $\mathcal{N}_j$ for each $j \in \{1, 2, \ldots, J\}$, where $K = [\alpha J]$ and the ratio α is predetermined, and calculate the local covariance matrix Γ_j defined in (9). Then, evaluate the local MD by (10). With the $J \times J$ affinity matrix W_x, the degree matrix D_x of size $J \times J$ is constructed as that in (12), and hence the DM Φ_t^x, where $t > 0$. As a result, the synchrosqueezed EEG spectral features $\mathcal{U}^x$ are converted into a set of new features $\Phi_t^x(\mathbf{u}^{(j)}) \in \mathbb{R}^{\hat{d}}$. We call them the *intrinsic sleep features* for the EEG signal x and denoted

$$\mathcal{F}^x := \{\Phi_t^x(\mathbf{u}^{(j)})\}_{j=1}^J \subset \mathbb{R}^{\hat{d}}. \tag{31}$$

See Figure 4 for an example of DM when $\hat{d} = 3$. It is clear that epochs of different sleep stages are clustered and separated, and this shows the reason we call $\mathcal{F}^x$ the intrinsic sleep features.

6.3 Step 3: Fuse Two Intrinsic Sleep Features to Common Intrinsic Sleep Feature

When we have two simultaneously recorded EEG channels x and y, for each channel, we obtain its intrinsic sleep features, denoted as $\mathcal{F}^x \subset \mathbb{R}^{\hat{d}_x}$ and $\mathcal{F}^y \subset \mathbb{R}^{\hat{d}_y}$, respectively, where $\hat{d}_x$ might be different from $\hat{d}_y$. Via AD, $A = A_x A_y$, we obtain the common metric between channels x and y. Then, run DM (30) with the common metric between the i-th epoch and the j-th epoch. Denote the first $\hat{d}$ nontrivial eigenvectors of the associated transition matrix by $\psi_2, \ldots, \psi_{\hat{d}+1}$, where $\hat{d}$ is chosen to be 10 in this work. For the co-clustering, we choose

$$\mathsf{M} = \begin{bmatrix} 0_{J\times J} & W_x W_y \\ W_y W_x & 0_{J\times J} \end{bmatrix} \in \mathbb{R}^{2J\times 2J}, \tag{32}$$

where W_x and W_y are affinity matrices defined in (30) and denote a diagonal matrix $\mathsf{D} \in \mathbb{R}^{2J\times 2J}$ so that its i-th diagonal entry is the sum of the i-th row of M. Denote $q_i \in \mathbb{R}^J$ to be the i-th left eigenvector of the transition matrix $\mathsf{D}^{-1}\mathsf{M}$. Since for each i, $q_i(l)$ and $q_i(J+l)$ correspond to the l-th epoch for each $l \in \{1, \ldots, J\}$, we could consider the $2\tilde{d}$ vector, $[q_2(j) \ \cdots \ q_{\tilde{d}+1}(j) \ q_2(J+j) \ \cdots \ q_{\tilde{d}+1}(J+j)]$, to be another set of features associated with the sleep stage of the j-th epoch, where $j \in \{1, \ldots, J\}$. With AD and co-clustering, call the $\hat{d} + 2\tilde{d}$ dimensional vector

$$v_j := [\psi_2(j), \ldots, \psi_{\hat{d}+1}(j), q_2(j), \cdots, q_{\tilde{d}+1}(j), q_2(J+j), \cdots, q_{\tilde{d}+1}(J+j)]^\top \tag{33}$$

the *common intrinsic sleep feature* associated with the j-th epoch. Denote

$$\mathcal{F}^{x,y} := \{v_j\}_{j=1}^J \subset \mathbb{R}^{\hat{d}+2\tilde{d}}.$$

An illustration of the result of AD with $\hat{d} = 3$ is shown in Figure 4.

6.4 Learning Step: Sleep Stage Classification by the Hidden Markov Model

To predict sleep stage, we choose the standard and widely used algorithm HMM for the classification purpose. HMM is particularly powerful if we want to model a sequence of variables that changes randomly in time. Although it is standard, to make the paper self-contained, we provide a summary of the HMM and its numerical implementation in this section.

In general, a HMM can be viewed as a doubly embedded stochastic sequence with a sequence that is not observable (hidden-state sequence) and can only be observed through another stochastic sequence (observable sequence). An HMM can be fully specified by the hidden-state space, the hidden-state transition matrix, the observation space, the emission probability matrix, and the initial status. From the training dataset, the HMM could be established as a prediction model for the testing dataset. Below, we provide a summary of the HMM and its numerical details.

The *hidden-state space* $\mathscr{S}$ consists of five sleep stages: Awake, REM, N1, N2, and N3. To simply the notation, we label Awake, REM, N1, N2, and N3 by 1, 2, 3, 4, and 5, respectively; that is, $\mathscr{S} = \{1, \ldots, 5\}$. The sleep stage on the j-th epoch is viewed as a random variable S_j, whose realization is denoted by $s_j \in \mathscr{S}$. To collect high-quality and reliable EEG signals, the calibration is generally carried out. During the calibration period, the testing subject is awake. Hence, the recording starts from an Awake epoch, which implies that $S_0 = 1$. Assume that the time series $\{S_0, S_1, S_2, \ldots\}$ is homogeneous and the hidden-state transition matrix $\mathfrak{M} := (m_{ij})_{1\leq i,j\leq 5}$ satisfies

$$m_{ij} = P(S_{t+1} = j \mid S_t = i). \tag{34}$$

By the homogeneous assumption, the probability of hidden-state transition on the right-hand side of (34) can be estimated by the number of transitions from state i to state j normalized by the number of transitions from state i, i.e., $\hat{m}_{ij} = \frac{\#\{t: s_t=i, s_{t+1}=j\}}{\#\{t: s_t=i\}}$.

The common intrinsic sleep features $\mathscr{F}^{x,y} = \{v_j\}_{j=1}^{J}$ can be viewed as our observation for the hidden-state sequence $\{s_j\}_{j=1}^{J}$. We create a codebook to quantize $\{v_j\}_{j=1}^{J}$ and define the *observation state space* by the Linde-Buzo-Gray (LBG) algorithm [52]. Denote the codebook as $\mathscr{O}$ for the observation state space, which is represented by symbols $\{1, 2, \ldots, |\mathscr{O}|\}$, where $|\mathscr{O}|$ is the cardinality of $\mathscr{O}$. Based on the above vector quantization, we have an observable time series $\{O_1, O_2, \ldots, O_J\}$, where O_j is the random variable describing the observation at the j-th epoch, which takes a value from the codebook $\mathscr{O}$. The realization of O_j is denoted as o_j.

Consider the emission probability $\mathsf{E} := (e_j(k))_{j\in\mathscr{S}, k\in\mathscr{O}}$ to quantify the probability of observing state k from the j-th hidden state, and assume the following relationship

$$e_j(k) = P(O_t = k \mid S_t = j). \tag{35}$$

Based on the time-homogeneous assumption, E can be estimated by the accumulated number of times in hidden state j and simultaneously observing symbol k normalized by the number of times in hidden state j; that is, $\hat{e}_j(k) = \frac{\#\{t: s_t=j, o_t=k\}}{\#\{t: s_t=j\}}$. With the initial distribution π_0, $\{m_{ij}\}$ and $\{e_j(k)\}$, we have

$$\begin{aligned} &P(S_0 = s_0, S_1 = s_1, \ldots, S_J = s_J, O_1 = o_1, \ldots, O_J = o_J) \\ &= \pi_0(s_0) m_{s_0 s_1} e_{s_1}(o_1) \ldots m_{s_{J-1} s_J} e_{s_J}(o_J). \end{aligned} \tag{36}$$

Also we have the Markov property for this chain.

Given the trained HMM, that is, the estimated hidden-state transition matrix, the estimated emission probability matrix, and the initial status, we now detail an algorithm to estimate the sleep stages of the testing subject $\mathscr{T} := \{v_1, \ldots, v_J\}$, where J is the number of epochs. By the codebook, $\mathscr{T}$ is vector-quantized into $\{o_1, \ldots, o_J\}$. By Markov property, our goal is to find the most possible sequence of hidden states for the testing subject i.e., a path $\left(s_1^*, \ldots, s_J^*\right)$ that maximizes the probability

$$P(S_1 = s_1, \ldots, S_J = s_J, O_1 = o_1, \ldots, O_J = o_J). \tag{37}$$

Here we assume that $s_0 = 1$ (awake) and $P(O_{J+1} = o_{J+1}, S_{J+1} = F) = 1$ for some F in the hidden state space and some o_{J+1} in the observation state space.

Define

$$\nu_t(j) = \max_{s_1, \ldots, s_{t-1}} P(s_1, \ldots, s_{t-1}, o_1, \ldots, o_t, S_t = j) \tag{38}$$

for $t = 1, 2, \ldots, J + 1$, which represents the maximal probability that the HMM is in state j at time t and $o_1, \ldots, o_t$ are the observations up to time t. The maximum in (38) is taken over all probable state sequence $s_1, \ldots, s_{t-1}$. Note that when $t = J + 1$,

$$\begin{aligned} \nu_{J+1}(F) &= \max_{s_1, \ldots, s_J} P(s_1, \ldots, s_J, o_1, \ldots, o_{J+1}, S_{J+1} = F) \\ &= \max_{s_1, \ldots, s_J} P(s_1, \ldots, s_J, o_1, \ldots, o_{J+1}) = P(s_1^*, \ldots, s_J^*, o_1, \ldots, o_{J+1}), \end{aligned} \tag{39}$$

where the last equality holds for some $s_1^*, \ldots, s_J^* \in \mathscr{S}$. This fact motivates us the following algorithm to find $\nu_t(j)$. First of all, for $t = 1$, $\nu_1(j) = P(o_1, S_1 = j) = m_{s_0 j} e_j(o_1) = m_{1j} e_j(o_1)$, where $j \in \mathscr{S}$. For $t = 2$,

$$\nu_2(j) = \max_{s_1} P(s_1, o_1, o_2, S_2 = j)$$

$$= \max_{s_1} m_{1s_1} e_{s_1}(o_1) m_{s_1 j} e_j(o_2) = \max_{s_1} \nu_1(s_1) m_{s_1 j} e_j(o_2). \quad (40)$$

Denote the maximizer of the right-hand side of (40) by $V(j, 2)$, that is,

$$V(j, 2) = \arg\max_{s_1} P(S_1 = s_1, o_1, o_2, S_2 = j) \in \mathscr{S}.$$

Note that since $V(j, 2) = \arg\max_{s_1} P(S_1 = s_1 \mid o_1, o_2, S_2 = j)$, $V(j, 2)$ is the most likely state at time index 1 when the state is in j at time index 2 and when o_1 and o_2 are observed. In general, we have

$$\nu_{t+1}(j) = \max_{s_t} \nu_t(s_t) m_{s_t j} e_j(o_{t+1}) \quad (41)$$

for $t = 2, \ldots, J$ and $j \in \mathscr{S}$. Denote the maximizer of the right-hand side of (41) by $V(j, t+1)$, which can be interpreted as the best *relay point* connecting the node $S_{t+1} = j$ with the most likely path that emits the symbols $o_1, \ldots, o_{t+1}$. With $\{V(\cdot, t)\}_{t=1,\ldots,J+1}$, we can find the optimal path from the state F at time index $J+1$ iteratively as follows:

$$s_J^* = V(F, J+1), \quad s_{J-1}^* = V(s_J^*, J) \ldots s_2^* = V(s_3^*, 3), \quad s_1^* = V(s_2^*, 2).$$

For details, see [20].

7 Material and Statistics

To evaluate the proposed algorithm, we consider a publicly available database and follow standard performance evaluation procedures.

7.1 Material

To evaluate the proposed algorithm, we consider the commonly considered benchmark database, Sleep-EDF Database [Expanded], from the public repository Physionet [21]. It contains two subsets (marked as SC* and ST*). The first subset SC* comes from healthy subjects without any sleep-related medication. The subset SC* contains Fpz-Cz/Pz-Oz EEG signals recorded from ten males and ten females without any sleep-related medication, and the age range is 25–34 year-olds. There are two approximately 20-hour recordings per subject, apart from a single subject for whom there is only a single recording. The EEG signals were recorded during two subsequent day-night periods at the subjects' home. The sampling rate is 100 Hz. The second subset ST* was obtained in a 1994 study of temazepam effects on the sleep of subjects with mild difficulty falling asleep. The subset ST* contains Fpz-Cz/Pz-Oz EEG signals recorded from 7 males and 15 females, who had mild difficulty falling asleep. Since this dataset is originally used for studying the effects of temazepam, the EEG signals were recorded in the hospital for two nights, one of which was after temazepam intake. Only their placebo nights can be downloaded from [21]. The sampling rate is 100 Hz. For both SC* and ST* sets, each 30s epoch of EEG data has been annotated into the classes Awake, REM, N1, N2, N3, and N4. The epochs corresponding to movement and unknown stages were excluded, and the epochs labeled by N4 are relabeled to N3 according to the AASM standard [2]. For more details of the database, we refer the reader to https://www.physionet.org/physiobank/database/sleep-edfx/.

7.2 Statistics

To evaluate the performance of the automatic sleep stage annotation, we shall distinguish two common cross-validation (CV) schemes. According to whether the training data and the testing data come from different subjects, the literature is divided into two groups, *leave-one-subject-out* and *non-leave-one-subject-out* CV. When the validation set and training set are determined *on the subject level*, that is, the training set and the validation set contain different subjects, we call it the leave-one-subject-out CV (LOSOCV) scheme; otherwise we call it the non-LOSOCV scheme. The main challenge of the LOSOCV scheme comes from the inter-individual variability, but this scheme is close to the real-world scenario – how to predict the sleep dynamics of a new arrival subject from a given annotated database. On the other hand, in the non-LOSOCV scheme, the training set and the testing set are dependent, and the performance might be overestimated. To better evaluate the performance of the proposed automatic sleep scoring algorithm, we choose the LOSOCV scheme. For each database, one subject is randomly chosen as the testing set and the other subjects form the training set. For the testing subject, we take the phenotype information to find the $\hat{K}$ most similar subjects to establish the HMM model. The impact of age on the sleep dynamics [53] and EEG signal [54, 55] is well-known, so the EEG information from subjects with similar age will provide more information. While the sleep dynamics is influenced by other phenotype information, since age is the common information among databases we consider, we determine the $\hat{K}$ most similar subjects by the age. Note that this approach imitates the real scenario – for a new-arriving subject, we can score its sleep stages by taking the existing database with annotation into account. Also note that this LOSOCV scheme helps prevent overfitting and fully takes the inter-individual variability into account in constructing the prediction model.

All performance measurements used in this paper are computed through the unnormalized confusion matrix $M \in \mathbb{R}^{5\times 5}$. For $1 \leq p, q \leq 5$, the entry M_{pq} represents the number of expert-assigned p-class epochs, which were predicted to the q-class. The precision (PR_p), recall (RE_p), and F1-score (F1_p) of the p-th class, where $p = 1, \ldots, 5$, are computed respectively through

$$\mathrm{PR}_p = \frac{M_{pp}}{\sum_{q=1}^{5} M_{qp}}, \quad \mathrm{RE}_p = \frac{M_{pp}}{\sum_{q=1}^{5} M_{pq}}, \quad \mathrm{F1}_p = \frac{2\mathrm{PR}_p \cdot \mathrm{RE}_p}{\mathrm{PR}_p + \mathrm{RE}_p}. \tag{42}$$

The overall accuracy (ACC), macro F1 score (Macro − F1), and kappa (κ) coefficient are computed respectively through

$$\mathrm{ACC} = \frac{\sum_{p=1}^{5} M_{pp}}{\sum_{p,q=1}^{5} M_{pq}}, \quad \mathrm{Macro} - \mathrm{F1} = \frac{1}{5}\sum_{p=1}^{5} \mathrm{F1}_p, \quad \kappa = \frac{\mathrm{ACC} - \mathrm{EA}}{1 - \mathrm{EA}}, \tag{43}$$

where EA means the expected accuracy, which is defined by

$$\mathrm{EA} = \frac{\sum_{p=1}^{5} \left(\sum_{q=1}^{5} M_{pq}\right) \times \left(\sum_{q=1}^{5} M_{qp}\right)}{\left(\sum_{p,q=1}^{5} M_{pq}\right)^2}. \tag{44}$$

To evaluate if two matched samples have the same mean, we apply the one-tail Wilcoxon signed-rank test under the null hypothesis that the difference between the pairs follows a symmetric distribution around zero. When we compare the variance, we apply the one-tail F-test under the null hypothesis that there is no difference between the variances. We consider the significance level of 0.05. To handle the multiple comparison issue, we consider the Bonferroni correction.

8 Results

We report the results of applying the proposed algorithm to the abovementioned two databases. The parameters in the numerical implementation are listed here. $\tau = 1/100$. For SST, we choose $H = 1001$ and $K = 4004$ in (26) and (27); that is, the bandwidth is $1001/100 = 10.01$ seconds, and we oversample the frequency domain by a factor of 4. For the local MD, we take $\alpha = 0.1$ in (30) and $d = 7$ in (10). For DM, the ϵ in (30) is chosen to be the 5% percentile of pairwise distances, the diffusion time is $t = 1$, and we choose $\hat{d} = 10$. For the co-clustering, $\tilde{d}$ is chosen to be 10. No systematic parameter optimization is performed to avoid overfitting. For the reproducibility purpose, the MATLAB code will be provided via request.

8.1 Sleep Dynamics Visualization

We start from showing the visualization of the intrinsic sleep features and the common intrinsic sleep features from 12 different subjects in the SC* database. See Figure 5 for a visualization by DM. See Figure 6 for a visualization by AD and co-clustering. Clearly, we see that Awake, REM, N2, and N3 stages are well clustered in all plots, while N1 is less clustered and tends to mixed up with other stages. Moreover, in AD and co-clustering, we further see a "circle" with a hole in the middle, and the sleep stages are organized on the circle and follow the usual sleep dynamics pattern. While the geometric organization of sleep dynamics can be easily visualized in Figure 6, it is not easy to visualize the temporal dynamics information. For this purpose, we show the final intrinsic features expanded in the time line in Figure 7. Another way to visualize the dynamics is via a video that encodes the temporal relationship among different sleep stages. See the video available in https://www.dropbox.com/s/21e8aw7scvo5kkb/dynamics_SC31.mp4?dl=0.

Next, see Figure 8 for a visualization of AD and co-clustering of the ST* database. While Awake, REM, N2, and N3 stages are still well clustered in all plots, compared with the normal subjects in SC* database, the separation and the "circle" are less clear.

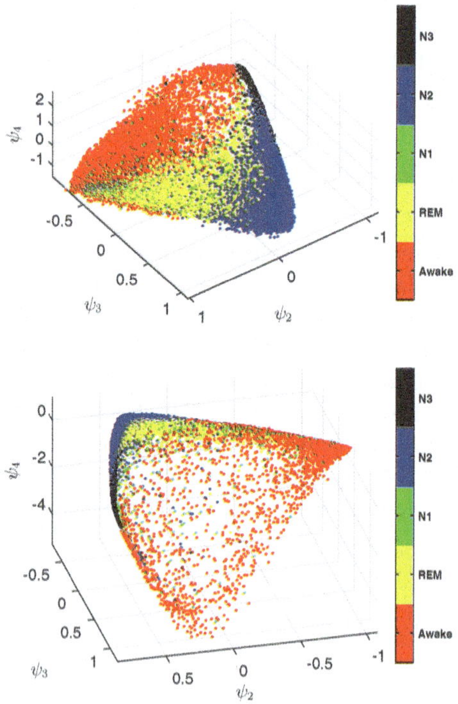

Fig. 5 A visualization of the intrinsic sleep features (from single channel) extracted from 12 different subjects from the Sleep-EDF database (SC*). The ratios of the stages Awake, REM, N1, N2, and N3 are 17.3%, 18.5%, 4.3%, 46.0%, and 13.9%, respectively. Each point corresponds to a 30-second epoch

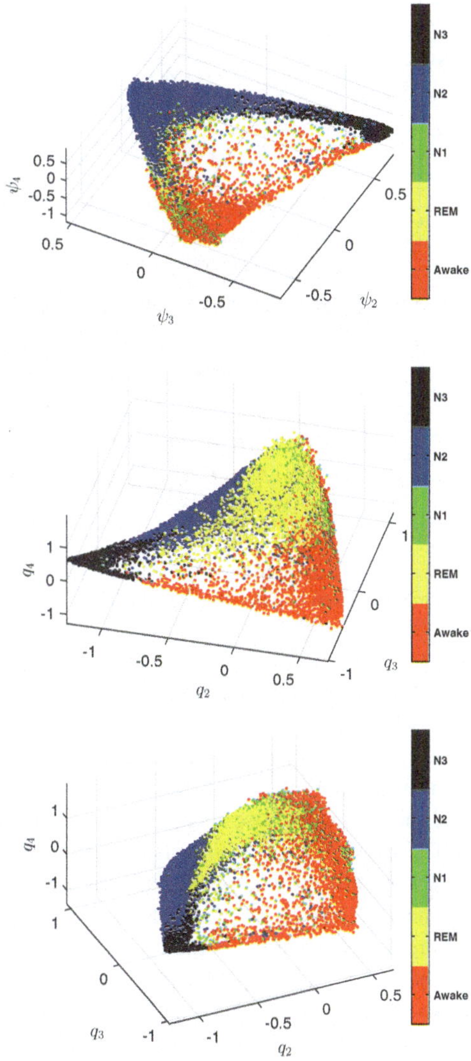

Fig. 6 A visualization of the common intrinsic sleep features (from two channels) extracted from 12 different subjects from the Sleep-EDF database (SC*). From top to bottom are ADM of Fpz-Cz & Pz-Oz, multiview DM of Fpz-Cz & Pz-Oz, and multiview DM of Fpz-Cz & Pz-Oz. In the middle subplot, we plot $\{[q_2(i), q_3(i), q_4(i)]\}_{i=1}^{J}$, and in the bottom subplot, we show $\{[q_2(i+J), q_3(i+J), q_4(i+J)]\}_{i=1}^{J}$. The ratios of the stages Awake, REM, N1, N2, and N3 are 17.3%, 18.5%, 4.3%, 46.0%, and 13.9%, respectively. Each point corresponds to a 30-second epoch

8.2 *Sleep Stage Prediction*

Since there are long periods of wakefulness at the start and the end of recordings, when a subject is not sleeping, [56] only includes 30 minutes of such periods just before and after the sleep periods. To have a fair comparison, we also follow this truncation rule. In the end, the labeled epochs are imbalanced, with 42.4% epochs labeled N2 and only 6.6% epochs labeled N1. Authors of [56], as well as [57, 58], handle the imbalanced data issue by setting the number of epochs per-stage per

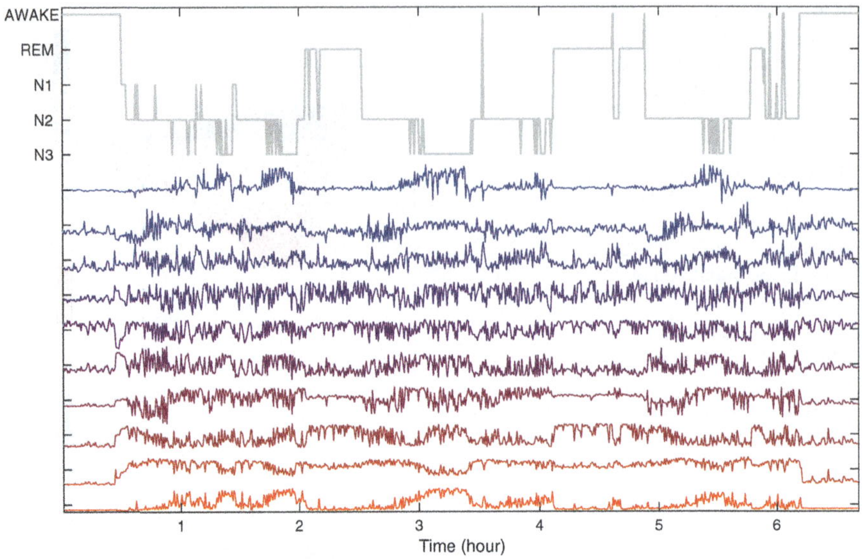

Fig. 7 A visualization of the final ten intrinsic features by ADM. The first is colored in red, and the tenth is colored in blue. The expert's labels are plotted on the top

recording belonging the training data (consisting of $\hat{K}$ subjects) equal to the number of epochs of the least represented stage, N1. To have a fair comparison and test the stability of the proposed algorithm, we apply the same scheme in the following way. We run the LOSOCV with $\hat{K} = 9$. For each subject in the testing set, we take the nine subjects with closest age and handle the imbalanced data like that of [56].

The averaged confusion matrix of the proposed algorithm over 20 subjects is shown in Table 1. The overall accuracy over 20 subjects is 82.49% ± 5.05%, and the macro F1 is 75.7% ± 5.2%, with Cohen's kappa 0.758 ± 0.07. Note that the N1 prediction accuracy is the lowest one, with 35% accuracy compared with other stages, and most N1 epochs are classified as REM or N2. This misclassification is related to the scattered N1 epochs in Figure 6 that can be visually observed, and it is the main reason to drag down the overall accuracy and macro F1. We also note that N3 is commonly classified wrongly as N2, Awake is commonly classified wrongly as N1, and REM is commonly classified wrongly as N2. To further examine the performance, the resulting hypnogram of one subject is shown in Figure 9. Note that the discrepancy between the experts' annotations and the prediction frequently happens when there is a "stage transition." Note that the sleep dynamics transition from one stage to another one often happens in the middle of one epoch. Thus, those epochs with sleep dynamics transition contain information that is not purely for one stage and hence harder to classify.

An ideal approach to handle the imbalanced data is collecting more data to enhance the prediction accuracy. In the SC* dataset, there were long periods of awake epochs before the start and after the end of sleep that we can use. To

Fig. 8 A visualization of the common intrinsic sleep features (from two channels) extracted from 12 different subjects from the Sleep-EDF database (ST*). From top to bottom are ADM of Fpz-Cz & Pz-Oz, first set of multiview DM of Fpz-Cz & Pz-Oz, and the second set of multiview DM of Fpz-Cz & Pz-Oz. In subplot 8.1, we plot $\{[q_2(i), q_3(i), q_4(i)]\}_{i=1}^{J}$, and in subplot 8.1, we show $\{[q_2(i+J), q_3(i+J), q_4(i+J)]\}_{i=1}^{J}$. The ratios of the stages Awake, REM, N1, N2, and N3 are 10.3%, 20.0%, 10.0%, 45.3%, and 14.3%, respectively. Each point corresponds to a 30-second epoch

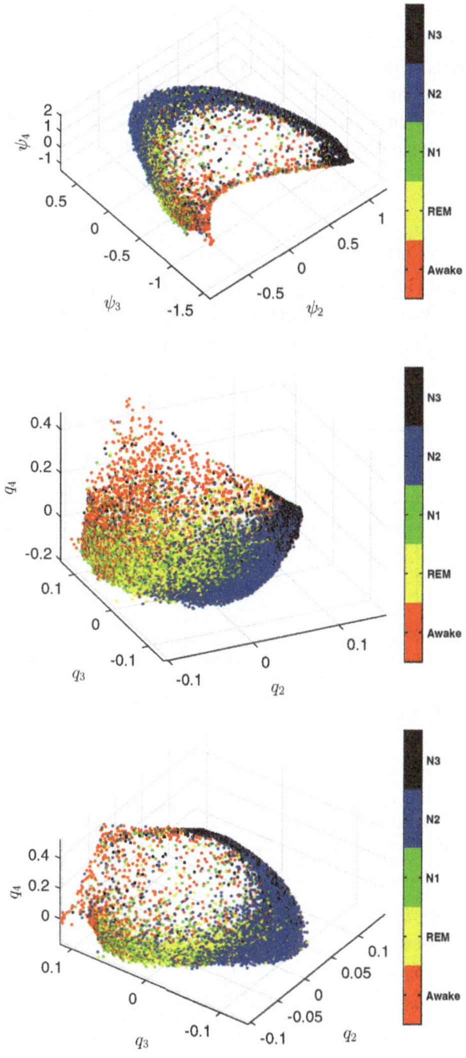

further evaluate the algorithm, we consider longer periods of wakefulness just before and after the sleep periods. Apart from the three recordings (sc4092e0, sc4191e0, sc4192e0), we included 90 minutes of awake periods before and after the sleep periods. For the sc4092e0, sc4191e0, and sc4192e0 recordings, we only included 60 minutes of awake periods just before and after the sleep periods due to the appearance of artifacts (labeled as MOVEMENT and UNKNOWN), which were at the start or the end of each recording. With more awake epochs, the corresponding comparison matrix with the same is shown in Table 2. All performance indices, including the overall accuracy, the macro F1 score, and the Cohen's kappa, are

Table 1 Comparison matrix obtained from 20-fold leave-one-subject-out cross-validation on Fpz-Cz and Pz-Oz channels from the Sleep-EDF SC* database. The common intrinsic sleep feature is used. The overall accuracy equals 82.57%, the macro F1 score equals 76.0%, and Cohen's kappa equals 0.763. If the classification accuracy, macro F1 score, and Cohen's kappa are computed for each night recording, the standard deviation of classification accuracy (resp. the macro F1 score and Cohen's kappa) for the 39-night recordings is 4.96% (resp. 5.15% and 0.068). We follow the class-balanced random sampling scheme used in [56–58]

	Predicted					**Per-class metrics**		
	Awake	REM	N1	N2	N3	PR	RE	F1
Awake (18%)	6943 (**88%**)	184 (**2%**)	625 (**8%**)	156 (**2%**)	19 (**0%**)	91	88	89
REM (18%)	112 (**1%**)	7063 (**92%**)	123 (**2%**)	419 (**5%**)	0 (**0%**)	73	92	81
N1 (7%)	378 (**13%**)	907 (**32%**)	967 (**35%**)	534 (**19%**)	18 (**1%**)	45	34	39
N2 (42%)	128 (**1%**)	1451 (**8%**)	412 (**2%**)	14557 (**82%**)	1251 (**7%**)	90	82	86
N3 (14%)	29 (**0%**)	16 (**0%**)	3 (**0%**)	545 (**10%**)	5110 (**90%**)	80	90	84

	Predicted					**Per-class Metrics**		
	Awake	REM	N1	N2	N3	PR	RE	F1
Awake (18%)	6943 (**88%**)	184 (**2%**)	625 (**8%**)	156 (**2%**)	19 (**0%**)	91	88	89
REM (18%)	112 (**1%**)	7063 (**92 %**)	123 (**2%**)	419 (**5%**)	0 (**0%**)	73	92	81
N1 (7%)	378 (**13%**)	907 (**32%**)	967 (**35%**)	534 (**19%**)	18 (**1 %**)	45	34	39
N2 (42%)	128 (**1%**)	1451 (**8%**)	412 (**2%**)	14557 (**82%**)	1251 (**7 %**)	90	82	86
N3 (14%)	29 (**0%**)	16 (**0%**)	3 (**0%**)	545 (**10 %**)	5110 (**90%**)	80	90	84

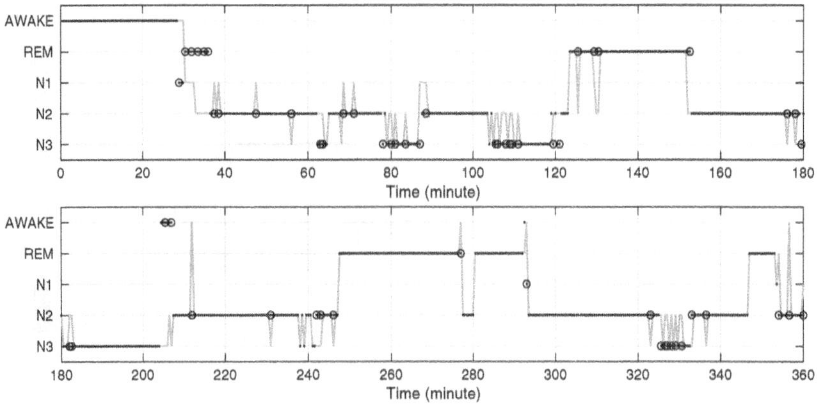

Fig. 9 The resulting hypnogram of one subject from SC*. The gray curve is the expert's label, and the black dots are the predicted sleep stages. The discrepancy is emphasized by the black circles

consistently higher than those from only including 30 minutes of awake periods just before and after the sleep periods reported in Table 1. The overall accuracy over 20 subjects is 84.21% ± 4.85%, and the macro F1 is 76.5% ± 5.18%, with Cohen's kappa 0.79 ± 0.06. Particularly, the accuracy of N1 prediction is increased to 42%.

Table 2 Confusion matrix obtained from 20-fold leave-one-subject-out cross-validation on Fpz-Cz and Pz-Oz channels from the Sleep-EDF SC* database with longer awake periods. The common intrinsic sleep feature is used. The overall accuracy equals 84.21%, the macro F1 score equals 76.5%, and Cohen's kappa equals 0.788. If the classification accuracy, macro F1 score, and Cohen's kappa are computed for each night recording, the standard deviation of classification accuracy (resp. the macro F1 score and Cohen's kappa) for the 39-night recordings is 4.85% (resp. 5.18% and 0.06). We apply the class-balanced random sampling scheme proposed in [56–58]

	Predicted					**Per-class metrics**		
	Awake	REM	N1	N2	N3	PR	RE	F1
Awake (34%)	15159 (**88%**)	339 (**2%**)	1572 (**9%**)	170 (**1%**)	45 (**0%**)	98	88	92
REM (15%)	24 (**1%**)	7162 (**93%**)	133 (**2%**)	395 (**5%**)	3 (**0%**)	75	93	83
N1 (5%)	232 (**8%**)	829 (**30%**)	1180 (**42%**)	544 (**19%**)	19 (**1%**)	33	42	37
N2 (35%)	89 (**0%**)	1196 (**7%**)	636 (**4%**)	14553 (**82%**)	1325 (**7%**)	90	82	86
N3 (11%)	19 (**0%**)	3 (**0%**)	21 (**0%**)	499 (**9%**)	5161 (**91%**)	79	91	84

Table 3 Comparison matrix obtained from 22-fold leave-one-subject-out cross-validation on Fpz-Cz and Pz-Oz channels from the Sleep-EDF ST* database. The common intrinsic sleep feature is used. The overall accuracy equals 77.01%, the macro F1 score equals 71.53%, and Cohen's kappa equals 0.6813. The standard deviation of classification accuracy (resp., the macro F1 score and Cohen's kappa) for the 22-night recordings is 6.63% (resp., 7.78% and 9.27%). We apply the class-balanced random sampling scheme proposed in [56–58]

	Predicted					**Per-class metrics**		
	Awake	REM	N1	N2	N3	PR	RE	F1
Awake (11%)	2008 (**88%**)	40 (**2%**)	178 (**8%**)	42 (**2%**)	16 (**0%**)	72	88	79
REM (20%)	59 (**1%**)	3489 (**85%**)	238 (**6%**)	334 (**8%**)	11 (**0%**)	79	84	81
N1 (10%)	548 (**27%**)	388 (**19%**)	697 (**34%**)	409 (**20%**)	2 (**0%**)	48	34	40
N2 (45%)	155 (**2%**)	514 (**5%**)	348 (**4%**)	7599 (**80%**)	877 (**9%**)	84	80	82
N3 (15%)	30 (**1%**)	2 (**0%**)	5 (**0%**)	654 (**21%**)	2454 (**79%**)	73	78	75

For the ST* database, we also run the LOSOCV with $\hat{K} = 9$. The averaged confusion matrix of the proposed algorithm over 22 subjects is shown in Table 3. The overall accuracy is 77.8% $\pm$ 5.77%, and the macro F1 is 71.5% $\pm$ 7.55%, with Cohen's kappa 0.69 $\pm$ 0.08. In this database, there are 10% epochs labeled as N1. Although it is slightly higher than that of SC* database, the prediction performance of N1 is 34%, which is still relatively low. Also, note that the prediction performance of N3 is lower and a significant portion of N3 is misclassified as N2.

8.3 *More Comparisons for Sleep Stage Prediction*

To appreciate the significance of the diffusion geometry-based sensor fusion framework, we report the results without two critical setups in the proposed algorithm – the sensor fusion and the local MD. First, we consider the case if we simply concatenate intrinsic sleep features of two channels, instead of taking the common intrinsic sleep features; that is, we concatenate $\Phi_t^x(\mathbf{u}^{(j)})$ and $\Phi_t^y(\mathbf{u}^{(j)})$ in (31) directly to replace (33) when we train the HMM model. Second, we consider the case if we do not use local MD to compare synchrosqueezed EEG spectral features but the ordinary L^2 distance; that is, $d^2(\mathbf{u}^{(i)}, \mathbf{u}^{(j)})$ in (30) is defined as $\|\mathbf{u}^{(i)} - \mathbf{u}^{(j)}\|_{\mathbb{R}^{10}}$ instead of $d^2_{\mathrm{LMD}}(\mathbf{u}^{(i)}, \mathbf{u}^{(j)})$. Third, we consider the single EEG channel; that is, we run HMM on the intrinsic sleep features extracted from Fpz-Cz or Pz-Oz channel.

The results of the above three combinations (confusion matrices not shown) for the SC* database are shown in Figure 10. Note that the mean and standard deviation of ACC, MF1, and Cohen's kappa are evaluated from all subjects, which are different from that shown in Table 1. It is clear that the averaged ACC, MF1, and Cohen's kappa are consistently downgraded in these three cases. In Figure 10, we see that compared with single channel or sensor fusion without local MD, the proposed sensor fusion of two channels consistently improves the result with

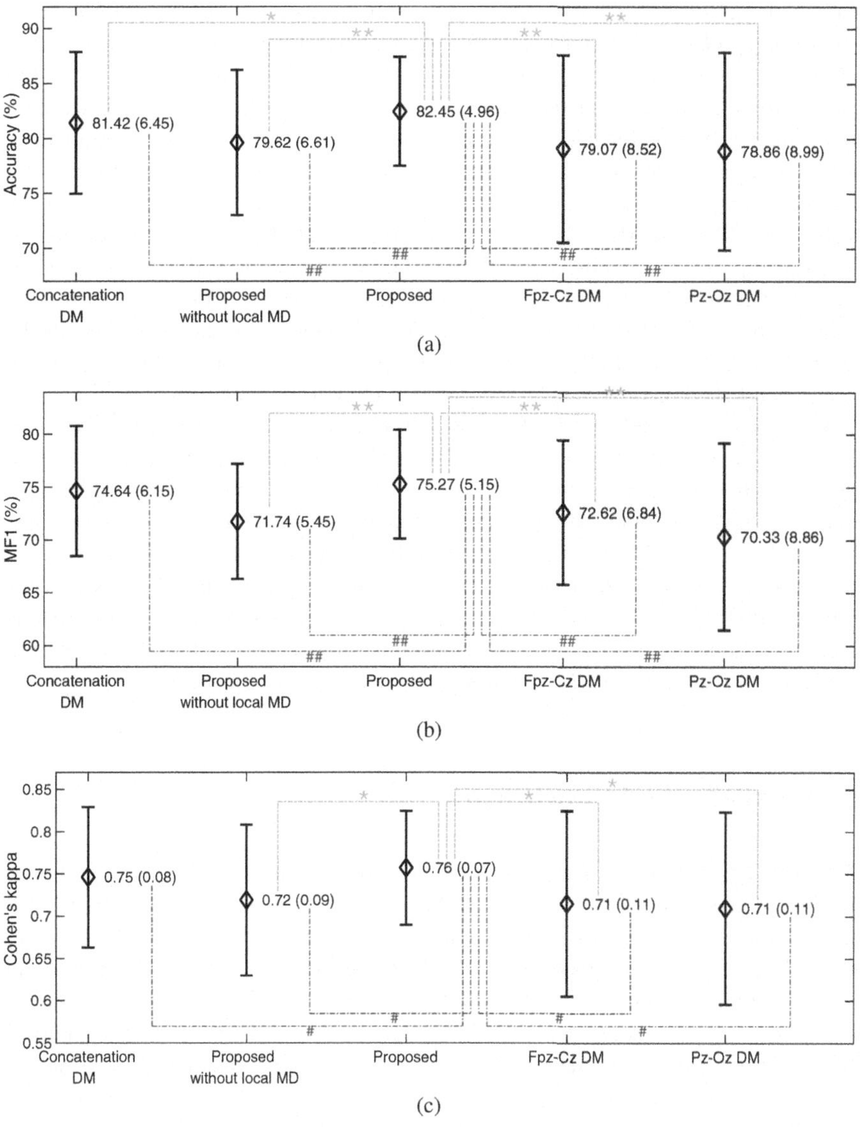

Fig. 10 Comparison between different information fusion methods in terms of the accuracy (ACC), macro F1-score (MF1), and Cohen's kappa for the SC* database. To evaluate if the mean is improved, the one-tail Wilcoxon signed-rank test is applied under the null hypothesis that the difference between the pairs follows a symmetric distribution around zero. To evaluate if the variance is smaller, we apply the one-tail F-test under the null hypothesis that there is no difference between the variances. ⋆ (respectively ⋆⋆) means statistical significance without (respectively with) the Bonferroni connection when the mean is compared; # (respectively ##) means statistical significance without (respectively with) the Bonferroni connection when the variance is compared. (**a**) Comparison of accuracy. (**b**) Comparison of macro-F1. (**c**) Comparison of Cohen's kappa

statistical significance, for both mean and variance of ACC and MF1. Although we can see a significant difference between the mean and variance of Cohen's kappas before the Bonferroni correction in these comparisons, the significance is not strong enough to stand the Bonferroni correction. Compared with a direct concatenation of intrinsic sleep features of two channels, we see that there is a significance between the averaged ACCs before the Bonferroni correction, but it disappears after the Bonferroni correction. However, we see a significant difference between the variance of ACC and MF1. This fact reflects the essential property of the diffusion-based unsupervised learning algorithms. Via the alternating diffusion process for the sensor fusion, the intrinsic sleep features are "stabilized" and hence the smaller variance. This comparison provides an empirical evidence of the usefulness of the sensor fusion and local Mahalanobis distance in the proposed algorithm, in addition to the established theoretical backup shown in the Online Supplementary.

We have considered LOSOCV to prevent overfitting. Here we further consider the fivefold leave-subject-out cross-validation (CV) to further evaluate the proposed algorithm; that is, we randomly divide all subjects into five non-overlapping groups, and for each group as the testing group, we train the model from the left four groups. The model is trained in the following way. For each subject in the testing set, we take epochs of the nine ($\hat{K} = 9$) subjects with closest age from the other four groups collected during the first night into account and balance classes by taking the epochs from the second night. The result is reported in Table 4. We see that the result with fivefold leave-subject-out CV is similar to the leave-one-subject-out CV.

9 Discussion and Conclusion

The unsupervised diffusion geometry-based sensor fusion framework is proposed to capture the geometric structure of the sleep dynamics. We take the spectral information of EEG signals as an example and test the framework on the publicly available benchmark database. With the learning algorithm HMM, we obtain an accurate prediction model, and the result is compatible with several state-of-the-art algorithms based on neural network (NN). In addition to the theoretical backup of the diffusion geometry framework provided in the online supplementary materials, a systematical examination of step in the diffusion geometry framework is provided. All these summarize the usefulness of the diffusion geometry framework for the sleep dynamics study. We mention that the proposed framework is flexible to study other physiological dynamics but not only for studying the sleep dynamics. For example, its variation has been applied to study f-wave from subjects with atrial fibrillation [59], intra-cranial EEG signal [60], etc.

Table 4 Comparison matrix obtained from fivefold cross-validation on Fpz-Cz and Pz-Oz channels from the Sleep-EDF SC* database. The common intrinsic sleep feature is used. The 20 subjects in the SC* database are divided into 5 groups. Each group contains four subjects. After one of the five groups is selected for testing, the remaining four groups are used for training purposes. The overall accuracy equals 82.25%, the macro F1 score equals 75.89%, and Cohen's kappa equals 0.7591. If the classification accuracy, macro F1 score, and Cohen's kappa are computed for each night recording, the standard deviation of classification accuracy (resp. the macro F1 score and Cohen's kappa) for the 39-night recordings is 4.99% (resp. 5.05% and 6.68%). The training set consist of 2-night recordings of the remaining 19 subjects with class-balanced random sampling, which is the same as the class balancing method used in [56–58]

	Predicted					**Per-class metrics**		
	Awake	REM	N1	N2	N3	PR	RE	F1
Awake (18%)	6857 (**86%**)	180 (**2%**)	732 (**9%**)	148 (**2%**)	10 (**0%**)	92	87	89
REM (18%)	124 (**2%**)	6965 (**90%**)	184 (**2%**)	443 (**6%**)	1 (**0%**)	75	90	82
N1 (7%)	327 (**12%**)	873 (**31%**)	1032 (**37%**)	550 (**20%**)	22 (**0%**)	42	37	39
N2 (42%)	131 (**1%**)	1301 (**7%**)	528 (**3%**)	14517 (**82%**)	1322 (**7%**)	90	82	85
N3 (14%)	33 (**1%**)	13 (**0%**)	11 (**0%**)	512 (**9%**)	5134 (**90%**)	79	90	84

9.1 Physiological Explanation of the Results

Although our overall prediction accuracy is compatible with the state-of-the-art prediction algorithm in the Sleep-EDF SC* database, like [56], we see that the prediction accuracy of N1 is relatively low by our algorithm (F1 is 39% by our method and 46.6% in [56]), and this low N1 accuracy downgrades the overall accuracy. This low prediction rate of N1 is also seen in the Sleep-EDF ST* database. This low prediction rate partially comes from the relatively small size of N1 epochs and partially comes from the algorithm and available channels.

Based on the AASM criteria [2], to distinguish N1 and REM, we need electrooculogram and electromyogram signals, which are not available in the dataset. The EEG backgrounds of N1 and N2 are the same, and experts distinguish N1 and N2 by the K-complex or spindle, as well as the *3-minute rule*. While the synchrosqueezed EEG spectral features capture the K-complex or spindle behavior, the 3-minute rule is not considered in the algorithm. In the proposed algorithm, in order to handle the inter-individual variability, the temporal information among epochs is not fully utilized when we design the intrinsic sleep feature but only used in the HMM. How to incorporate the temporal information into the diffusion geometry framework will be explored in the future. Furthermore, there are other information in addition to the spectral information discussed in this paper. We do not extensively explore all possible information but focus on the diffusion geometry and sensor fusion framework. For example, while the vertex sharp is a common "landmark" indicating transition from N1 to N2, we do not take it into account since this feature is not always present and a rule-based approach is needed to include this temporally transient feature. Another interesting reasoning that it is possible to improve the N1 accuracy is the deep neural network (DNN) result. This suggests that by taking experts' labels into account, some distinguishable EEG structure of N1 that is not sensible by spectral information can be efficiently extracted by the DNN framework proposed in [56]. In conclusion, since the proposed features depend solely on the spectral information, we may need features of different categories to capture this unseen N1 structure. On the other hand, it is well-known that different EEG leads provide different information for N1. For example, the occipital lead has a stronger alpha wave strength, compared with the central lead, when transiting from wakefulness to N1. When there are multiple EEG leads, this lead information could be taken into account to further improve the accuracy.

Note that beside N1, the prediction performance of N3 is also lower in the Sleep-EDF ST* database, where the subjects take temazepam before data recording. It has been well-known that in general, benzodiazepine hypnotics [61] reduces the low-frequency activity and enhances spindle frequency. Since our features are mainly based on the spectral information, a N3 epoch might look more like N2 epochs and hence the confusion and the lower performance. This outcome emphasizes the importance of the drug history when designing the algorithm.

9.2 *Visualization for Sleep Dynamics Exploration*

In Figures 6 and 8, we show the underlying geometric structure of the sleep dynamics captured by the proposed algorithm – the Awake, REM, N2, and N3 are well clustered, with N1 scattered around Awake, REM, and N2. Furthermore, in Figures 6 and 8, we can even visualize a "circle." An interesting physiological finding from these plots is the close relationship between N3 and Awake. Note that the same result is also shown in Figure 4. This geometric relationship indicates the similarity between the common intrinsic sleep features of N3 and Awake stages. This similarity comes from the well-known fact that before arousal, particularly across the descending part of sleep cycles and in the first cycle, we can observe the "delta-like burst" that mimics the delta wave specific for N3 stages [62, 63]. Note that epochs from 1 subject are used to generate Figure 4 and 12 different subjects are pooled together to generate Figures 6 and 8 and we see the same geometric structure. This finding exemplifies our observation that this distribution is consistent across subjects. On the other hand, due to the sleep apnea disturbance, this "circle" is in general less obvious. This indicates the interruption of sleep dynamics by sleep apnea and hence the frequent random transition from one sleep stage to another. A possible direction is applying the topological data analysis tools to quantify the existence of circle and hence a quantification of sleep apnea disturbance.

9.3 *Sleep Stage Classification and Comparison with Related Work*

There have been many proposed algorithms for the sake of automatic sleep stage scoring. Since we focus on the LOSOCV scheme and predict five different sleep stages, here we only mention papers considering the LOSOCV scheme and predicting five different sleep stages from single- or two- EEG channels.

In [57], the performance of the stacked sparse autoencoder was evaluated in Sleep-EDF SC*, and the overall accuracy was 78.9%. Instead of extracting features based on the domain knowledge, features in [58] are automatically learned by the convolutional neural networks (CNNs). The overall accuracies was 74.8% for the Sleep-EDF SC* database. In [56], the authors proposed a deep learning model, called DeepSleepNet, which reaches the state-of-the-art 82.0% of overall accuracy on the Sleep-EDF SC* database. In [64], a similar approach based on the deep CNN with modifications is considered and achieves a compatible result. All the above studies focus on the single-channel EEG signal.

Compared with the state-of-the-art DNN approach [56], which is supervised in nature, our approach is unsupervised in nature. Recall that the main difference between the supervised learning and unsupervised learning is that the label information is taken into account in the supervised learning approach. The success of

DNN in many fields is well-known [65], and it is not surprising that it has a great potential to help medical data analysis.

While DNN is in general a useful tool for the engineering purpose, it is often criticized of working as a black box. For medical problems and datasets, when interpretation is needed, a mathematically solid and interpretable tool would be useful. The algorithm we proposed, on the other hand, has a clear interpretation with solid mathematical supports. Moreover, a peculiar property of medical databases, the "uncertainty," deserves more discussion. Take the sleep dynamics studied in this paper as an example. It is well-known that the inter-expert agreement rate is only about 80% for normal subjects, not to say for subjects with sleep problems [19]. With this uncertainty, a supervised learning algorithm *might* learn both good and bad labels. On the other hand, the proposed unsupervised approach is independent of the provided labels, and the chosen spectral features all come from the EEG signal and speak solely for the sleep dynamics but not the labels. To some extent, the "uncertainty" issue is less critical via the unsupervised approach, since the uncertain labels are not taken into account in the feature extraction step.

Since both supervised and unsupervised approaches have their own merits, it is natural to seek for a way to combine both. We are exploring the possibility of combining DNN and the proposed unsupervised techniques, and the result will be reported in the future work.

9.4 "Self-Evolving" Artificial Intelligence System

Due to the advance of the technology and computational power, in the past decades, a lot of effort has been devoted to establish an artificial intelligence (AI) system for the automatic sleep stage annotation purpose. In addition to being able to accurately score sleep stages, an ideal AI system should also be able to "self-learn" or accumulate knowledge from the historical database. Note that despite the well-accepted homeostasis assumption of physiological system, physiological dynamics vary from subject to subject. Therefore, the main challenge of this self-learning capability is handling the inter-individual variability. This challenge is actually ubiquitous – for a new-arriving subject, how may one utilize the existing database with the expert annotations?

This challenge is actually empirically handled in this article. Recall that for each given subject, we take the age into account to find "similar subjects" to train the model to automatically annotate the sleep stage of the given subject. This idea is a special case of the general picture commonly encountered in the clinical situation. The inter-individual variability is inevitable, and this variability is the main obstacle toward a self-evolving system that can self-learn like a human being. Our solution is respecting the physicians' decision-making process and clinical experience to build up the system – when a physician sees a subject the first time, an automatic "classification" of this subject is established. This "decision tree" is mostly based on the physician's knowledge. Although it varies from physician to physician, the overall structure of the decision tree "should be" relatively stable, ideally. For example, a physician will not consider any menopause-related diseases if the patient is a five year boy. In the sleep dynamics problem studied in this article, we take the impact of age on the sleep dynamics [53] and EEG signal [54, 55] into account. Note that we only consider age since information provided in the available databases is limited. In general, this approach can be understood as a high-level filtering based on the *phenotype* information [66]. Another benefit of this phenotype-based approach is its flexibility for the growing database. While it is widely believed that the larger the database is, the more accurate model we can establish, we should take the limited computational resource into account. By only choosing those subjects sharing similar phenotype to establish the prediction model, computational efficiency can be achieved.

This phenotype-based idea allows us to establish a "self-evolving" AI system for the automatic annotation purpose. Armed with the above ideas, the system can accumulate experience/wisdom from each new subject – after applying the existing system with n subjects with experts' annotations to the new subject to alleviate the physician's load, the physician can update/train the system by providing his/her feedback. The updated system with $n + 1$ subjects is then more "knowledgeable." This close loop forms the self-evolving or self-learning part of the artificial intelligence system. See Figure 11 for an illustration of the general framework.

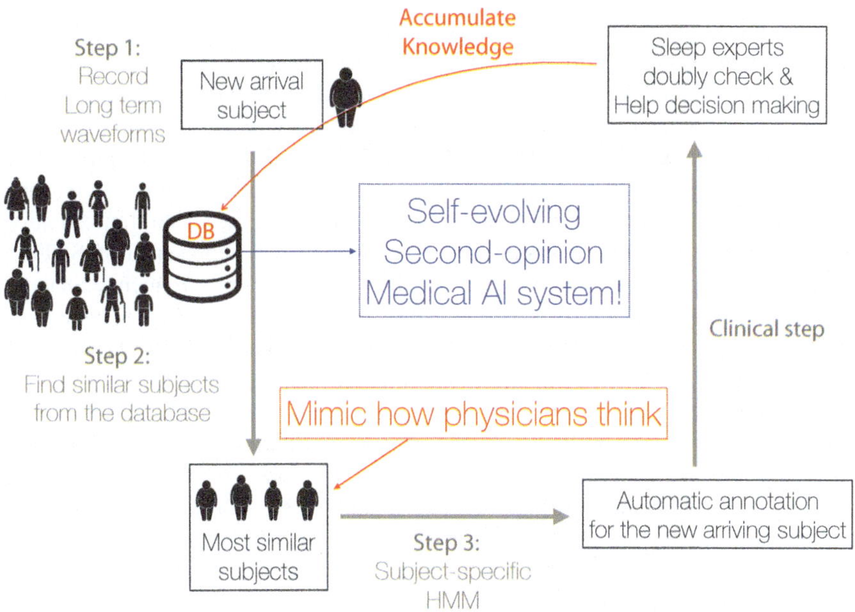

Fig. 11 An illustration of the "self-evolving" artificial intelligence system

To further validate and develop this proposed self-evolving system, in our future study, we will include more health-related information and mimic physicians' decision-making rules to determine the similarity between subjects. Moreover, we will establish a statistical model to better quantify the decision tree and better handle the "inter-physician" variability of their decision trees. The "reward" idea from the reinforcement learning will also be considered.

One main clinical application of this self-evolving AI system is establishing a more accurate sleep apnea screening. With the sleep stage information, the apnea-hypopnea index could be calculated. For that purpose, it is a common consensus to consider fewer channels for the patient to wear. While more channels provide more information, they may disturb the sleep. The respiratory flow is a common channel that people consider to screen the sleep apnea at home. In addition to providing the sleep apnea information, it has been known that the respiratory flow signal also contains abundant information about the sleep stage and is based on different physiological mechanisms compared with the EEG signal [67]. In this scenario, the proposed sensor fusion algorithm has the potential to incorporate the information hidden in the flow signal and design a more accurate prediction system. An exploration of this direction will be postponed to our future work.

9.5 Limitation and Future Work

Despite the strength of the proposed method, the discussion is not complete without mentioning its limitations. While we test the algorithm on the publicly available benchmark database and compare our results with those of state-of-the-art algorithms, those databases are small. To draw a conclusion and confirm its clinical applicability, a large-scale and prospective study is needed. We focus only on the spectral information in the EEG signals. There are other features, for example, [68], we can consider to further improve the performance. While the spectral information is mainly determined by the nonlinear-type time-frequency representation SST for the purpose of preventing energy leakage, there are other time-frequency analysis tools that we can consider, for example, the scattering transform [69]. A systematic study of other possibilities will be explored in the future work. While with two channels our algorithm is compatible with that reported in [56] which depends on only one channel, when we have only one channel, our algorithm does not perform better (for Fpz-Cz, the accuracy and F1 of our algorithm are 78.5% and 67.9%, while the accuracy and F1 reported in [56] are 82% and 76.9%. For Pz-Oz, the accuracy and F1 of our algorithm are 79.3% and 70.3%, while the accuracy and F1 reported in [56] are 79.8% and 73.1%). As discussed above, this limitation comes from the poor features for N1 classification. We need to find features that can better quantify N1 dynamics and better understand how and why the deep neural network achieves the accuracy. Another related open problem is how to further take the temporal information into account when we deal with the inter-individual prediction. Note that in the current algorithm, although the temporal relationship of epochs is considered in the HMM model, it is not taken into account to design the feature. The abovementioned limitations will be studied and reported in the future work.

Although extended theoretical understandings of applied algorithms have been established in the past decade, there are still open problems we need to explore from the theoretical perspective. In general, we know that by taking the phase information into account, we obtain a sharper time-frequency representation. Although we do empirically find that the classification performance is better with the sharpened time-frequency representation determined by SST, we should be careful that a sharper time-frequency representation is not equivalent to the "correct" time-frequency representation. A mathematical question to ask is when there is no obvious oscillatory pattern in the EEG signal, like the "mixed frequency" property of the theta wave in N1, how does SST behave, and what is the mathematical property of the time-frequency representation determined by SST. While AD and co-clustering look similar, they are developed under different motivations, and the consequence and relationship are never discussed. Understanding this relationship might allow us to further improve diffusion-based sensor fusion algorithms.

10 Funding

G.-R. Liu is supported by Ministry of Science and Technology (MOST) grant `MOST 106-2115-M-006 -016 -MY2`. Y.-L. Lo is supported by MOST grant `MOST 101-2220-E-182A-001, 102-2220-E-182A-001, 103-2220-E-182A-001 and MOST 104-220-E-182-002`. Y.-C. Sheu is supported by MOST grant `MOST 106-2115-M-009-006` and NCTS, Taiwan.

References

1. A. Rechtschaffen, A. Kales, *A Manual of Standardized Terminology, Techniques and Scoring System for Sleep Stages of Human Subjects* (Public Health Service, US Government Printing Office, Washington, 1968)
2. C. Iber, S. Ancoli-Isreal, A. Chesson, S. Quan, *The AASM Manual for Scoring of Sleep and Associated Events-Rules: Terminology and Technical Specification* (American Academy of Sleep Medicine, 2007)
3. C.B. Saper, The neurobiology of sleep. Continuum **19**(1), 19–31 (2013)
4. T. Kanda, N. Tsujino, E. Kuramoto, Y. Koyama, E.A. Susaki, S. Chikahisa, H. Funato, Sleep as a biological problem: an overview of frontiers in sleep research. J. Physiol. Sci. **66**(1), 1–13 (2016)
5. A. Karni, D. Tanne, B.S. Rubenstein, J.J. Askenasy, D. Sagi, Dependence on REM sleep of overnight improvement of a perceptual skill. Science **265**(5172), 679–682 (1994)
6. F. Roche Campo, X. Drouot, A.W. Thille, F. Galia, B. Cabello, M.-P. D'Ortho, L. Brochard, Poor sleep quality is associated with late noninvasive ventilation failure in patients with acute hypercapnic respiratory failure. Crit. Care Med. **38**(2), 477–485 (2010)

7. J.-E. Kang, M.M. Lim, R.J. Bateman, J.J. Lee, L.P. Smyth, J.R. Cirrito, N. Fujiki, S. Nishino, D.M. Holtzman, Amyloid-b Dynamics are regulated by Orexin and the sleep-wake cycle. Science **326**, 1005–1007 (2009)
8. D. Leger, V. Bayon, J. Laaban, P. Philip, Impact of sleep apnea on economics. Sleep Med. Rev. **16**(5), 455–462 (2012)
9. I.G. Campbell, Eeg recording and analysis for sleep research. Curr. Protocols Neurosci. **49**(1), 10–2 (2009)
10. I. Daubechies, J. Lu, H.-T. Wu, Synchrosqueezed wavelet transforms: an empirical mode decomposition-like tool. Appl. Comput. Harmon. Anal. **30**, 243–261 (2011)
11. H.-T. Wu, Adaptive Analysis of Complex Data Sets. Ph.D. thesis, Princeton University (2011)
12. B. Ricaud, B. Torresani, A survey of uncertainty principles and some signal processing applications. Adv. Comput. Math. **40**(3), 629–650 (2014)
13. A. Singer, R.R. Coifman, Non-linear independent component analysis with diffusion maps. Appl. Comput. Harmon. Anal. **25**(2), 226–239 (2008)
14. R. Talmon, R. Coifman, Empirical intrinsic geometry for nonlinear modeling and time series filtering. Proc. Nat. Acad. Sci. **110**(31), 12535–12540 (2013)
15. R.R. Coifman, S. Lafon, Diffusion maps. Appl. Comput. Harmon. Anal. **21**(1), 5–30 (2006)
16. R.R. Lederman, R. Talmon, Learning the geometry of common latent variables using alternating-diffusion. Appl. Comput. Harmon. Anal. **44**(3), 509–536 (2015)
17. I.S. Dhillon, Co-clustering documents and words using bipartite spectral graph partitioning, in *Proceedings of the Seventh ACM SIGKDD International Conference on Knowledge Discovery and Data Mining* (ACM, 2001), pp. 269–274
18. B.W. Rotenberg, C.F. George, K.M. Sullivan, E. Wong, Wait times for sleep apnea care in Ontario: a multidisciplinary assessment. Can. Respir. J. **17**(4), 170–174 (2010)
19. R.G. Norman, I. Pal, C. Stewart, J.A. Walsleben, D.M. Rapoport, Interobserver agreement among sleep scorers from different centers in a large dataset. Sleep **23**(7), 901–908 (2000)
20. A.M. Fraser, *Hidden Markov Models and Dynamical Systems* (SIAM, 2008)
21. A. Goldberger, L. Amaral, L. Glass, J. Hausdorff, P. Ivanov, R. Mark, J. Mietus, G. Moody, C.-K. Peng, H. Stanley, Physiobank, physiotoolkit, and physionet: components of a new research resource for complex physiologic signals. Circulation **101**(23), e215–e220 (2000)
22. A. Berrian, N. Saito, Adaptive synchrosqueezing based on a quilted short-time Fourier transform. Int. Soc. Opt. Photon. (SPIE) **10394**, 1039420 (2017)
23. Y.-C. Chen, M.-Y. Cheng, H.-T. Wu, Nonparametric and adaptive modeling of dynamic seasonality and trend with heteroscedastic and dependent errors. J. R. Stat. Soc. B **76**(3), 651–682 (2014)
24. O. Katz, R. Talmon, Y.-L. Lo, H.-T. Wu, Diffusion-based nonlinear filtering for multimodal data fusion with application to sleep stage assessment. Inform. Fusion **45**, 346–360 (2019)
25. R. Talmon, R.R. Coifman, Empirical intrinsic geometry for nonlinear modeling and time series filtering. Proc. Natl. Acad. Sci. U. S. A. **110**(31), 12535–12540 (2013)
26. J. Malik, C. Shen, N. Wu, H.-T. Wu, Connecting dots – from covariance to geodesics, empirical intrinsic geometry, and locally linear embedding. Pure Appl. Anal. accepted for publication
27. P. Bérard, G. Besson, S. Gallot, Embedding Riemannian manifolds by their heat kernel. Geom. Funct. Anal. **4**, 373–398 (1994)
28. M. Belkin, P. Niyogi, Towards a theoretical foundation for Laplacian-based manifold methods, in *Proceedings of the 18th Conference on Learning Theory (COLT)* (2005), pp. 486–500
29. M. Hein, J. Audibert, U. von Luxburg, From graphs to manifolds – weak and strong pointwise consistency of graph Laplacians, in *COLT* (2005), pp. 470–485
30. A. Singer, From graph to manifold Laplacian: the convergence rate. Appl. Comput. Harmon. Anal. **21**(1), 128–134 (2006)
31. A. Singer, H.-T. Wu, Spectral convergence of the connection Laplacian from random samples. Inform. Inference: J. IMA **6**(1), 58–123 (2017)
32. N.G. Trillos, M. Gerlach, M. Hein, D. Slepcev, Error estimates for spectral convergence of the graph Laplacian on random geometric graphs towards the Laplace–Beltrami operator. Found. Comput. Math. **20**(4), 827–887 (2020)
33. X. Wang, Spectral Convergence Rate of Graph Laplacian, ArXiv:1510.08110 in (2015)

34. E. Giné, V. Koltchinskii, Empirical graph Laplacian approximation of Laplace-Beltrami operators: large sample results, in *IMS Lecture Notes*, vol. 51, ed. by A. Bonato, J. Janssen. Monograph Series (The Institute of Mathematical Statistics, 2006), pp. 238–259
35. P.W. Jones, M. Maggioni, R. Schul, Manifold parametrizations by eigenfunctions of the Laplacian and heat kernels. Proc. Natl. Acad. Sci. U. S. A. **105**(6), 1803–1808 (2008)
36. J. Bates, The embedding dimension of Laplacian eigenfunction maps. Appl. Comput. Harmon. Anal. **37**(3), 516–530 (2014)
37. J.W. Portegies, Embeddings of Riemannian manifolds with heat kernels and eigenfunctions. Commun. Pure Appl. Math. **69**(3), 478–518 (2016)
38. N. El Karoui, H.-T. Wu, Connection graph Laplacian methods can be made robust to noise. Ann. Stat. **44**(1), 346–372 (2016)
39. I. Gel'fand, N.Y. Vilenkin, *Generalized Function Theory*, vol. 4 (Academic Press, 1964)
40. P. Bérard, *Spectral Geometry: Direct and Inverse Problems* (Springer, 1986)
41. A. Singer, H.-T. Wu, Vector diffusion maps and the connection Laplacian. Commun. Pure Appl. Math. **65**(8), 1067–1144 (2012)
42. N. El Karoui, On information plus noise kernel random matrices. Ann. Stat. **38**(5), 3191–3216 (2010)
43. R. Talmon, H.-T. Wu, Discovering a latent common manifold with alternating diffusion for multimodal sensor data analysis. Appl. Comput. Harmon. Anal. In press (2018)
44. O. Lindenbaum, A. Yeredor, M. Salhov, A. Averbuch, Multi-View diffusion maps. Inf fusion. **55**, 127–149 (2020)
45. W. Hardle, *Canonical Correlation Analysis* (Springer, Berlin/Heidelberg, 2007), pp. 321–330
46. D. Lahat, T. Adali, C. Jutten, Multimodal data fusion: an overview of methods, challenges, and prospects. Proc. IEEE **103**(9), 1449–1477 (2015)
47. N.F. Marshall, M.J. Hirn, Time coupled diffusion maps. Appl. Comput. Harmon. Anal. **45**(3), 709–728 (2018)
48. T. Michaeli, W. Wang, K. Livescu, Nonparametric canonical correlation analysis, in *International Conference on Machine Learning* (2016), pp. 1967–1976
49. T. Shnitzer, M. Ben-Chen, L. Guibas, R. Talmon, H.-T. Wu, Recovering hidden components in multimodal data with composite diffusion operators. SIAM Journal on Mathematics of Data Science **1**(3), 588–616 (2019)
50. F. Chung, *Spectral Graph Theory* (American Mathematical Society, 1996)
51. J.R. Lee, S.O. Gharan, L. Trevisan, Multiway spectral partitioning and higher-order Cheeger inequalities. J. ACM **61**(6), 37:1–37:30 (2014)
52. A. Buzo, A. Gray, R. Gray, J. Markel, Speech coding based upon vector quantization. IEEE Trans. Acoust. Speech Signal Process. **28**(5), 562–574 (1980)
53. M.V. Vitiello, L.H. Larsen, K.E. Moe, Age-related sleep change. J. Psychosom. Res. **56**(5), 503–510 (2004)
54. M. Boselli, L. Parrino, A. Smerieri, M.G. Terzano, Effect of age on EEG arousals in normal sleep. Sleep **21**(4), 361–367 (1998)
55. E. Van Cauter, R. Leproult, L. Plat, Age-related changes in slow wave sleep and rem sleep and relationship with growth hormone and cortisol levels in healthy men. JAMA **284**(7), 861–868 (2000)
56. A. Supratak, H. Dong, C. Wu, Y. Guo, DeepSleepNet: a model for automatic sleep stage scoring based on raw single-channel EEG. IEEE Trans. Neural. Syst. Rehabil. Eng. **25**, 1998–2008 (2017)
57. O. Tsinalis, P.M. Matthews, Y. Guo, Automatic sleep stage scoring using time-frequency analysis and stacked sparse autoencoders. Ann. Biomed. Eng. **44**(5), 1587–1597 (2016)
58. O. Tsinalis, P.M. Matthews, Y. Guo, S. Zafeiriou, Automatic sleep stage scoring with single-channel EEG using convolutional neural networks, arXiv:1610.01683 in (2016)
59. J. Malik, N. Reed, C.-L. Wang, H.-T. Wu, Single-lead f-wave extraction using diffusion geometry. Physiol. Meas. **38**, 1310–1334 (2017)
60. S. Alagapan, H.W. Shin, F. Frohlich, H.-T. Wu, Diffusion geometry approach to efficiently remove electrical stimulation artifacts in intracranial electroencephalography. J. Neural Eng. (2019). https://doi.org/10.1088/1741-2552/aaf2ba

61. A. Borbély, P. Mattmann, M. Loepfe, I. Strauch, D. Lehmann, Effect of benzodiazepine hypnotics on all-night sleep EEG spectra. Hum. Neurobiol. **4**(3), 189–194 (1985)
62. M. Bonnet, D. Carley et al., EEG arousals: Scoring rules and examples. A preliminary report from the Sleep Disorders Atlas Task Force of the American Sleep Disorders Association. Sleep **15**(2), 173–184 (1992)
63. P. Halasz, M. Terzano, L. Parrino, R. Bodizs, The nature of arousal in sleep. J. Sleep Res. **13**, 1–23 (2004)
64. A. Vilamala, K.H. Madsen, L.K. Hansen, Deep convolutional neural networks for interpretable analysis of EEG sleep stage scoring, in *2017 IEEE International Workshop on Machine Learning for Signal Processing* (2017)
65. Y. LeCun, Y. Bengio, G. Hinton, Deep learning. Nature **521**, 436–444 (2015)
66. H.-T. Wu, J.-C. Wu, P.-C. Huang, T.-Y. Lin, T.-Y. Wang, Y.-H. Huang, Y.-L. Lo, Phenotype-based and self-learning inter-individual sleep apnea screening with a level IV-like monitoring system. Front. Physiol. **9**, 723 (2018)
67. S.R. Thompson, U. Ackermann, R.L. Horner, Sleep as a teaching tool for integrating respiratory physiology and motor control. Adv. Physiol. Educ. **25**(2), 29–44 (2001)
68. S. Motamedi-Fakhr, M. Moshrefi-Torbati, M. Hill, C.M. Hill, P.R. White, Signal processing techniques applied to human sleep EEG signals – a review. Biomed. Signal Process. Control **10**, 21–33 (2014)
69. S. Mallat, Group invariant scattering. Commun. Pure Appl. Math. **65**(10), 1331–1398 (2012)

Harmonic Functions in Slabs and Half-Spaces

W. R. Madych

Abstract The usual solution to the Dirichlet problem for the Laplace equation $\Delta u = 0$ in the slab $\mathbb{R}^n \times (a, b)$, where $-\infty < a < b < \infty$, and the half-space $\mathbb{R}^n \times (0, \infty)$ involves convolution of the data with a Poisson kernel. Interestingly, the class of distributions which is convolvable with the natural Poisson kernel Q for the slab is considerably wider than that which is convolvable with the classical Poisson kernel P for the half-space. We investigate this curious phenomenon and observe that arbitrary tempered distributions can be convolved with Q, resulting in functions harmonic in the slab with no greater than polynomial growth in the interior and distributionally bounded on hyperplanes parallel to the boundary. Conversely, we show that all harmonic functions in the slab which enjoy no greater than polynomial growth in the interior and are distributionally bounded on hyperplanes parallel to the boundary can be characterized as Poisson integrals of tempered distributions. In the case of the half-space we observe that the classical Poisson kernel can be modified so that the result is applicable to all tempered distributions and gives rise to harmonic functions in the half-space with the prescribed boundary values. In both cases if the boundary data is given by polynomials then so is the resulting harmonic function. In the appendix we record some additional properties of the kernel Q and offer several pertinent comments and observations.

Subject Classification: 31B05, 31B10, 31B25, 46F99

W. R. Madych (✉)
Department of Mathematics, University of Connecticut, Storrs, CT, USA
e-mail: madych@math.uconn.edu

M. Th. Rassias (ed.), *Harmonic Analysis and Applications*, Springer Optimization and Its Applications 168, https://doi.org/10.1007/978-3-030-61887-2_12

1 Introduction

1.1 Background

Consider the Dirichlet problem associated with functions u harmonic in the domain Ω in $\mathbb{R}^{n+1}$.

$$\begin{aligned} \Delta u &= 0 \quad \text{in } \Omega \\ u &= f \quad \text{on } \partial\Omega . \end{aligned} \tag{1}$$

Here Δ is the Laplace operator and $\partial\Omega$ is the boundary of Ω. The boundary values are to be taken on in some sense appropriate to the nature of f. For instance, if f is continuous, then the solution u should be continuous in the closure of Ω.

If $\Omega = \mathbb{R}^{n+1}_+ = \mathbb{R}^n \times (0, \infty)$ is the upper half-space and f is a sufficiently well-behaved function on $\partial\Omega = \mathbb{R}^n \times \{0\} = \mathbb{R}^n$, then a solution of (1) is given by the convolution-type integral

$$u(x, y) = \int_{\mathbb{R}^n} P(x - t, y) f(t) dt . \tag{2}$$

Here $P(x - t, y)$ is the classical Poisson kernel for the upper half-space defined by

$$P(x, y) = c_n \frac{y}{(|x|^2 + y^2)^{(n+1)/2}} \quad \text{where} \quad c_n = \frac{\Gamma((n + 1)/2)}{\pi^{(n+1)/2}} \tag{3}$$

and $(x, y) = (x_1, \ldots, x_n, y) \in \mathbb{R}^n \times (0, \infty)$, $t = (t_1, \ldots, t_n) \in \mathbb{R}^n$, and $|x|^2 = x_1^2 + \cdots + x_n^2$.

If $\Omega = \mathbb{R}^n \times (a, b)$, where $-\infty < a < b < \infty$, is a slab in $\mathbb{R}^{n+1}$ and the restrictions of f to $\mathbb{R}^n \times \{a\}$ and $\mathbb{R}^n \times \{b\}$ are respectively the functions f_a and f_b defined on $\mathbb{R}^n$, then, in analogy to (2), the solution to (1) is given by

$$u(x, y) = \int_{\mathbb{R}^n} Q(x - t, b - y; c) f_a(t) dt + \int_{\mathbb{R}^n} Q(x - t, y - a; c) f_b(t) dt \tag{4}$$

where $c = b - a$ and for $0 \le y < c$

$$Q(x, y; c) = (2\pi)^{-n} \int_{\mathbb{R}^n} \frac{\sinh y|\xi|}{\sinh c|\xi|} e^{i\langle x,\xi\rangle} d\xi . \tag{5}$$

The pair $\big(Q(x - t, b - y; c), Q(x - t, y - a; c)\big)$ may be regarded as the Poisson kernel for the slab.

Note that for fixed (x, y) the right-hand side of (2) may be viewed as a linear functional evaluated at f. It may also be viewed as the linear functional represented by f evaluated at $P(x - t, y)$ as a function of t or, what is the same thing, as f

evaluated at a family of "test" functions parameterized by (x, y). The expression on the right-hand side of (4) may be viewed similarly. It is this view we adopt in what follows.

Notice that for $c = b - a > 0$ and $(x, y) \in \mathbb{R}^n \times (a, b)$, the test functions $Q(x - t, b - y; c)$ and $Q(x - t, y - a; c)$ as functions of t are in the Schwartz class $\mathscr{S}(\mathbb{R}^n)$ of infinitely differentiable rapidly decreasing functions. As a consequence, the right-hand side of (4) can be extended in a natural way to include any pair of tempered distributions f_a and f_b resulting in $u(x, y)$ which is harmonic in the slab, $(x, y) \in \mathbb{R}^n \times (a, b)$.

On the other hand, while $P(x - t, y)$ as a function of t is infinitely differentiable, it is only $O(|t|)^{-n-1}$ as $|t| \to \infty$ and thus is not in the Schwartz class $\mathscr{S}(\mathbb{R}^n)$. As a consequence the class of functionals f which give rise to harmonic functions u via (2) requires additional restrictions at infinity and cannot be extended to the whole class of tempered distributions. Indeed, from (2) it is not immediately clear whether arbitrary continuous functions $f(x)$ which grow faster than $|x|$ as $|x| \to \infty$ can be the boundary values of functions $u(x, y)$ harmonic in $\mathbb{R}^n \times (0, \infty)$.

This issue can be dealt with in certain cases by modifying the classical Poisson kernel P appropriately; examples can be found in [3, 4, 14, 15, 17, 24, 32] and elsewhere. Nevertheless the difference in behavior of the kernels P and Q is a curious phenomenon which deserves further study. In this note we record pertinent observations concerning these kernels and the behavior of harmonic functions in the slab and half-space.

The theory of harmonic functions is classical and has been exhaustively studied. There are several texts devoted to the subject including the classic [19] and the more recent examples [4, 6]; specific studies for the half-space, the slab, and more general unbounded domains include [3, 7, 11, 14, 15, 20, 31] and [16], respectively. Nevertheless our observations outlined in the next subsection seem to be new.

1.2 Contents

In Section 2 we characterize harmonic functions u which enjoy representation (4) in the slab $\mathbb{R}^n \times (a, b)$ with tempered distributions f_a and f_b. In the case when f_a and f_b are polynomials of degree k and m, respectively, we show that such a solution u is a harmonic polynomial of degree no greater than $1 + \max\{k, m\}$.

Section 3 is primarily devoted to providing substitute kernels for $P(x - t, y)$ that are suitable for use with tempered distributions f. Specifically we exhibit a family of kernels which are convolvable with any tempered distribution f but may result in harmonic functions with faster than polynomial growth as $y \to \infty$. As a curious observation, we show that taking $f_a = f$ and $f_b = 0$ in (4) and letting $b \to \infty$ do not necessarily lead to a harmonic function in the half-space.

In the Appendix, Section 4, we provide several alternate representations and other significant properties of the kernel $Q(x, y; c)$. In addition, we record several pertinent comments and observations.

The remainder of this section is devoted to collecting further background material that is germane to our development.

1.3 Notation

We use standard mathematical notation and terminology.

Recall that the components ν_j, $j = 1, \ldots, n$, of a multi-index $\nu = (\nu_1, \ldots, \nu_n)$ are all non-negative integers and its length $|\nu|$ is defined by $|\nu| = \nu_1 + \cdots + \nu_n$. For ordinary elements x in $\mathbb{R}^n$, $|x|$ denotes the usual Euclidean norm of x. Also, $x^\nu = x_1^{\nu_1} \cdots x_2^{\nu_n}$ and, if $D_j = \partial/\partial x_j$, then D^ν denotes the derivative of order $|\nu|$ defined by $D^\nu = D_1^{\nu_1} \cdots D_n^{\nu_n}$.

The Schwartz space $\mathscr{S} = \mathscr{S}(\mathbb{R}^n)$ consists of infinitely differentiable rapidly decreasing functions ϕ equipped with the semi-norms

$$\|\phi\|_{M,N} = \sum_{|\nu| \le M} \sup_{x \in \mathbb{R}^n} |(1 + |x|)^N D^\nu \phi(x)| \,.$$

Thus $\phi \in \mathscr{S}(\mathbb{R}^n)$ means that $\|\phi\|_{M,N}$ is finite for every $M = 0, 1, 2, \ldots$ and $N = 0, 1, 2, \ldots$, and $\lim_{k\to\infty} \phi_k = \phi$ in $\mathscr{S}(\mathbb{R}^n)$ means that $\lim_{k\to\infty} \|\phi_k - \phi\|_{M.N} = 0$ for every M and N.

Its dual $\mathscr{S}' = \mathscr{S}'(\mathbb{R}^n)$ is the space of tempered distributions that consists of continuous linear functionals on $\mathscr{S}(\mathbb{R}^n)$. We use the notation $\langle \phi, f \rangle$ to denote the evaluation of a tempered distribution $f \in \mathscr{S}'(\mathbb{R}^n)$ at $\phi \in \mathscr{S}(\mathbb{R}^n)$. Thus $f \in \mathscr{S}'(\mathbb{R}^n)$ means that

$$\lim_{k\to\infty} \langle \phi_k, f \rangle = \langle \phi, f \rangle \quad \text{whenever } \lim_{k\to\infty} \phi_k = \phi \text{ in } \mathscr{S}(\mathbb{R}^n),$$

which is equivalent to the existence of M and N and a constant C such that

$$|\langle \phi, f \rangle| \le C \|\phi\|_{M,N}$$

for all ϕ in $\mathscr{S}(\mathbb{R}^n)$.

In the case when both ϕ and ψ are in $\mathscr{S}(\mathbb{R}^n)$

$$\langle \phi, \psi \rangle = \int_{\mathbb{R}^n} \phi(x)\psi(x)dx \,.$$

Thus $\langle \phi, f \rangle$ may make sense even when ϕ fails to be in $\mathscr{S}(\mathbb{R}^n)$, for example, when both ϕ and f are in $L^2(\mathbb{R}^n)$. Using the notation $P(x - \cdot, y)$ to denote $P(x - t, y)$ as a function of t for fixed (x, y) relation (2) reduces to

$$u(x, y) = \langle P(x - \cdot, y), f \rangle \,.$$

Δ is used to denote the Laplace operator in the (x, y) variable. Thus

$$\Delta = \frac{\partial^2}{\partial x_1^2} + \cdots + \frac{\partial^2}{\partial x_n^2} + \frac{\partial^2}{\partial y^2} .$$

The Fourier transform, $f \to \widehat{f}$, is always taken only in the x variable, and, unless it is meaningful in a stronger sense, it should be interpreted in the distributional sense. Thus $\widehat{f}$ is well defined for every tempered distribution f via $\langle \widehat{f}, \phi \rangle = \langle f, \widehat{\phi} \rangle$. The corresponding frequency variable in $\mathbb{R}^n$ is denoted by ξ. The normalization we use gives

$$\widehat{f}(\xi) = \int_{\mathbb{R}^n} e^{-i\langle x,\xi \rangle} f(x)dx$$

whenever f is in $L^1(\mathbb{R}^n)$.

A remarkably complete and succinct exposition of distribution theory that includes multi-index notation, the spaces $\mathscr{S}(\mathbb{R}^n)$ and $\mathscr{S}'(\mathbb{R}^n)$, and basic facts concerning the Fourier transform can be found in [18, Chapter 1]. Other accessible sources include [10, 27, 29], and [30] among others.

For convenience we often use notation such as $f(x)$ or $f(\xi)$ to denote a distribution in the x or ξ variable, respectively, even when its values need not be defined pointwise. As is customary, we use the symbol C to denote generic constants whose value depends on the context.

2 Functions Harmonic in a Slab

2.1 Some Basic Formulas

One way of obtaining a representation for the solution u of

$$\begin{aligned} \Delta u(x, y) = 0 \quad & \text{for } (x, y) \text{ in } \mathbb{R}^n \times (a, b) \\ \text{with} \quad u(x, a) = f_a(x) \quad & \text{and} \quad u(x, b) = f_b(x) \end{aligned} \tag{6}$$

is to take the Fourier transform in the x variable. This leads to the univariate boundary value problem parametrized by the frequency variable $\xi \in \mathbb{R}^n$

$$\begin{aligned} & \frac{d^2}{dy^2}\widehat{u}(\xi, y) - |\xi|^2\widehat{u}(\xi, y) = 0, \quad a < y < b, \\ & \text{with} \quad \widehat{u}(\xi, a) = \widehat{f_a}(\xi) \quad \text{and} \quad \widehat{u}(\xi, b) = \widehat{f_b}(\xi). \end{aligned} \tag{7}$$

If $c = b - a$ and

$$\widehat{Q}(\xi, y; c) = \frac{\sinh y|\xi|}{\sinh c|\xi|} \quad \text{for } 0 \le y \le c$$

then

$$\widehat{u}(\xi, y) = \widehat{Q}(\xi, b - y; c)\widehat{f_a}(\xi) + \widehat{Q}(\xi, y - a; c)\widehat{f_b}(\xi) \tag{8}$$

is a solution of (7) for every $\xi \in \mathbb{R}^n$ and hence (8) is, formally at least, the Fourier transform of a solution to (6).

Note that if $0 \le y < c$, then for fixed y, the variable $\widehat{Q}(\xi, y; c)$, as a function of ξ, is in the Schwartz class $\mathscr{S}(\mathbb{R}^n)$. Hence its inverse Fourier transform

$$Q(x, y; c) = (2\pi)^{-n} \int_{\mathbb{R}^n} \widehat{Q}(\xi, y; c)\, e^{i\langle x, \xi\rangle} d\xi\,, \tag{9}$$

as a function of x, is also in $\mathscr{S}(\mathbb{R}^n)$.

Next, observe that

$$\widehat{Q}(\xi, y; c) = \frac{\sinh y|\xi|}{\sinh c|\xi|} = \frac{\sinh \frac{y}{c}|c\xi|}{\sinh |c\xi|}.$$

The final expression in the above string implies that

$$Q(x, y; c) = c^{-n} Q(x/c, y/c; 1). \tag{10}$$

This allows us, when convenient, to simplify our notation to

$$Q(x, y) = Q(x, y; 1). \tag{11}$$

In the case $n = 1$, it is known that $Q(x, y)$ can also be expressed as

$$Q(x, y) = \frac{1}{2} \frac{\sin \pi y}{\cosh \pi |x| + \cos \pi y} \qquad \text{for } (x, y) \in \mathbb{R}^n \times [0, 1). \tag{12}$$

For example, see [28, p. 510, item 10], [13, p. 31, item (14)], [31], [30, page 208], and [7]. For a derivation of (12) and analogous expressions for Q when $n > 1$, see the Appendix, Section 4, where alternate representations of Q can also be found.

The inverse Fourier transform of (8) leads to the convolution-type expression

$$u(x, y) = \langle Q(x - \cdot, b - y; c), f_a\rangle + \langle Q(x - \cdot, y - a; c), f_b\rangle \tag{13}$$

which, because $Q(x - t, y; c)$ is in $\mathscr{S}(\mathbb{R}^n)$ as a function of t for all $(x, y) \in \mathbb{R}^n \times (0, c)$, makes sense for any pair of distributions f_a and f_b in $\mathscr{S}'(\mathbb{R}^n)$.

In what follows, it will sometimes be convenient to use

$$u(x, y) = (2\pi)^{-n}\langle \widehat{Q}(\cdot, b - y; c)e^{i\langle x,\cdot\rangle}, \widehat{f_a}\rangle + (2\pi)^{-n}\langle \widehat{Q}(\cdot, y - a; c)e^{i\langle x,\cdot\rangle}, \widehat{f_b}\rangle \tag{14}$$

which is equivalent to (13).

To avoid tedious repetition of restrictions such as $0 < y < 1$, $0 \le y \le c$, or $a \le y \le b$ when considering the expressions such as $Q(x.y)$, $Q(x, y; c)$ or $Q(x, b - y; c)$, unless otherwise indicated, we will always automatically assume that such restrictions are satisfied.

2.2 Properties of $\widehat{Q}$

The following properties of $\widehat{Q}$ can be verified directly:

(i) For $(x, y) \in \mathbb{R}^n \times (0; c)$

$$\Delta\left(\widehat{Q}(\xi, y; c)\, e^{i\langle x,\xi\rangle}\right) \text{ exists and is equal to 0 in } \mathscr{S}(\mathbb{R}^n).$$

That is, the limit operations which define the result of applying Δ to $\widehat{Q}(\xi, y; c)\, e^{i\langle x,\xi\rangle}$ in the (x, y) variables converge in $\mathscr{S}(\mathbb{R}^n)$ as a function of ξ. In particular, this means that

$$\Delta\left\langle\widehat{Q}(\cdot, y; c)e^{i\langle x,\cdot\rangle}, f\right\rangle = 0 \text{ for } (x, y) \in \mathbb{R}^n \times (0, c) \text{ and all distributions } f \text{ in } \mathscr{S}'(\mathbb{R}^n).$$

(ii) For every M, N, and y_0, with $0 < y_0 < c$, there is a constant C such that

$$\|\widehat{Q}(\cdot, y; c)\, e^{i\langle x,\cdot\rangle}\|_{M,N} \le Cy(1 + |x|)^M \quad \text{whenever } 0 \le y \le y_0.$$

(iii) $\lim_{y\to 0} \widehat{Q}(\xi, y; c) = 0$ in $\mathscr{S}(\mathbb{R}^n)$. In other words, for every M and N

$$\lim_{y\to 0} \|\widehat{Q}(\xi, y; c)\|_{M,N} = 0.$$

(iv) For every test function $\phi \in \mathscr{S}(\mathbb{R}^n)$

$$\lim_{y\to c} \widehat{Q}(\xi, y; c)\, \phi(\xi) = \phi(\xi) \quad \text{in } \mathscr{S}(\mathbb{R}^n).$$

(v) For every pair M and N, there is a constant C independent of ϕ such that

$$\|\widehat{Q}(\cdot, y; c)\phi\|_{M,N} \le C\|\phi\|_{M,N} \quad \text{for } 0 \le y \le c.$$

2.3 Harmonic Functions of Polynomial Growth in the Slab

The following theorems provide a characterization of harmonic functions $u(x, y)$ that enjoy no greater than polynomial growth in the slab $\mathbb{R}^n \times (a, b)$ as $|x|$ tends to ∞.

Theorem 1 *Suppose $u(x, y)$ is defined by* (13) *with f_a and f_b in $\mathscr{S}'$. Then $u(x, y)$ is harmonic in $\mathbb{R}^n \times (a, b)$ and satisfies the following:*

(A1) *For every test function ϕ in $\mathscr{S}$*

$$\lim_{y\to a} \langle u(\cdot, y), \phi\rangle = \langle f_a, \phi\rangle \quad \textit{and} \quad \lim_{y\to b} \langle u(\cdot, y), \phi\rangle = \langle f_b, \phi\rangle. \tag{15}$$

(A2) *There is a number N such that if y_0 and y_1 is any pair of numbers that satisfy $a < y_0 < y_1 < b$ and y satisfies $y_0 \le y \le y_1$, then*

$$|u(x, y)| \le C(1 + |x|)^N, \tag{16}$$

where C is independent of $x \in \mathbb{R}^n$.

(A3) *There are constants M, N, and C such that for every test function ϕ in $\mathscr{S}$ and all y satisfying $a < y < b$*

$$|\langle u(\cdot, y), \phi\rangle| \le C\|\phi\|_{M,N}\,. \tag{17}$$

Theorem 2 *Conversely, every function $u(x, y)$ that is harmonic in $\mathbb{R}^n \times (a, b)$ and satisfies properties (A2) and (A3) listed above enjoys representation* (13) *for some unique pair of distributions f_a and f_b in $\mathscr{S}'$.*

Some restrictions on the growth of $u(x, y)$ as $|x|$ tends to ∞ are necessary for uniqueness. Evidence for this is provided, in the case $n = 1$, by the example

$$u(x, y) = e^{\pi x/c} \sin\big(\pi(y - a)/c)\big).$$

Proof (of Theorem 1) The fact that u is harmonic in $\mathbb{R}^n \times (a, b)$ follows from item 2.2(i).

To see (A1) assume $f_b = 0$ and write

$$\langle u(\cdot, y), \phi\rangle = \langle \widehat{Q}(\cdot, b - y; c)\check{\phi}, \widehat{f_a}\rangle \to \begin{cases} \langle \check{\phi}, \widehat{f_a}\rangle & \text{as } y \to a \\ 0 & \text{as } y \to b \end{cases} \tag{18}$$

where the limiting values are consequences of items 2.2 (iii) and (iv). Here $\check{\phi}$ denotes the inverse Fourier transform of ϕ. Using the fact that $\langle \check{\phi}, \widehat{f} \rangle = \langle \phi, f \rangle$ for all $\phi \in \mathscr{S}$ and all $f \in \mathscr{S}'$, relation (18) implies the desired result in this case.

The analogous result is valid when $f_a = 0$, *mutatis mutandis*, and the general case follows as a consequence.

To see (A2) note, there are constants C, M, and N such that

$$|\langle \phi, \widehat{f_a} \rangle| \leq C \|\phi\|_{M,N} \tag{19}$$

for all ϕ in $\mathscr{S}$. With the choice $\phi(\xi) = \widehat{Q}(\xi, b - y; c)\, e^{i\langle x, \xi \rangle}$, item 2.2 (ii) allows us to conclude that there is a constant C

$$\|\phi\|_{M,N} \leq C(1 + |x|)^M$$

whenever $a < y_0 \leq y \leq b$. This implies the desired result when $f_b = 0$.

A similar conclusion holds when $f_a = 0$, *mutatis mutandis*. Together both cases imply the general result.

To see (A3) compute as above using item 2.2(v). QED

Proof (of Theorem 2) It suffices to consider the case $a = 0$ and $b = c$. We do so to avoid excessively cumbersome notation.

Let $a_k = c/k$ and $b_k = c - c/k$, $k = 3, 4, 5, \ldots$. Let $f_{a_k}(x) = u(x, a_k)$ and $f_{b_k}(x) = u(x, b_k)$. Then in view of (A3), we may use a weak compactness-type argument, for example see [27, p. 68], to conclude that there is a subsequence, which we also index with k, and distributions f_0 and f_c in $\mathscr{S}'$ such that

$$\lim_{k \to \infty} \langle \phi, f_{a_k} \rangle = \langle \phi, f_0 \rangle \quad \text{and} \quad \lim_{k \to \infty} \langle \phi, f_{b_k} \rangle = \langle \phi, f_c \rangle.$$

Let

$$v(x, y) = \langle Q(x - \cdot, c - y; c), f_0 \rangle + \langle Q(x - \cdot, y; c), f_c \rangle .$$

To see that $v(x, y) = u(x, y)$ for all $(x, y) \in \mathbb{R}^n \times (0, c)$, let

$$u_k(x, y) = \langle Q(x - \cdot, b_k - y; b_k - a_k), f_{a_k} \rangle + \langle Q(x - \cdot, y - a_k; b_k - a_k), f_{b_k} \rangle ,$$

$$v_k(x, y) = \langle Q(x - \cdot, c - y; c), f_{a_k} \rangle + \langle Q(x - \cdot, y; c), f_{b_k} \rangle ,$$

and, for the moment, assume that

$$u_k(x, y) = u(x, y) \quad \text{for all} \quad (x, y) \in \mathbb{R}^n \times (a_k, b_k). \tag{20}$$

Fix the point $(x, y) \in \mathbb{R}^n \times (0, c)$, and write

$$u - v = u - u_k + u_k - v_k + v_k - v.$$

Now for k sufficiently large, say $k \geq k_1$, the value of y will satisfy $a_k < y < b_k$ and in view of our assumption

$$u(x, y) - u_k(x, y) = 0. \tag{21}$$

Next, if

$$\phi(t) = Q(x - t, c - y; c) \quad \text{and} \quad \psi(t) = Q(x - t, y; c),$$

then both ϕ and ψ are in $\mathscr{S}$, and, in view of the way f_0 and f_c were obtained, for sufficiently large k, say $k \geq k_2$,

$$|\langle \phi, f_{a_k} - f_0 \rangle| < \epsilon/2 \quad \text{and} \quad |\langle \psi, f_{b_k} - f_c \rangle| < \epsilon/2.$$

This last pair of inequalities implies that

$$|v_k(x, y) - v(x, y)| \leq \epsilon \quad \text{whenever } k > k_2. \tag{22}$$

Finally, let ϕ and ψ be as above and let

$$\phi_k(t) = Q(x - t, b_k - y; b_k - a_k) \quad \text{and} \quad \psi_k(t) = Q(x - t, y - a_k; b_k - a_k).$$

Then in view of (17)

$$|\langle \phi_k - \phi, f_{a_k} \rangle| \leq C\|\phi_k - \phi\|_{M,N} \quad \text{and} \quad |\langle \psi_k - \psi, f_{b_k} \rangle| \leq C\|\phi_k - \phi\|_{M,N}.$$

By choosing k sufficiently large, say $k \geq k_3$, the right-hand side of each of the last two inequalities is less than $\epsilon/2$. It follows that

$$|u_k(x, y) - v_k(x, y)| < \epsilon \quad \text{whenever } k \geq k_3. \tag{23}$$

Hence by choosing $k > \max\{k_1, k_2, k_3\}$ relations (21), (22), and (23), imply that

$$|u(x, y) - v(x, y)| < 2\epsilon.$$

In view of the fact that ϵ is arbitrary, we may conclude that

$$u(x, y) = v(x, y).$$

It remains to prove (20).

In the argument that follows, we use the fact that if the tempered distributions f_a and f_b are continuous functions, then the function $u(x, y)$ defined by (13) is harmonic in $\mathbb{R}^n \times (a.b)$ and continuous in its closure $\mathbb{R}^n \times [a.b]$. The fact that u is harmonic, of course, follows from Theorem 1, while the fact that it's continuous is a consequence of Theorem 3 below.

In view of the above announcement, the function $w(x, y) = u(x, y) - u_k(x, y)$ is harmonic in $\mathbb{R}^n \times (a_k, b_k)$, continuous in its closure $\mathbb{R}^n \times [a_k, b_k]$, and identically 0 on the boundary $(\mathbb{R}^n \times \{a_k\}) \cup (\mathbb{R}^n \times \{b_k\})$. The strategy is to show that for any fixed point (t, y_0) with $t = (t_1, \dots, t_n)$ in $\mathbb{R}^n$ and $a_k \le y_0 \le b_k$

$$w(t, y_0) = 0. \tag{24}$$

This of course implies (20).

Next, without loss of generality, assume w is real valued, and for any fixed positive ϵ, let

$$w_\epsilon(x, y) = w(x, y) + \epsilon \left\{ \prod_{j=1}^{n} \cosh\left(\frac{\pi}{c\sqrt{n}}(x_j - t_j)\right) \right\} \sin\left(\frac{\pi}{c} y\right).$$

Then w_ϵ is harmonic in $\mathbb{R}^n \times (a_k, b_k)$ and, since $0 < a_k < b_k < c$, positive on the boundary $(\mathbb{R}^n \times \{a_k\}) \cup (\mathbb{R}^n \times \{b_k\})$. Furthermore, in view of (A2) when $|x - t|$ is sufficiently large, say $|x - t| > r$, $w_\epsilon(x, y)$ is positive for all y in the range $a_k \le y \le b_k$. The maximum principle for harmonic functions now implies that $w_\epsilon \ge 0$ on all of $\mathbb{R}^n \times [a_k, b_k]$. In particular this means that $w(t, y_0) \ge -\epsilon$.

The same argument with $w(x, y)$ replaced with $-w(x, y)$ in the definition of w_ϵ shows that $w(t, y_0) \le \epsilon$. Since ϵ is arbitrary (24) follows. QED

2.4 More Properties of *Q* and *u*

Theorem 1(A1) and its proof suggest that the kernel $Q(x, y; c)$ acts like an approximation of the identity convolution kernel in the x variable as $y \to c$. In the case $n = 1$, this was verified in [31] by use of (12). To verify this in the general case, it suffices to check that

$$\lim_{y \to c} \int_{\mathbb{R}^n} Q(x, y; c) dx = 1\,, \tag{25}$$

$$\|Q(\cdot, y; c\|_{L^1(\mathbb{R}^n)} \le C \quad \text{for all } y,\ 0 \le y < c, \tag{26}$$

and that

$$\lim_{y \to c} \int_{|x| > \epsilon} Q(x, y; c) dx = 0 \quad \text{for every positive } \epsilon. \tag{27}$$

Items (25) and (26) follow from the fact that

$$\|Q(\cdot, y; c)\|_{L^1(\mathbb{R}^n)} = \int_{\mathbb{R}^n} Q(x, y; c)dx = \widehat{Q}(0, y; c) = \frac{y}{c}$$

where the first equality is a consequence of Lemma 1 below, which implies that Q is non-negative.

Lemma 1

$$Q(x, y; c) > 0 \quad \textit{for all} \quad (x, y) \in \mathbb{R}^n \times (0, c).$$

Proof For positive σ let

$$\widehat{Q}_\sigma(\xi, y; c) = \widehat{Q}(\xi, y; c)\, e^{-\sigma|\xi|^2}.$$

Then $Q_\sigma(x, y; c)$ as a function of (x, y) is harmonic in the slab $\mathbb{R}^n \times (0, c)$ and, in view of 2.2(iii) and (iv), continuous in the closure $\mathbb{R}^n \times [0, c]$ with

$$Q_\sigma(x, 0; c) = 0 \quad \text{and} \quad Q_\sigma(x, c; c) = (2\pi\sigma)^{-n/2}\, e^{-|x|^2/\sigma}.$$

The maximum principle for harmonic functions allows us to conclude that

$$Q_\sigma(x, y; c) > 0 \quad \text{for } (x, y) \text{ in } \mathbb{R}^n \times (0, c).$$

Since this is true for all positive σ, the desired result follows. QED

The validity of condition (27) is a consequence of the somewhat stronger property $Q(x, y; c)$ is asserted to have in the Lemma 2 below. This property will also be useful in what follows.

Lemma 2 *Suppose ϕ is a function in $C^\infty(\mathbb{R}^n)$ which satisfies*

$$\phi(x) = \begin{cases} 0 & \textit{if } |x| \le \epsilon_0 \\ 1 & \textit{if } |x| \ge \epsilon_1 \end{cases}$$

for some ϵ_0 and ϵ_1 that satisfy $0 < \epsilon_0 < \epsilon_1 < \infty$. Then

$$\lim_{y \to c} \phi(x)\, Q(x, y; c) = 0 \quad \textit{in } \mathscr{S}(\mathbb{R}^n).$$

Proof Since

$$\|(1 + |x|)^N D^\nu \phi(x) Q(x, y; c)\|_{L^\infty(\mathbb{R}^n)} \le \frac{1}{\epsilon_0^k}\|(1 + |x|)^N |x|^k D^\nu \phi(x) Q(x, y; c)\|_{L^\infty(\mathbb{R}^n)},$$

it suffices to show that for each ν

$$\lim_{y \to c} \|x^{\mu} D^{\nu} Q(x, y; c)\|_{L^{\infty}(\mathbb{R}^n)} = 0 \quad \text{whenever } |\mu| \text{ is sufficiently large.} \tag{28}$$

Identity (28) follows from the fact, which can be verified directly, that for every multi-index ν

$$\lim_{y \to c} \|D^{\mu}\big(\xi^{\nu} \widehat{Q}(x, y; c)\big)\|_{L^{1}(\mathbb{R}^n)} = 0 \quad \text{whenever } |\mu| \text{ is sufficiently large.} \tag{29}$$

For more details and a more precise variant of (29), see Lemma 10 in Section 4. QED

Lemma 3 *Suppose f is a distribution in $\mathscr{S}'$ that is continuous in a neighborhood of a point x_0 in $\mathbb{R}^n$. If $(x, y) \in \mathbb{R}^n \times (0, c)$, then*

$$\lim_{\substack{(x,y) \to (x_0,c) \\ (x,y) \in \mathbb{R}^n \times (0,c)}} \langle Q(x - \cdot, y; c), f \rangle = f(x_0)\,.$$

Proof First assume that $f = 0$ in a neighborhood $\mathscr{N}_0 = \{x : |x - x_0| < \epsilon\}$ of x_0. For $0 < \epsilon_0 < \epsilon_1 < \epsilon$, let ϕ be a function in $C^{\infty}(\mathbb{R}^n)$ with the property that

$$\phi(x) = \begin{cases} 0 & \text{if } |x - x_0| < \epsilon_0 \\ 1 & \text{if } |x - x_0| > \epsilon_1 . \end{cases}$$

Write

$$\langle Q(x - \cdot, y; c), f \rangle = \langle Q(x - \cdot, y; c), (1 - \phi) f \rangle + \langle Q(x - \cdot, y; c), \phi f \rangle$$

and notice that

$$\langle Q(x - \cdot, y; c), (1 - \phi) f \rangle = 0$$

while Lemma 2 implies that

$$\lim_{\substack{(x,y) \to (x_0,c) \\ (x,y) \in \mathbb{R}^n \times (0,c)}} \langle Q(x - \cdot, y; c), \phi f \rangle = 0.$$

So we may conclude that

$$\lim_{\substack{(x,y) \to (x_0,c) \\ (x,y) \in \mathbb{R}^n \times (0,c)}} \langle Q(x - \cdot, y; c), f \rangle = 0.$$

Now, if f is merely continuous in the same neighborhood $\mathscr{N}_0$ of x_0, then the desired result follows by essentially the same calculation as above except for the term involving $(1 - \phi) f$. In this case we get the conclusion

$$\lim_{\substack{(x,y)\to(x_0,c)\\(x,y)\in\mathbb{R}^n\times(0,c)}} \langle Q(x-\cdot,y;c),(1-\phi)f\rangle = f(x_0)$$

by the use of properties (25), (26), and (27). QED

Lemma 4 *For every distribution f in $\mathcal{S}'$*

$$\lim_{\substack{(x,y)\to(x_0,0)\\(x,y)\in\mathbb{R}^n\times[0,c)}} \langle Q(x-\cdot,y;c),f\rangle = 0\,.$$

uniformly on compact subsets of the variable $x_0 \in \mathbb{R}^n$.

Proof Since

$$\langle Q(x-\cdot,y;c),f\rangle = (2\pi)^{-n}\langle \widehat{Q}(\cdot,y;c)e^{i\langle x,\cdot\rangle},\widehat{f}\rangle$$

and

$$|\langle \widehat{Q}(\cdot,y;c)e^{i\langle x,\cdot\rangle},\widehat{f}\rangle| \le C\|\widehat{Q}(\cdot,y;c)e^{i\langle x,\cdot\rangle}\|_{M,N}$$

for some M and N, the desired result is a consequence of Property 2.2(ii). QED

Lemmas 3 and 4 taken together imply the following:

Theorem 3 *Suppose $u(x,y)$ is defined by* (13) *with f_a and f_b in $\mathcal{S}'$. If f_a is continuous in a neighborhood of a point x_0 in $\mathbb{R}^n$, then*

$$\lim_{\substack{(x,y)\to(x_0,a)\\(x,y)\in\mathbb{R}^n\times(a,b)}} u(x,y) = f_a(x_0)\,.$$

A similar statement involving f_b is also valid.

2.5 Harmonic Polynomials

It is well-known that if $\Omega = B$ is a solid spherical ball in $\mathbb{R}^n$ and f is the restriction of a polynomial $\sum_{|\nu|\le m} x^\nu$ to its boundary $\partial\Omega = S$, the surface of the sphere, then there is a harmonic polynomial u of degree no greater than m which solves the Dirichlet problem (2). In other words, in $\mathbb{R}^n$ the restriction of any n variate polynomial $p(x)$ to the surface S of a sphere is equal to the restriction of a harmonic polynomial u to S. See, for example, [30, Corollary 2.2, page 140].

If $n \ge 2$, the analogue of this is not necessarily true for other regions Ω even if the boundary is very regular, like the zero set of some n variate polynomial. For example, if $\Omega = \{x : 1 < |x| < 2\}$ and $p(x) = |x|^2$, then the harmonic function u which is equal to p on the boundary of Ω is defined by

$$u(x) = \begin{cases} \dfrac{3\log|x|}{\log 2} + 1 & \text{when } n = 2 \\[2ex] \dfrac{4\cdot 2^{n-2} - 1 - 3(2/|x|)^{n-2}}{2^{n-2} - 1} & \text{when } n \geq 3. \end{cases}$$

In the case Ω is the slab $\mathbb{R}^n \times (a, b)$ in $\mathbb{R}^{n+1}$, we have the following:

Theorem 4 *Suppose $u(x, y)$ is defined by* (13) *and both f_a and f_b are polynomials, $f_a = p_a$ and $f_b = p_b$.*

(i) *Then u is a harmonic polynomial. Furthermore u is the unique harmonic polynomial which satisfies*

$$u(x, a) = p_a(x) \quad \text{and} \quad u(x, b) = p_b(x)\,.$$

(ii) *If the degrees p_a and p_b are k and m, respectively, then the degree of $u(x, y)$ is no greater than $1 + \max\{k, m\}$.*

(iii) *If p_a and p_b are both harmonic, then*

$$u(x, y) = \frac{b - y}{b - a}\, p_a(x) + \frac{y - a}{b - a}\, p_b(x)\,.$$

Proof The fact that u is a polynomial follows by a direct calculation which is outlined below. Uniqueness follows from Theorem 2.

Suppose f is the monomial,

$$f(x) = p_\nu(x) = x^\nu\,.$$

To see the nature of $\langle Q(x - \cdot, y; c), p_\nu\rangle$, note that

$$\widehat{Q}(\xi, y; c) = \frac{y}{c}\,\frac{\sum_{k=0}^{\infty} \frac{(y|\xi|)^{2k}}{(2k+1)!}}{\sum_{k=0}^{\infty} \frac{(c|\xi|)^{2k}}{(2k+1)!}} \tag{30}$$

which, when $|\xi|$ is sufficiently small, can be expressed as

$$\widehat{Q}(\xi, y; c) = \frac{y}{c}\left(1 + \sum_{k=1}^{\infty} q_{2k}(y; c)(-|\xi|^2)^k\right) \tag{31}$$

where

$$q_{2k}(y; c) = \sum_{j=0}^{k} a_j c^{2j} y^{2(k-j)} \tag{32}$$

are polynomials homogeneous of degree $2k$ in (y, c) and, because $\widehat{Q}(\xi, c; c) = 1$, satisfy

$$q_{2k}(c; c) = 0 \quad \text{for } k = 1, 2, \dots . \tag{33}$$

In particular,

$$q_2(y, c) = \frac{c^2 - y^2}{3!} \tag{34}$$

and

$$q_4(y; c) = \frac{c^4}{(3!)^2} - \frac{c^4}{5!} - \frac{c^2 y^2}{(3!)^2} + \frac{y^4}{5!} .$$

Let

$$\widehat{Q}_m(\xi, y; c) = \frac{y}{c}\left(1 + \sum_{0<2k\le m} q_{2k}(y; c)(-|\xi|^2)^k\right),$$

write

$$\widehat{Q}(\xi, y; c) = \widehat{Q}_m(\xi, y; c) + \left\{\widehat{Q}(\xi, y; c) - \widehat{Q}_m(\xi, y; c)\right\},$$

and note that

$$\left\{\widehat{Q}(\xi, y; c) - \widehat{Q}_m(\xi, y; c)\right\}(iD_\xi)^\nu \delta(\xi) = 0 \quad \text{whenever } m \ge |\nu|.$$

Hence

$$\widehat{Q}(\xi, y; c)(iD_\xi)^\nu \delta(\xi) = \widehat{Q}_m(\xi, y; c)(iD_\xi)^\nu \delta(\xi),$$

and using the fact that the inverse Fourier transform of $(2\pi)^n(-|\xi|^2)^k(iD_\xi)^\nu\delta(\xi)$ is $\Delta^k x^\nu$, it follows that

$$\langle Q(x - \cdot, y; c), p_\nu\rangle = \frac{y}{c}\left(p_\nu(x) + \sum_{0<2k\le|\nu|} q_{2k}(y; c)\, \Delta^k p_\nu(x)\right). \tag{35}$$

Because $\Delta^k p_\nu(x) = 0$ for $2k > |\nu|$, the upper bound on the index of summation on the right-hand side of (35) can be extended to ∞ without changing the value of the sum. Since (35) is valid for all multi-indexes ν, we may summarize the above observations in the following lemma.

Lemma 5 *For any polynomial*

$$p(x) = \sum_{|\nu| \leq m} a_\nu x^\nu$$

on $\mathbb{R}^n$, *the function*

$$u(x, y) = \langle Q(x - \cdot, y; c), p \rangle$$

initially defined on the slab $\mathbb{R}^n \times [0, c]$ *can be extended to all of* $\mathbb{R}^{n+1}$ *as*

$$u(x, y) = \frac{y}{c} \left(p(x) + \sum_{k=1}^{\infty} q_{2k}(y; c)\, \Delta^k p(x) \right)$$

and is the unique harmonic polynomial on $\mathbb{R}^{n+1}$ *with the property that*

$$u(x, 0) = 0 \quad \text{and} \quad u(x, c) = p(x)\,.$$

Here $q_{2k}(y; c)$ *is a polynomial of degree* $2k$ *in* y *defined by* (30), (31), *and* (32).

The remaining conclusions of the Theorem follow as a corollary. QED

Note that the polynomials $q_{2k}(y; c)$ can be defined iteratively as follows:

$$y\, q_0(y; c) = y\,,$$

and for $k = 1, 2, \ldots$

$$\frac{d^2}{dy^2}\{y\, q_{2k}(y; c)\} = -y\, q_{2(k-1)}(y; c)$$

with the constraints $y\, q_{2k}(y; c)|_{y=0} = y\, q_{2k}(y; c)|_{y=c} = 0$.

Before closing this section, we mention that, in view of the phrase "no greater than," the formulation of item (ii) of the Theorem may seem somewhat unwieldy. However, this phrase is necessary in view of the possibility that the degree of $u(x, y)$ may be strictly less than $1 + \max\{k, m\}$. This is conveniently illustrated by taking $p_a = p_b = p$ in item (iii) of the Theorem, which results in $u(x, y) = p(x)$.

We also mention that there is a larger literature concerning the nature of bounded domains Ω where the Dirichlet problem with polynomial or entire data necessarily leads to a polynomial or entire solution. For a representative sampling, see [9, 21–23, 25, 26].

3 Functions Harmonic in a Half-Space

3.1 Basic Formulas

There are several ways of obtaining the classical Poisson kernel for the upper half-space $\mathbb{R}^n \times (0, \infty)$. One way is to proceed as in Subsection 2.1. Then formally, for any scalar θ,

$$\hat{u}(\xi, y) = \left\{(1-\theta)e^{-y|\xi|} + \theta e^{y|\xi|}\right\} \hat{f}(\xi) \tag{36}$$

is the Fourier transform of a solution u to Laplace's equation in $\mathbb{R}^n \times (0, \infty)$ with $u(x, 0) = f(x)$. (Note that u is in fact well defined, for example, if $\hat{f}(\xi)$ is a distribution with compact support and is sufficiently regular in a neighborhood of the origin.)

If θ is not zero, then, in addition to issues at the origin, because of the rapid growth of $e^{y|\xi|}$ as $|\xi| \to \infty$, the above solution does not make sense for tempered distributions $\hat{f}$ in general. This and reasons related to uniqueness lead to the standard solution

$$\hat{u}(\xi, y) = e^{-y|x|} \hat{f}(\xi) \tag{37}$$

which makes sense for all $\hat{f}$ in $\mathscr{S}'$ that satisfy an appropriate restriction in some neighborhood of the origin.

For fixed positive y, the function $e^{-y|\xi|}$ is continuous and decays exponentially as $|\xi| \to \infty$, so its inverse Fourier transform

$$P(x, y) = (2\pi)^{-n} \int_{R^n} e^{-y|\xi|} e^{i\langle x,\xi\rangle} d\xi \,.$$

is well defined and can be simplified to

$$P(x, y) = c_n \frac{y}{(|x|^2 + y^2)^{(n+1)/2}} \quad \text{where} \quad c_n = \frac{\Gamma((n+1)/2)}{\pi^{(n+1)/2}} \,. \tag{38}$$

For details see, for example, [29, page 61].

The inverse Fourier transform of (37) leads to the convolution-type expression

$$u(x, y) = \langle P(x - \cdot, y), f \rangle \tag{39}$$

which makes sense for all $(x, y) \in \mathbb{R}^n \times (0, \infty)$ and all distributions f in $\mathscr{S}'$ that satisfy a restriction on their growth at ∞. For example, (39) makes sense for all distributions f with compact support as well as for all locally integrable f such that

$$\int_{\mathbb{R}^n} \frac{|f(x)|}{1+|x|^{n+1}} dx < \infty .$$

The harmonic function u in (39) may also be expressed as

$$u(x, y) = (2\pi)^{-n} \langle e^{-y|\cdot|+i\langle x,\cdot\rangle}, \hat{f}\rangle . \tag{40}$$

The function $u(x, y)$ is well defined via (40) for all $(x, y) \in \mathbb{R}^n \times (0, \infty)$ and all $\hat{f}$ in $\mathscr{S}'$ that enjoy an appropriate restriction in some neighborhood of the origin. For example, the restriction that $\hat{f}$ be a measure in such a neighborhood does the job.

Characterization and behavior of harmonic functions u of the form (39) involving various classes of data f have been exhaustively studied in the literature. For example, see [6, Chapter 7], [29, Chapters 3 and 7], or [30, Chapter 2]. These works deal with data f for which u is well defined via (39).

Modifications to the classical Poisson kernel (38) that can be used with continuous functions $f(x)$ that enjoy no greater than polynomial growth as $|x|$ tends to ∞ can be found in [3, 4, 14, 15, 17, 24, 32] where the rate of growth of f determines the necessary modifications. Such kernels can also be used with more general initial data f; for example, various classes of measures and locally integrable functions are treated in most of the abovementioned works. A related formula providing an analytic function which represents an arbitrary distribution in $\mathscr{D}'$ is recorded in [8, Section 5.9]; the resulting expression is a convergent series generated by subtracting appropriate terms not unlike the meromorphic function with prescribed poles generated by the Mittag-Leffler theorem [2, page 185].

Here we provide a modification of (39) which gives rise to harmonic functions u in the half-space with any initial data f from $\mathscr{S}'$. The formula is independent of the growth or any other specific behavior of f.

3.2 Harmonic Polynomials

A solution to the Dirichlet problem in half-space $\Omega = \mathbb{R}^n \times (0, \infty)$ when f is a polynomial can be obtained by "inspection." Indeed, if f is the polynomial,

$$p(x) = \sum_{|\nu|\le m} a_\nu x^\nu ,$$

it is not difficult to see that by adding polynomial terms to p each of which contains a factor of some positive power of y, one can construct a harmonic polynomial $u(x, y)$ such that $u(x, 0) = p(x)$. A direct calculation shows that

$$\sum_{k=0}^{\infty}(-1)^k \frac{y^{2k}}{(2k)!} \Delta^k p(x)$$

does the job. Notice that if $2k > m$, where m is the degree of p, then $\Delta_x^k p(x) = 0$ so the above sum has only a finite number of non-zero terms.

This polynomial adjustment is not unique since one can always add any harmonic polynomial that vanishes on $\mathbb{R}^n \times \{0\}$.

We summarize this as follows:

Proposition 1 *Suppose*

$$p(x) = \sum_{|\nu| \leq m} a_\nu x^\nu ,$$

is a polynomial on $\mathbb{R}^n$.

(i) *Then*

$$u(x, y) = \sum_{k=0}^{\infty} (-1)^k \frac{y^{2k}}{(2k)!} \Delta^k p(x) \tag{41}$$

is a harmonic polynomial of degree no greater than the degree of p such that

$$u(x, 0) = p(x) . \tag{42}$$

(ii) *Any harmonic polynomial u which satisfies* (42) *is not unique. If the degree of p is greater than one, then u is not even unique in the sense of having minimal degree.*

(iii) *The polynomial*

$$v(x, y) = \sum_{k=0}^{\infty} (-1)^k \frac{y^{2k+1}}{(2k+1)!} \Delta^k p(x) \tag{43}$$

is a harmonic polynomial which satisfies $v(x, 0) = 0$.

3.3 A Modified Poisson Kernel

To obtain a replacement for $P(x, y)$ which will permit the application of an arbitrary tempered distribution f in the corresponding analog of (39), reconsider (36). If $\theta = 1/2$, then the term in braces reduces to $\cosh y|\xi|$ which, as a function of ξ, is real analytic on all of $\mathbb{R}^n$. In this case, the right-hand side of (36) makes distributional sense in every neighborhood of the origin. The difficulty is in neighborhoods of ∞. However θ need not be constant as a function of ξ.

With this in mind, let $\theta(\xi) = \frac{1}{2}\lambda(\xi)$ where $\lambda(\xi)$ is an infinitely differentiable compactly supported function which is identically 1 in a neighborhood of the origin. The term in braces on the right-hand side of (36) reduces to

$$\widehat{P_\lambda}(\xi, y) = \lambda(\xi)\cosh y|\xi| + (1 - \lambda(\xi))e^{-y|\xi|}. \tag{44}$$

This function has the following properties that can be directly verified:

(i) For fixed $y > 0$, $\widehat{P_\lambda}(\xi, y)$ is in $\mathscr{S}$ as a function of ξ.
(ii) For fixed M, N, and positive y_0, there is a constant C, independent of x and y, such that

$$\|\widehat{P_\lambda}(\cdot, y)\, e^{-i\langle x,\cdot\rangle}\|_{M,N} \le Ce^{ay}(1 + |x|)^M \quad \text{for } y > y_0$$

where $a > \rho = \min\{r : |\lambda(\xi)| = 0 \text{ for all } \xi \text{ that satisfy } |\xi| \ge r\}$.
(iii) For every pair M and N, there is a constant C, independent of $\phi \in \mathscr{S}$, such that

$$\|\widehat{P_\lambda}(\cdot, y)\,\phi\|_{M,N} \le Ce^{ay}\|\phi\|_{M,N}\,.$$

where a is as in item (ii) directly above.
(iv) For all $\phi \in \mathscr{S}$

$$\lim_{y\to 0} \widehat{P_\lambda}(\xi, y)\phi(\xi) = \phi(\xi) \quad \text{in } \mathscr{S}.$$

(v) For $(x, y) \in \mathbb{R}^n \times (0, \infty)$, $\Delta\widehat{P_\lambda}(\xi, y)e^{i\langle x,\xi\rangle}$ exists and is equal to 0 in $\mathscr{S}$.
(vi) For each multi-index ν

$$\lim_{y\to 0^+} \|D_\xi^\mu\left(\xi^\nu\widehat{P_\lambda}(\xi, y)\right)\|_{L^1(\xi\in\mathbb{R}^n)} = 0$$

for all multi-indexes μ whose length $|\mu|$ is sufficiently large.

Its inverse Fourier transform

$$P_\lambda(x, y) = (2\pi)^{-n}\int_{\mathbb{R}^n} \widehat{P_\lambda}(\xi, y)\, e^{i\langle x,\xi\rangle} d\xi \tag{45}$$

has the following properties:

$$\int_{\mathbb{R}^n} P_\lambda(x, y)dx = 1 \tag{46}$$

If $0 < y \le y_1 < \infty$, then there is a constant C such that

$$\|P_\lambda(\cdot, y)\|_{L^1(\mathbb{R}^n)} \le C\,. \tag{47}$$

If ϕ is a function in $C^\infty(\mathbb{R}^n)$ which satisfies

$$\phi(x) = \begin{cases} 0 & \text{if } |x| \leq \epsilon_0 \\ 1 & \text{if } |x| \geq \epsilon_1 \end{cases}$$

for some $0 < \epsilon_0 < \epsilon_1 < \infty$, then

$$\lim_{y\to 0} \phi(x)\, P_\lambda(x, y) = 0 \quad \text{in } \mathscr{S}. \tag{48}$$

These properties of the modified kernel $P_\lambda(x, y)$ give rise to the following theorem; the proofs of the various items in its statement are analogous to the proofs of the corresponding items in the statements of Theorems 1, 3, and 4, *mutatis mutandis*.

Theorem 5 *Let λ be a compactly supported infinitely differentiable function on $\mathbb{R}^n$ which is $= 1$ in a neighborhood of the origin and define P_λ via* (44) *and* (45). *Suppose $f \in \mathscr{S}'$ and*

$$u(x, y) = \langle P_\lambda(x - \cdot, y), f \rangle . \tag{49}$$

Then $u(x, y)$ is harmonic in $\mathbb{R}^n \times (0, \infty)$ and satisfies the following:

(i) *For every test function ϕ in $\mathscr{S}$*

$$\lim_{y\to 0} \langle u(\cdot, y), \phi \rangle = \langle f, \phi \rangle.$$

(ii) *There is a number N such that if y_0 and y_1 is any pair of numbers that satisfy $0 < y_0 < y_1 < \infty$ and y satisfies $y_0 \leq y \leq y_1$, then*

$$|u(x, y)| \leq C(1 + |x|)^N$$

where C is a constant independent of x.

(iii) *There are constants M, N, and C such that for every test function ϕ in $\mathscr{S}$*

$$|\langle u(\cdot, y), \phi \rangle| \leq C\|\phi\|_{M,N}$$

whenever $0 < y \leq y_1 < \infty$.

(iv) *If f is continuous in a neighborhood of a point x_0 in $\mathbb{R}^n$, then*

$$\lim_{\substack{(x,y)\to(x_0,0) \\ (x,y)\in\mathbb{R}^n\times(0,\infty)}} u(x, y) = f(x_0) .$$

(v) *If f is a polynomial, $f = p$, then u is the harmonic function described by* (41).

3.4 Remarks

(i) The harmonic function u defined by (49) depends on both f and λ. As such, it is not uniquely determined by the boundary distribution f. For example, if $f(x) = \cos x$ and $\lambda(\xi) = 1$ when $|\xi| = 1$, then the harmonic function defined by (49) is $u(x, y) = \cos x \cosh y$. However, if λ is such that $\lambda(\xi) = 0$ when $|\xi| = 1$, then the harmonic function defined by (49) is $u(x, y) = \cos x \exp(-y)$.

(ii) We remind the reader that a distribution f in $\mathscr{S}'$ can have arbitrarily fast growth as $|x| \to \infty$. In the case $n = 1$, this is illustrated by the continuous function, and also a tempered distribution, f defined by

$$f(x) = g(x) \cos G(x) = \frac{d}{dx} \sin G(x)$$

where

$$G(x) = G(0) + \int_0^x g(t)\, dt$$

and $g(t)$ is any continuous function on $\mathbb{R}$ that can have arbitrarily fast growth as t tends to $\pm\infty$.

(iii) It is somewhat ironic to think that if f is a distribution of the sort described by the univariate example mentioned in item (ii) above, then the harmonic function $u(x, y)$ is essentially well defined in terms of f and the classical Poisson kernel via (39). Namely, if $f(x) = \frac{d}{dx}F(x)$ where $F(x)$ is a continuously differentiable function that is bounded, then

$$u(x, y) = \lim_{r\to\infty} \int_{-r}^{r} P(x - t, y)\frac{d}{dt}F(t)dt = -\int_{-\infty}^{\infty} F(t)\frac{d}{dt}P(x - t, y)dt.$$

In other words, $u(x, y) = \frac{\partial}{\partial x}v(x, y)$ where $v(x, y)$ is a bounded harmonic function on the upper half-plane $\mathbb{R} \times (0, \infty)$.

On the other hand, if $f(x) = x^2$, which has relatively mild growth at $\pm\infty$, then (39) makes no sense.

(iv) It is tempting to construct harmonic functions in the half-space $\mathbb{R}^n \times (0, \infty)$ with boundary values f as limits of harmonic functions in the slabs $\mathbb{R}^n \times (0, c)$ as $c \to \infty$ with boundary values f at $y = 0$ and 0 at $y = c$, respectively. In view of the fact that for each $\xi \in \mathbb{R}^n$

$$\lim_{c\to\infty} \widehat{Q}(\xi, c - y; c) = e^{-y|\xi|},$$

this should work if $\hat{f}$ is sufficiently well behaved in a neighborhood of the origin. In general, however, such a construction will fail. This is nicely illustrated by the case when f is a polynomial, say $f(x) = |x|^2$. Recall that the harmonic polynomial $u_c(x, y)$ that satisfies $u_c(x, 0) = p(x)$ and $u_c(x, c) = 0$ is given by

$$u_c(x, y) = \frac{c-y}{c}\left(p(x) + \sum_{k=1}^{\infty} q_{2k}((c-y); c)\, \Delta^k p(x)\right)$$

and that

$$q_2(c-y, c) = \frac{c^2 - (c-y)^2}{3!} = \frac{2cy - y^2}{3!}.$$

See Lemma 5 and formula (34). Hence if $p(x) = |x|^2$

$$u_c(x, y) = \frac{c-y}{c}\left\{|x|^2 + \frac{2cy - y^2}{3!} 2n\right\}$$

and, if $y > 0$,

$$\lim_{c\to\infty} |u_c(x, y)| = \infty\,.$$

4 Appendix

4.1 Properties of the Kernel Q(x,y)

In this section we present several alternate expressions for $Q(x, y)$ and its Fourier transform $\widehat{Q}(\xi, y)$ and record several useful properties. Recall that these functions are defined by

$$\widehat{Q}(\xi, y) = \frac{\sinh y|\xi|}{\sinh |\xi|} \tag{50}$$

and

$$Q(x, y) = (2\pi)^{-n} \int_{\mathbb{R}^n} \frac{\sinh y|\xi|}{\sinh |\xi|}\, e^{i\langle x,\xi\rangle} d\xi. \tag{51}$$

Note that $Q(x, y)$ is well defined for $(x, y) \in \mathbb{R}^n \times (-1, 1)$ and is odd as a function of y, namely, $Q(x, -y) = -Q(x, y)$.

Furthermore, in view of (10), the Poisson kernel $Q(x, y; c)$ for the slab $\mathbb{R}^n \times (a, b)$, $c = b - a$, is given by

$$Q(x, y; c) = c^{-n} Q(x/c, y/c). \tag{52}$$

Next, let $P(x, y)$ be the Poisson kernel for the upper half-space $\mathbb{R}^n \times (0, \infty)$. Namely,

$$P(x, y) = c_n \frac{y}{(|x|^2 + y^2)^{(n+1)/2}} \quad \text{where} \quad c_n = \frac{\Gamma((n+1)/2)}{\pi^{(n+1)/2}}. \tag{53}$$

In what follows, we use the fact that $P(x, y)$ is well defined on $\mathbb{R}^{n+1} \setminus \{0\}$. Let X be the discrete set consisting of points (x, y) in $\mathbb{R}^n \times \mathbb{R} = \mathbb{R}^{n+1}$ where $x = 0$ and y is an odd integer; in other words, $X = \{(x, y) \in \mathbb{R}^n \times \mathbb{R} : x = 0 \text{ and } y = 2k+1,\ k = 0, \pm 1, \pm 2, \ldots\}$.

Lemma 6 *$Q(x, y)$ enjoys the representation*

$$Q(x, y) = \sum_{k=-\infty}^{\infty} P(x, 2k + (1 - y)) \tag{54}$$

where $P(x, y)$ is the Poisson kernel (53). *In the case $n \geq 2$, this series converges absolutely for all $(x, y) \in \mathbb{R}^{n+1} \setminus X$. In the case $n = 1$, it converges in the sense that*

$$Q(x, y) = \lim_{N \to \infty} \sum_{k=-N}^{N-1} P(x, 2k + 1 - y).$$

Proof To see (54) write

$$\begin{aligned}
\frac{\sinh y|\xi|}{\sinh |\xi|} &= \frac{e^{y|\xi|} - e^{-y|\xi|}}{e^{|\xi|} - e^{-|\xi|}} \\
&= e^{-|\xi|}\{e^{y|\xi|} - e^{-y|\xi|}\} \frac{1}{1 - e^{-2|\xi|}} \\
&= \{e^{-(1-y)|\xi|} - e^{-(1+y)|\xi|}\} \sum_{k=0}^{\infty} e^{-2k|\xi|} \\
&= \sum_{k=0}^{\infty} \{e^{-(2k+1-y)|\xi|} - e^{-(2k+1+y)|\xi|}\}
\end{aligned}$$

or, more succinctly,

$$\widehat{Q}(\xi, y) = \sum_{k=0}^{\infty}\{e^{-(2k+1-y)|\xi|} - e^{-(2k+1+y)|\xi|}\}. \tag{55}$$

Taking the inverse Fourier transform and using the fact that $\widehat{P}(\xi, y) = e^{-y|\xi|}$ result in

$$Q(x, y) = \sum_{k=0}^{\infty}\{P(x, 2k+1-y) - P(x, 2k+1+y)\}. \tag{56}$$

Since $P(x, y)$ is odd in the y variable, write

$$-P(x, 2k+1+y) = P(x, -2k-1-y) = P(x, -2k-2+1-y),$$

substitute this into (56), and apply an appropriate change of summation variables to produce the desired result (54). QED

Identity (54) can be used to obtain other useful expressions for $Q(x, y)$. For example:

Lemma 7 *$Q(x, y)$ also enjoys the representation*

$$Q(x, y) = c_n|x|^{1-n}\sum_{k=1}^{\infty}(-1)^{k+1}F(k\pi|x|)\sin(k\pi y) \tag{57}$$

where

$$F(\tau) = \int_{-\infty}^{\infty}\frac{z\sin(\tau z)}{(1+z^2)^{(n+1)/2}}\,dz.$$

Proof Identity (57) follows from an application of an appropriate variant of Poisson's summation formula to the right-hand side of (54).

More specifically, if $r = |x|$, $z = 1-y$, and $f(z) = \dfrac{z}{(z^2+1)^{(n+1)/2}}$, then

$$P(x, 1-y) = c_n\frac{z}{(z^2+r^2)^{(n+1)/2}} = \frac{c_n}{r^n}\frac{z/r}{\left((z/r)^2+1\right)^{(n+1)/2}} = \frac{c_n}{r^n}f(z/r)$$

and

$$\sum_{k=-\infty}^{\infty}P(x, 2k+1-y) = \frac{c_n}{r^n}\sum_{k=-\infty}^{\infty}f\left((z+2k)/r\right). \tag{58}$$

The last expression is 2 periodic in the z variable. Hence

$$\sum_{k=\infty}^{\infty} f\big((z+2k)/r\big) = \sum_{k=-\infty}^{\infty} \gamma_k e^{i\pi kz} \tag{59}$$

where

$$\gamma_k = \frac{1}{2}\int_0^2 \left\{ \sum_{k=\infty}^{\infty} f\big((z+2k)/r\big) \right\} e^{-i\pi kz} dz$$
$$= \frac{1}{2}\int_{-\infty}^{\infty} f(z/r) e^{-i\pi kz} dz = \frac{r}{2}\int_{-\infty}^{\infty} f(z) e^{-i\pi krz} dz \tag{60}$$

In view of (54), identities (58), (59), and (60) imply that

$$Q(x,y) = \frac{c_n}{2r^{n-1}} \sum_{k=-\infty}^{\infty} \left\{ \int_{-\infty}^{\infty} f(z) e^{-i\pi krz} dz \right\} e^{i\pi k(1-y)}.$$

Simplifying the last expression for $Q(x, y)$ while using the fact that $f(z)$ is odd results in (57). QED

The function $Q(x, y)$, $(x, y) \in \mathbb{R}^n \times [0, 1)$ is radial in the x variable. That is, for fixed y, we may write $Q(x, y) = h(|x|)$ where $h(r)$ is a rapidly decaying infinitely differentiable function of the variable r, $0 \le r < \infty$. The exact nature of the function $h(r)$, which depends on n, is the subject of the next lemma, which is also a consequence of representation (54).

Lemma 8 *Let $h_n(r, y)$ be the function of two variables $(r, y) \in [0, \infty) \times (-1, 1)$ such that for $x \in \mathbb{R}^n$*

$$h_n(|x|, y) = Q(x, y), \qquad n = 1, 2, \ldots .$$

Then

$$h_{n+2}(r, y) = \frac{-1}{2\pi r} \frac{\partial}{\partial r} h_n(r, y) . \tag{61}$$

Furthermore

$$h_1(r, y) = \frac{1}{2} \frac{\sin \pi y}{\cosh \pi r + \cos \pi y}, \tag{62}$$

$$h_3(r, y) = \frac{1}{4} \frac{\sin \pi y \, \dfrac{\sinh \pi r}{r}}{(\cosh \pi r + \cos \pi y)^2}, \tag{63}$$

and, more generally, for $k = 0, 1, 2, \ldots$

$$h_{2k+1}(r, y) = \frac{1}{2}\left(\frac{-1}{2\pi\, r}\frac{\partial}{\partial r}\right)^k \frac{\sin \pi y}{\cosh \pi r + \cos \pi y}\,. \tag{64}$$

Proof Let

$$P_n(r, y) = c_n \frac{y}{(r^2 + y^2)^{(n+1)/2}}$$

and observe that (61) follows from the fact that

$$P_{n+2}(r, y) = \frac{-1}{2\pi\, r}\frac{\partial}{\partial r} P_n(r, y)$$

and identity (54).

The fact that $h_1(|x|, y)$ is essentially the Poisson kernel for the slab in the case $n = 1$ is well-known; see the remark after (12).

Identity (62) can also be verified directly by using relation (54). To see this, set $2z = 1 - y$ and $2s = r$ and write

$$P_n(r, 2k + 1 - y) = P_n(2s, 2k + 2z) = \frac{c_n}{2^n}\frac{z + k}{\left((z + k)^2 + s^2\right)^{(n+1)/2}}\,.$$

Thus, the sum in (54) in the case $n = 1$ can be reduced to

$$\frac{1}{2\pi}\sum_{k=-\infty}^{\infty}\frac{z + k}{(z + k)^2 + s^2} = \frac{1}{2\pi}\lim_{N\to\infty}\sum_{k=-N}^{N-1}\frac{z + k}{(z + k)^2 + s^2}.$$

To simplify the last expression, note that

$$\frac{z}{z^2 + s^2} = \frac{1}{2}\left\{\frac{1}{z + is} + \frac{1}{z - is}\right\}$$

and use the classical summation technique, as in, for example, [2, page 187], to write

$$\sum_{k=-\infty}^{\infty}\frac{1}{z + is + k} = \pi \cot \pi(z + is) \quad \text{and} \quad \sum_{k=-\infty}^{\infty}\frac{1}{z - is + k} = \pi \cot \pi(z - is)\,.$$

Hence

$$\sum_{k=-\infty}^{\infty} \frac{z+k}{(z+k)^2+s^2} = \frac{\pi}{2}\{\cot \pi(z+is) + \cot \pi(z-is)\}$$

$$= \frac{\pi}{2} \frac{2\sin 2\pi z}{\cos 2\pi is - \cos 2\pi z}$$

$$= \frac{\pi \sin 2\pi z}{\cosh 2\pi s - \cos 2\pi z}.$$

Finally, multiply by $\frac{1}{2\pi}$, use the substitutions $1-y = 2z$ and $r = 2s$, and apply (54), to get (62).

Identities (63) and (64) are immediate consequences of (61) and (62). QED

Note that (64) allows us to conclude that in the case when n is odd

$$Q(x, y) = O(|x|^{(1-n)/2} e^{-\pi|x|}) \quad \text{as } |x| \to \infty$$

or, somewhat more precisely,

$$Q(x, y) \leq C|x|^{(1-n)/2} e^{-\pi|x|} \sin \pi y \quad \text{when } |x| \geq \epsilon > 0, \tag{65}$$

where C may depend on n and ϵ but is otherwise independent of x and y.

Another way of obtaining (65), which is valid in all the cases $n \geq 2$, involves estimates on both the function $F(\tau)$ in (57) and the sum of the resulting series. The following argument provides more details:

Proof Assume $n \geq 2$ and write

$$F(\tau) = \int_{-\infty}^{\infty} \frac{z\sin(\tau z)}{(1+z^2)^{(n+1)/2}} dz$$

$$= \frac{1}{1-n} \int_{-\infty}^{\infty} \left\{ \frac{d}{dz} \frac{1}{(1+z^2)^{(n-1)/2}} \right\} \sin(\tau z)\, dz$$

$$= \frac{\tau}{n-1} \int_{-\infty}^{\infty} \frac{\cos(\tau z)}{(1+z^2)^{(n-1)/2}} dz$$

and note that the last expression in the above string is a multiple of the classical Bessel potential $g_{n-1}(\tau)$ that, for $\alpha > 0$, is defined as the inverse Fourier transform of $(1+z^2)^{-\alpha}$, namely,

$$g_\alpha(\tau) = \frac{1}{2\pi} \int_{-\infty}^{\infty} \frac{e^{i\tau z}}{(1+z^2)^{\alpha/2}} dz.$$

Hence

$$F(\tau) = \frac{2\pi\tau}{n-1} g_{n-1}(\tau)$$

and substituting this in (57) yields

$$Q(x, y) = \frac{2\pi^2 c_n}{n-1} |x|^{2-n} \sum_{k=1}^{\infty} (-1)^{k+1} k\, g_{n-1}(k\pi |x|) \sin(k\pi y). \tag{66}$$

The potentials g_α can be expressed in terms of the special functions $K_{(\alpha-1)/2}$, also known as modified Bessel functions of the third kind [1, p. 376, formula 9.6.25], [5, p. 414, formula (2.8)], [12, p. 83, formula (27)]; such special functions have known asymptotics [1, p. 378, formula 9.7.2], [5, p. 415, formula (3,6)], [12, p. 23, formula (1)] that can be used to estimate g_α, [5, p. 417, formula (4,3)].

Alternatively, one can estimate g_α directly by using the representation

$$g_\alpha(\tau) = \frac{1}{(2\sqrt{\pi})\, \Gamma(\alpha/2)} \int_0^\infty \left\{ \exp\left(-\left(\frac{|\tau|^2}{4t} + t \right) \right) \right\} t^{(\alpha-1)/2} \frac{dt}{t}, \tag{67}$$

that is a consequence of taking the inverse Fourier transform of

$$(1 + z^2)^{-\alpha/2} = \frac{1}{\Gamma(\alpha/2)} \int_0^\infty e^{-(1+|z|^2)t}\, t^{\alpha/2} \frac{dt}{t}.$$

If $|\tau| \geq \epsilon > 0$, an estimate of the right-hand side of (67) leads to

$$C_0 |\tau|^{(\alpha-2)/2} e^{-|\tau|} \leq g_\alpha(\tau) \leq C_1 |\tau|^{(\alpha-2)/2} e^{-|\tau|}. \tag{68}$$

where C_0 and C_1 are positive constants independent of τ.

Hence if $|x| \geq \epsilon$, we may use (66) and (68) to write

$$\begin{aligned} Q(x, y) &\leq C|x|^{2-n} \sum_{k=1}^{\infty} k\, |k\pi x|^{(n-3)/2} e^{-|k\pi x|} \sin(k\pi y) \\ &\leq C|x|^{(1-n)/2} \left\{ \sum_{k=1}^{\infty} k^{(n-1)/2} e^{-(k-1)\pi\epsilon} \right\} e^{-|\pi x|} \sin(\pi y), \end{aligned}$$

where we use the inequality $|\sin(k\pi y)| \leq k|\sin(\pi y)|$, which can be verified by induction. We may conclude that

$$Q(x, y) \leq C|x|^{(1-n)/2} e^{-|\pi x|} \sin(\pi y) \quad \text{when } |x| \geq \epsilon > 0, \tag{69}$$

where the constant C depends on n and ϵ but is otherwise independent of x and y. QED

We summarize these observations thusly:

Lemma 9 *When $n \geq 2$, the series expansion* (57) *can be re-expressed as* (66). *Furthermore, $Q(x, y)$, which is non-negative, enjoys the bound* (69) *for all n, $n = 1, 2, 3, \ldots$.*

Finally, for completeness, we provide a proof of item (29) in Subsection 2.4.

As a preface to what follows, we remind the reader that if $\nu = (\nu_1, \ldots, \nu_n)$ is a multi-index, then $|\nu| = \nu_1 + \cdots + \nu_n$. Otherwise, for elements x in $\mathbb{R}^n$, $|x|$ denotes the Euclidean norm of x. Also, for two multi-indexes, $\mu \leq \nu$ means that $\mu_j \leq \nu_j$ for $j = 1, \ldots, n$; $\mu < \nu$ means that $\mu \leq \nu$ and $\mu_j < \nu_j$ for at least one index j.

Lemma 10 *Given any multi-index ν*

$$\lim_{y \to 1} \int_{\mathbb{R}^n} \left| D^{\mu}\left(\xi^{\nu} \widehat{Q}(\xi, y)\right) \right| d\xi = 0 \tag{70}$$

whenever the multi-index μ satisfies $|\mu| \geq |\nu| + n + 1$.

Proof To get a feel for what's involved, one may verify the relatively straightforward case $n = 1$ with $\nu = 0$. To see the general case, write

$$\int_{\mathbb{R}^n} \left| D^{\mu}\left(\xi^{\nu} \widehat{Q}(\xi, y)\right) \right| d\xi = I_0 + I_1 \tag{71}$$

where

$$I_0 = \int_{|\xi| \leq 1} \left| D^{\mu}\left(\xi^{\nu} \widehat{Q}(\xi, y)\right) \right| d\xi \quad \text{and} \quad I_1 = \int_{|xi| > 1} \left| D^{\mu}\left(\xi^{\nu} \widehat{Q}(\xi, y)\right) \right| d\xi .$$

To estimate I_0 express $\widehat{Q}$ as

$$\widehat{Q}(\xi, y) = y \frac{\phi(y\xi)}{\phi(\xi)} = y \left\{ \frac{\phi(y\xi) - \phi(\xi)}{\phi(\xi)} + 1 \right\}$$

where

$$\phi(\xi) = \frac{\sinh |\xi|}{|\xi|} .$$

Note that, among other things, ϕ is infinitely differentiable on $\mathbb{R}^n$ and for any non-zero multi-index β

$$\lim_{y \to 1} D^{\beta} \left\{ \frac{\phi(y\xi) - \phi(\xi)}{\phi(\xi)} + 1 \right\} = \lim_{y \to 1} D^{\beta} \left\{ \frac{\phi(y\xi) - \phi(\xi)}{\phi(\xi)} \right\} = 0. \tag{72}$$

If μ is such that $D^{\mu} \xi^{\nu} = 0$, in particular if $|\mu| > |\nu|$, then

$$D^\mu\left(\xi^\nu \widehat{Q}(\xi,y)\right) = y \sum_{0<\beta\le\mu} c_\beta D^{\mu-\beta}\xi^\nu\, D^\beta \left\{\frac{\phi(y\xi)-\phi(\xi)}{\phi(\xi)}\right\}$$

where the sum contains no terms with $\beta = 0$. Hence, in view of (72) and the bounded convergence theorem

$$\lim_{y\to 1} I_0 = 0\,. \tag{73}$$

Estimating I_1 is a bit more involved. First express $\widehat{Q}$ as

$$\widehat{Q}(\xi,y) = \widehat{Q}(\xi,y) - e^{(y-1)|\xi|} + e^{(y-1)|\xi|},$$

note that

$$\widehat{Q}(\xi,y) - e^{(y-1)|\xi|} = -e^{-|\xi|}\widehat{Q}(\xi,1-y),$$

and write

$$I_1 \le I_{1,1} + I_{1,2} \tag{74}$$

where

$$I_{1,1} = \int_{|\xi|>1} \left|D^\mu\xi^\nu\left(e^{-|\xi|}\widehat{Q}(\xi,1-y)\right)\right| d\xi \quad\text{and}\quad I_{1,2} = \int_{|\xi|>1} \left|D^\mu\xi^\nu e^{(y-1)|\xi|}\right| d\xi.$$

The fact that

$$\lim_{y\to 1} I_{1,1} = 0 \tag{75}$$

follows immediately from the fact that

$$\lim_{y\to 0} \widehat{Q}(\xi,y) = 0 \quad \text{in } \mathscr{S}(\mathbb{R}^n).$$

To estimate $I_{1,2}$, use the substitution $s = 1-y$ to simplify notation, write

$$D^\mu\left(\xi^\nu e^{-s|\xi|}\right) = \sum_{\substack{\beta\\ \beta\le\mu}} c_\beta \left\{D^{\mu-\beta}\xi^\nu\right\} D^\beta e^{-s|\xi|}, \tag{76}$$

and notice that $D^{\mu-\beta}\xi^\nu$ will be identically zero when $|\mu-\beta| > |\nu|$ and equal to a constant multiple of $\xi^{\nu-(\mu-\beta)}$ when $\nu-(\mu-\beta)\ge 0$. Since $|\beta| < |\mu|-|\nu|$ implies that $|\mu-\beta| > |\nu|$, we may write

$$\left|D^{\mu}\left(\xi^{\nu}e^{-s|\xi|}\right)\right| \leq C \sum_{\substack{\beta \\ |\mu|-|\nu|\leq|\beta|\leq|\mu|}} |\xi|^{|\nu|-|\mu|+|\beta|}\left|D^{\beta}e^{-s|\xi|}\right|. \tag{77}$$

Next, note that

$$D^{\beta}e^{-s|\xi|} = \sum_{k=0}^{|\beta|-1} s^{|\beta|-k}H_k(\xi)e^{-s|\xi|}$$

where H_k is homogeneous of degree $-k$ and infinitely differentiable on $\mathbb{R}^n \setminus \{0\}$ so that $|H_k(\xi)| \leq C|\xi|^{-k}$. (In the case $n = 1$, $H_0(\xi) = (\operatorname{sgn}\xi)^{|\beta|}$ while $H_k(\xi)$ is identically zero on $\mathbb{R} \setminus \{0\}$. In general the exact expression for H_k depends on β.) Hence

$$|\xi|^p|D^{\beta}e^{-s|\xi|}| \leq C \sum_{k=0}^{|\beta|-1} s^{|\beta|-k}|\xi|^{p-k}e^{-s|\xi|} \tag{78}$$

and it follows that

$$\int_{|\xi|>1} |\xi|^p|D^{\beta}e^{-s|\xi|}|d\xi \leq C\sum_{k=0}^{|\beta|-1} J_k \tag{79}$$

where

$$J_k = s^{|\beta|-k}\int_1^{\infty} r^{p-k+n-1}e^{-sr}dr,$$

which satisfies the bounds

$$J_k \leq C\begin{cases} s^{|\beta|-k} & \text{if } p+n+1 \leq k, \\ s^{|\beta|-(p+n)}\left(1+\log(1/s)\right) & \text{if } k = p+n, \\ s^{|\beta|-(p+n)} & \text{if } k \leq p+n-1. \end{cases}$$

In view of the assumption that $|\mu| \geq |\nu|+1$, taking $p = |\nu|-|\mu|+|\beta|$ implies that $|\beta|-(p+n) = |\mu|-|\nu|-n \geq 1$ and,

$$\text{when } 0 \leq k \leq |\beta|-1, \quad \lim_{s\to 0} J_k = 0. \tag{80}$$

Finally, setting $p = |\nu|-|\mu|+|\beta|$, items (78), (79), and (80) together with (76) and (77) imply that

$$\lim_{y\to 1} I_{1,2} = 0. \tag{81}$$

Lemma 10 now follows from identities (71), (73), (74), (75), and (81). QED

4.2 Comments and Observations

(i) In view of (52) and (69), we may conclude that the convolution

$$u(x, y) = \langle Q(x - \cdot, y; c), f \rangle \tag{82}$$

is well defined and makes sense for a class of distributions that is significantly wider than $\mathscr{S}'(\mathbb{R}^n)$. For example, (82) makes sense whenever the product

$$f(x)\, |x|^{(1-n)/2}\, e^{-\pi |x|/c}$$

is integrable over $\mathbb{R}^n$. This includes smooth functions such as

$$f(x) = (1 + |x|^2)^{p/2} \cosh(\pi |x|/c), \quad \text{where } p < (n-3)/2,$$

that are not in $\mathscr{S}'(\mathbb{R}^n)$. In the case $n = 1$, this phenomenon has been considered to some extent in [31].

(ii) In this article, we characterized functions $u(x, y)$ harmonic in the slab $\mathbb{R}^n \times (a, b)$ with boundary data in the class of tempered distributions $\mathscr{S}'(\mathbb{R}^n)$. The characterization of such harmonic functions with boundary data in $L^p(\mathbb{R}^n)$ is similar to that what is known for the upper half-space; for example, see [30, Chapter 2]. Analogous results in the cases when the boundary data are in certain natural classes of distributions suggested by item (i) above are not so clear.

(iii) If the functions $f_a(x)$ and $f_b(x)$ are entire, then according to a theorem in the recently published article [20], there is a harmonic function $u(x, y)$ on $\mathbb{R}^{n+1}$ that satisfies $u(x, a) = f_a(x)$, $u(x, b) = f_b(x)$, and can be extended to be harmonic on all of $\mathbb{R}^{n+1}$. Thus statement (i) of Theorem 4 is a special case that is a consequence of a constructive argument that is both direct and accessible. It might be an interesting exercise to check if the general result also follows via a similar argument.

References

1. M. Abramowitz, I.A. Stegun (eds.), *Handbook of Mathematical Functions, with Formulas, Graphs, and Mathematical Tables* (Dover Publications, Inc., New York, 1972), pp. xiv+1046
2. L.V. Ahlfors, *Complex Analysis*, 2nd edn. (McGraw-Hill, New York, 1966)
3. D.H. Armitage, Representations of harmonic functions in half-spaces. Proc. Lond. Math. Soc. **18**(3), 53–71 (1979)
4. D.H. Armitage, S.J. Gardiner, *Classical Potential Theory* (Springer, London, 2001)
5. N. Aronszajn, K.T. Smith, Theory of Bessel potentials. I. Ann. Inst. Fourier (Grenoble) **11**, 385–475 (1961)
6. S. Axler, P. Bourdon, W. Ramey, *Harmonic Function Theory* (Springer, New York, 1992)

7. F.T. Brawn, The Green and Poisson kernels for the strip $\mathbb{R}^n \times]0, 1[$. J. Lond. Math. Soc. **2**(2), 439–454 (1970)
8. H. Bremermann, *Distributions, Complex Variables, and Fourier Transforms* (Addison-Wesley, Reading, 1965)
9. M. Chamberland, D. Siegel, Polynomial solutions to Dirichlet problems. Proc. Am. Math. Soc. **129**(1), 211–217 (2001)
10. W.F. Donoghue Jr., *Distributions and Fourier transforms*. Pure and Applied Mathematics, vol. 32 (Academic Press, New York, 1969)
11. W. Durand, On some boundary value problems on a strip in the complex plane. Rep. Math. Phys. **52**(1), 1–23 (2003)
12. A. Erdelyi, W. Magnus, F. Oberhettinger, F.G. Tricomi, *Higher Transcendental Functions*, vol. II (McGraw-Hill, New York, 1953)
13. A. Erdelyi, W. Magnus, F. Oberhettinger, F.G. Tricomi, *Tables of Integral Transforms*, vol. I (McGraw-Hill, New York, 1954)
14. M. Finkelstein, S. Scheinberg, Kernels for solving problems of Dirichlet type in a half-plane. Adv. Math. **18**, 108–113 (1975)
15. S.J. Gardiner, The Dirichlet and Neumann problems for harmonic functions in half-spaces. J. Lond. Math. Soc. **24**(2), 502–512 (1981)
16. S.J. Gardiner, The Dirichlet problem with noncompact boundary. Math. Z. **213**(1), 163–170 (1993)
17. S. Gergün, I.V. Ostrovskii, On the Poisson representation of a function harmonic in the upper half-plane. Comput. Methods Funct. Theory **2**(1), 191–213 (2002)
18. L. Hörmander, *Linear partial differential operators*. Third revised printing. Die Grundlehren der mathematischen Wissenschaften, Band 116 (Springer-Verlag New York Inc., New York, 1969)
19. O.D. Kellogg, *Foundations of Potential Theory*, reprint of 1929 ed (Dover, New York, 1929)
20. D. Khavinson, E. Lundberg, H. Render, The Dirichlet problem for the slab with entire data and a difference equation for harmonic functions. Can. Math. Bull. **60**(1), 146–153 (2017)
21. D. Khavinson, H.S. Shapiro, Dirichlet's problem when the data is an entire function. Bull. Lond. Math. Soc. **24**(5), 456–468 (1992)
22. D. Khavinson, E. Lundberg, *Linear Holomorphic Partial Differential Equations and Classical Potential Theory*. Mathematical Surveys and Monographs, vol. 232 (American Mathematical Society, Providence, 2018)
23. E. Lundberg, H. Render, The Khavinson-Shapiro conjecture and polynomial decompositions. J. Math. Anal. Appl. **376**(2), (2011), 506–513
24. L. Qiao, Dirichlet problems for harmonic functions in half spaces. Ukr. Math. J. **66**(10), 1530–1543 (2015)
25. H. Render, Real Bargmann spaces, Fischer decompositions, and sets of uniqueness for polyharmonic functions. Duke Math. J. **142**(2), 313–352 (2008)
26. H. Render, A characterization of the Khavinson-Shapiro conjecture via Fischer operators. Potential Anal. **45**(3), 539–543 (2016)
27. W. Rudin, *Functional Analysis*, McGraw-Hill Series in Higher Mathematics (McGraw-Hill Book Co., New York-Düsseldorf-Johannesburg, 1973)
28. S.M. Selby (ed.), *CRC Standard Mathematical Tables*, 18th edn. (The Chemical Rubber Co., Cleveland, 1970)
29. E.M. Stein, *Singular Integrals and Differentiability Properties of Functions* (Princeton University Press, Princeton, 1970)
30. E.M. Stein, G. Weiss, *Introduction to Fourier Analysis on Euclidean Spaces* (Princeton University Press, Princeton, 1971)
31. D.V. Widder, Functions harmonic in a strip. Proc. Am. Math. Soc. **12**, 67–72 (1961)
32. Y.H. Zhang, G.T. Deng, T. Qian, Integral representations of a class of harmonic functions in the half space. J. Differ. Equ. **260**(2), 923–936 (2016)

The manufacturer's authorised representative in the EU is Springer Nature Customer Service Centre GmbH, Europaplatz 3, 69115 Heidelberg, Germany. If you have any concerns regarding our products, please contact ProductSafety@springernature.com

Printed and bound by CPI Group (UK) Ltd, Croydon, CR0 4YY

30/07/2026

02182012-0009